Dielectric Insulation in Medium- and High-Voltage Power Equipment—Degradation and Failure Mechanism, Diagnostics, and Electrical Parameters Improvement

Dielectric Insulation in Medium- and High-Voltage Power Equipment—Degradation and Failure Mechanism, Diagnostics, and Electrical Parameters Improvement

Editor

Tomasz Norbert Kołtunowicz

MDPI

Basel • Beijing • Wuhan • Barcelona • Belgrade • Novi Sad • Cluj • Manchester

Editor
Tomasz Norbert Kołtunowicz
Lublin University of
Technology
Lublin
Poland

Editorial Office
MDPI AG
Grosspeteranlage 5
4052 Basel, Switzerland

This is a reprint of articles from the Special Issue published online in the open access journal *Energies* (ISSN 1996-1073) (available at: https://www.mdpi.com/journal/energies/special_issues/0AOPUX89EG).

For citation purposes, cite each article independently as indicated on the article page online and as indicated below:

Lastname, A.A.; Lastname, B.B. Article Title. *Journal Name* **Year**, *Volume Number*, Page Range.

ISBN 978-3-7258-1973-7 (Hbk)
ISBN 978-3-7258-1974-4 (PDF)
doi.org/10.3390/books978-3-7258-1974-4

Contents

energies MDPI

Editorial

Dielectric Insulation in Medium- and High-Voltage Power Equipment—Degradation and Failure Mechanism, Diagnostics, and Electrical Parameters Improvement

Tomasz N. Koltunowicz

Department of Electrical Devices and High Voltage Technology, Lublin University of Technology, 38a, Nadbystrzycka Street, 20-618 Lublin, Poland; t.koltunowicz@pollub.pl; Tel.: +48-81-538-47-13

The proper operation of medium- and high-voltage power equipment is greatly affected by the degradation of its insulation. There are five main factors that cause insulation degradation: electrical, mechanical, and thermal exposures, as well as chemical aggression and environmental pollution. Therefore, it is very important to be able to control the development of insulation material degradation and thus reduce the possibility of medium- and high-voltage power equipment failures. The diagnostic tests carried out to this end help to eliminate damage to the insulating components of power equipment. This effectively reduces or even eliminates catastrophic failures and thus reduces the risk of environmental pollution. This Special Issue addresses topics related to the measurement of and the use of sensors and other solutions to monitor the condition of power equipment components. Issues related to the determination of degradation and damage mechanisms of dielectric insulation are presented, as are ways to monitor insulating elements of power equipment, including the condition of solid, liquid, and gaseous insulation.

The Special Issue "*Dielectric Insulation in Medium- and High-Voltage Power Equipment—Degradation and Failure Mechanism, Diagnostics, and Electrical Parameters Improvement*" has received good responses. Of 16 submissions, 12 were accepted, which gives a 75% acceptance rate. This Special Issue does not include review papers; all accepted papers are original research papers.

Z. Pu et al. in paper [1] present a solution to the problem of winding turn-to-turn breakdown faults in dry-type transformers on wind farms under overvoltage conditions. For this purpose, a simulation model based on the structural dimensions and material parameters of transformer windings is developed. The winding distribution parameters are calculated using the finite element method. The simulation results show that the maximum overvoltage between turns of the transformer winding under lightning shock is 5.282 kV; the maximum overvoltage between turns of the winding under very fast transient overvoltage is 11.6 kV and occurs between the first two and three layers of the section, close to the insulation breakdown margin. The optimization of the proposed process is carried out. The accuracy of the winding structure optimization simulation study is verified by testing the transformer's impulse voltage before and after optimization, providing a reference for the stable operation of dry-type transformers in practical wind farm applications.

Paper [2] focuses on issues related to the problems generated when the interruption of a vacuum circuit breaker on a wind farm causes high amplitude and rapid overvoltages, damaging the transformer's inter-winding insulation. Based on the physical process of dielectric recovery during the opening of the vacuum circuit breaker, a model of dielectric strength recovery is built to simulate arc reignition. The simulation results show that the overvoltage amplitude and reignition times calculated by the model are closer to the measured values. Compared with the traditional linear curve reignition model, the accuracy was increased by 24% and 51.2%, respectively. A suitable suppression scheme is proposed by installing a combined arrester on the high-voltage side of the transformer and

Citation: Koltunowicz, T.N. Dielectric Insulation in Medium- and High-Voltage Power Equipment—Degradation and Failure Mechanism, Diagnostics, and Electrical Parameters Improvement. *Energies* **2024**, *17*, 2704. https://doi.org/10.3390/en17112704

Received: 7 May 2024
Revised: 24 May 2024
Accepted: 28 May 2024
Published: 3 June 2024

connecting a choke coil in series, which can limit the phase-to-ground and phase-to-phase voltages and reduce voltage steepness.

X. Ge et al. [3] present four insulation resistance (IR) degradation models for cross-linked polyethylene-insulated cables under thermal ageing. IR is an essential metric indicating the insulation conditions of extruded power cables. In the paper, the influences of methodologies and temperature profiles on IR simulation are evaluated. Cable cylindrical insulation is first divided into sufficiently small segments whose temperatures are simulated by jointly using a finite volume method and an artificial neural network to model the thermal ageing experiment conditions. The thermal degradation of IR is then simulated by dichotomy models that randomly sample fully degraded segments based on an overall insulation ageing condition estimation and discretization models that estimate the gradual degradation of individual segments, respectively. The insulation resistance simulation results are not only compared between different models, but also discussed in terms of the sensitivity of insulation resistance simulation to segment sizes and degradation rates.

X. Yi et al. [4] investigate the deterioration of and the abnormal temperature rise in the GFRP core rod material of compact V-string composite insulators subjected to prolonged alternating flexural loading under wind-induced stresses. The axial stress on the GFRP (Glass-Fiber-Reinforced Plastic) core rod, resulting from transverse wind loads, is a focal point of the examination. By establishing a stress model and damage model, the paper simulates and computes the evolution of damage in the outer arc material of composite insulator core rods subjected to alternating flexural loads. The research study underscores the significance of understanding the ageing and decay-like fracture process of compact line V-string composite insulators and provides crucial insights for future research aimed at enhancing the material properties of composite insulator core rods.

In paper [5], M. Florkowski and M. Kuniewski present information about partial discharges (PDs) and the subsequent deterioration of electrical insulation caused by the high electric-field stress and high-frequency switching phenomena as well as the impact of environmental conditions. The authors describe a novel combined approach based on surface resistance and potential mapping to reveal the effects of internal processes and the deterioration of insulating material due to the actions of partial discharges. The following two-step approach is proposed: multi-point resistance mapping is used to identify the spots of discharge channels, pointing to surface resistance that is a few orders of magnitude lower as compared to untreated areas, whereas high spatial and temporal resolution allows for the precise mapping and tracing of decay patterns. Experiments were carried out on polyethylene (PE) and Nomex specimens that contained embedded voids. The presented methodology and experimental results expand our understanding of PD mechanisms and internal surface processes.

P. Mikrut and P. Zydroń in paper [6] present the conditions for the formation of PD pulses in gaseous voids located in the XLPE insulation of an HVDC cable. The MATLAB® procedure and the coupled electro-thermal simulation model implemented in COMSOL Multiphysics® software are used in the analysis. The FEM model was used to study the effect of the applied voltage, the temperature field, and the location of the gaseous void in the distribution and values of the electric field in the cable insulation. In the numerical simulation procedure, the time sequences of PDs arising in the gaseous defects of the HVDC cable insulation were analyzed by observing changes caused by the increase in the temperature of the cable core. The model was used to study the conditions for PD formation in models of three HVDC cables for DC voltages from 150 kV to 500 kV. The critical dimensions of gaseous voids— which, if exceeded, make voids sources of PDs—were also estimated for each of the analyzed cables.

K. Meresch et al. [7] present a study on the use of acoustic inspection to detect partial discharges. Ultrasonic sensors have made detecting partial discharges through acoustic sensing increasingly feasible. Interpreting acoustic signals can pose challenges, as it requires extensive expertise and knowledge of equipment configuration. To solve this problem, a technique based on zero-crossing rate and fundamental frequency estimation is proposed to standardize insulator diagnosis. In the experiment proposed by the authors, a database of 72 raw acoustic signals with frequencies from 0 to 128 kHz was used, and various types of contamination and defects were introduced into the chain of insulators. The proposed technique allowed for the detection of partial discharges and their classification according to type, such as corona or surface discharges. This method simplifies the process while providing valuable insights into the severity of the phenomena observed in the field.

Paper [8] by A. Cichon and M. Wlodarz presents measurements of the technical condition of an on-load tap changer (OLTC). The methods used require the transformer to be taken out of service for the duration of the diagnostic procedure for the sake of precision. This authors create an online OLTC diagnostics method based on acoustic emission (AE) signals. An extensive measurement database containing many frequently occurring OLTC defects is used, and a method of feature extraction from AE signals based on wavelet decomposition is developed. Several machine learning models for classifying OLTC defects are created, and the most effective model is selected.

E. Taine et al. [9] present research on silicone elastomers, which are commonly used in high-voltage engineering. They are used in outdoor insulation as coatings or structural elements, or at interfaces between network elements, such as cable sealing ends (CSEs). Developing reliable systems that operate under high electric fields and variable repeated strains requires a thorough understanding of the mechanisms behind electrical breakdown and its coupling to mechanical cycling. The effect of Mullins damage and mechanical fatigue on silicone elastomers is investigated in the paper. An electro-mechanical instability model that considers cyclic softening allows for predicting the evolution of breakdown strength depending on loading history.

P. Zukowski et al. in paper [10] present studies of the site percolation phenomenon for square matrixes with dimensions L = 55, 101, and 151 using the Monte Carlo computer simulation method. The study features an in-depth analysis of the test results using a metrological approach, which determines the uncertainty of estimating the iteration results with statistical methods. The authors establish that the statistical distribution of the percolation threshold value is a normal distribution. The average value of the percolation threshold for relatively small numbers of iterations varies in a small range. For large numbers of iterations, this value stabilizes and practically does not depend on the dimensions of the matrix. The value of the standard deviation of the percolation threshold for small numbers of iterations also fluctuates to a small extent. For a large number of iterations, the standard deviation values reach a steady state. Along with the increase in the dimensions of the matrix, there is a clear decrease in the value of the standard deviation. The application of the metrological approach to the analysis of the percolation phenomenon simulation results allows for the development of a new method of optimizing the determination and reducing the uncertainty of percolation threshold estimation. It consists in selecting the dimensions of the matrix and the number of iterations in order to obtain the assumed uncertainty in determining the percolation threshold. Such a procedure can be used to simulate the percolation phenomenon and to estimate the value of the percolation threshold and its uncertainty in matrices with other matrix shapes than square ones.

Paper [11] focuses on determining the percolation phenomenon for square matrices using the Monte Carlo simulation method. The spatial distributions of the coordinates of the nodes creating the percolation channel are determined, and maps of the density distribution of these nodes are created. It is established that in matrices with finite dimensions, an edge phenomenon occurs, consisting of a decrease in the concentration of nodes creating a percolation channel as one approaches the edge of the matrix. As the matrix dimensions increase, the intensity of this phenomenon decreases. Clusters whose dimensions are close to half of the matrix dimensions are most likely to occur. The research shows that both the values of the standard deviation of the percolation threshold and the intensity of the edge phenomenon are clearly related to the dimensions of the matrices and decrease as matrix dimensions increase.

In paper [12], an in-depth analysis of the percolation phenomenon for square matrices with dimensions from $L = 50$ to 600 for a sample number of 5×10^4 is performed using Monte Carlo computer simulations. The percolation threshold value is defined as the number of conductive nodes remaining in the matrix before drawing the node interrupting the last percolation channel, in connection with the overall count of nodes within the matrix. The dependencies of the expected value of the percolation threshold and the standard deviation of the dimensions of the matrix are determined. It is established that the standard deviation decreases with the increase in matrix dimensions. The analysis involves not only the spatial distributions of nodes interrupting the percolation channels but also the overall patterns of node interruption in the matrix. The distributions reveal an edge phenomenon within the matrices, characterized by the maximum concentration of nodes interrupting the final percolation channel occurring at the center of the matrix. Increasing the dimensions of the matrix slows down the rate of decrease in the number of nodes towards the edge. In doing so, the area in which values close to the maximum occur is expanded. The approximation of the experimental results allows for the determination of formulas describing the spatial distributions of the nodes interrupting the last percolation channel and the values of the standard deviation from the matrix dimensions.

Conflicts of Interest: The author declares no conflicts of interest.

References

1. Pu, Z.; Yu, X.; Wang, Y.; Liu, H.; Feng, Z. Inter-Turn Breakdown Fault Analysis and Winding Structure Optimisation of Winding of Dry-Type Transformers in Wind Farms. *Energies* **2023**, *16*, 2012. [CrossRef]
2. Pu, Z.; Liu, H.; Wang, Y.; Yu, X.; Wu, T. Simulation and Protection of Reignition Overvoltage in Wind Farm Considering Microscopic Dielectric Recovery Process of Vacuum Circuit Breaker. *Energies* **2023**, *16*, 2070. [CrossRef]
3. Ge, X.; Fan, F.; Given, M.J.; Stewart, B.G. Insulation Resistance Degradation Models of Extruded Power Cables under Thermal Ageing. *Energies* **2024**, *17*, 1062. [CrossRef]
4. Yi, X.; Ding, L.; Liu, H.; Zhang, J.; Liu, J.; Li, Z.; Zhang, Y.; Wang, P.; Geng, J. Study on the Mechanism Effect of Bending Loads on the Decay-like Degradation of Composite Insulator GFRP Core Rod. *Energies* **2024**, *17*, 423. [CrossRef]
5. Florkowski, M.; Kuniewski, M. Partial Discharge-Originated Deterioration of Insulating Material Investigated by Surface-Resistance and Potential Mapping. *Energies* **2023**, *16*, 5973. [CrossRef]
6. Mikrut, P.; Zydroń, P. Numerical Modeling of PD Pulses Formation in a Gaseous Void Located in XLPE Insulation of a Loaded HVDC Cable. *Energies* **2023**, *16*, 6374. [CrossRef]
7. Maresch, K.; Freitas-Gutierres, L.F.; Oliveira, A.L.; Borin, A.S.; Cardoso, G., Jr.; Damiani, J.S.; Morais, A.M.; Correa, C.H.; Martins, E.F. Advanced Diagnostic Approach for High-Voltage Insulators: Analyzing Partial Discharges through Zero-Crossing Rate and Fundamental Frequency Estimation of Acoustic Raw Data. *Energies* **2023**, *16*, 6033. [CrossRef]
8. Cichoń, A.; Włodarz, M. OLTC Fault detection Based on Acoustic Emission and Supported by Machine Learning. *Energies* **2024**, *17*, 220. [CrossRef]
9. Taine, E.; Andritsch, T.; Saeedi, I.A.; Morshuis, P.H.F. Dielectric Breakdown Strength of PDMS Elastomers after Mechanical Cycling. *Energies* **2023**, *16*, 7424. [CrossRef]
10. Zukowski, P.; Okal, P.; Kierczynski, K.; Rogalski, P.; Borucki, S.; Kunicki, M.; Koltunowicz, T.N. Investigations into the Influence of Matrix Dimensions and Number of Iterations on the Percolation Phenomenon for Direct Current. *Energies* **2023**, *16*, 7128. [CrossRef]
11. Zukowski, P.; Okal, P.; Kierczynski, K.; Rogalski, P.; Bondariev, V. Analysis of Uneven Distribution of Nodes Creating a Percolation Channel in Matrices with Translational Symmetry for Direct Current. *Energies* **2023**, *16*, 7647. [CrossRef]
12. Zukowski, P.; Okal, P.; Kierczynski, K.; Rogalski, P.; Bondariev, V.; Pogrebnjak, A.D. Monte Carlo Simulation of Percolation Phenomena for Direct Current in Large Square Matrices. *Energies* **2023**, *16*, 8024. [CrossRef]

energies

MDPI

Article

Inter-Turn Breakdown Fault Analysis and Winding Structure Optimisation of Winding of Dry-Type Transformers in Wind Farms

Ziheng Pu [1], Xinyun Yu [1,*], Yaoqiang Wang [2], Hao Liu [1] and Zihao Feng [1]

[1] College of Electrical Engineering & New Energy, China Three Gorges University, Yichang 443002, China
[2] Hainan Jinpan Electric Research Institute Co., Ltd., Wuhan 430074, China
* Correspondence: xinyun142924@163.com

Abstract: To address the problem of winding turn-to-turn breakdown faults in 35 kV dry-type transformers in wind farms under overvoltage conditions, this paper establishes a simulation model based on the structural dimensions and material parameters of the transformer windings. The winding distribution parameters were calculated using the finite element method. The transient processes inside the high-voltage coil were calculated by constructing a multi-conductor transmission line model (MTL) that took into account the influence of the secondary winding. The voltage distribution of the winding was analysed for both lightning shock and extra-fast transient overvoltage conditions. The simulation results show that the maximum overvoltage between turns of the transformer winding under lightning shock is 5.282 kV; the maximum overvoltage between turns of the winding under very fast transient overvoltage is 11.6 kV, which occurs between the first 2–3 layers of the section, close to the insulation breakdown margin. On this basis, the transformer winding structure was optimised and the maximum inter-turn overvoltage after optimisation was 9.104 kV, reducing the likelihood of insulation breakdown by 24.1%. Finally, the accuracy of the winding structure optimisation simulation study was verified by testing the transformer's impulse voltage before and after optimisation, providing a reference for the stable operation of 35 kV dry-type transformers in wind farm practical applications.

Keywords: dry type transformer; ultra-fast transient overvoltage; distributed parameter; voltage distribution; winding structure optimisation

Citation: Pu, Z.; Yu, X.; Wang, Y.; Liu, H.; Feng, Z. Inter-Turn Breakdown Fault Analysis and Winding Structure Optimisation of Winding of Dry-Type Transformers in Wind Farms. *Energies* **2023**, *16*, 2012. https://doi.org/10.3390/en16042012

Academic Editor: Tomasz Norbert Kołtunowicz

Received: 21 December 2022
Revised: 10 January 2023
Accepted: 10 February 2023
Published: 17 February 2023

1. Introduction

In recent years, with the large number of wind farms in operation, 35 kV dry-type transformers have been used in large numbers in the field. Due to the differences between the electrical network structure of wind farms and the general distribution system, some of the conventional 35 kV dry-type transformers generated for commissioning in recent years have experienced winding turn-to-turn breakdown faults during operation due to the extra-fast transient overvoltages generated by the frequent operation of vacuum circuit breakers [1–4]. Although 35 kV dry-type transformers are less expensive and have a smaller failure rate, the high cost of shutdown and transport to replace them after a breakdown fault seriously affects the benefits. Therefore, it is important to improve and optimise the transformer winding structure to minimise the probability of this type of fault, in combination with the calculation and analysis of the fault location and winding inter-turn voltage of dry-type transformers, for the safe and stable operation of wind farms and to reduce maintenance costs.

Considering that inter-turn voltage measurement is more difficult, most of the existing studies on the distribution of winding inter-turn transient overvoltage in lightning overvoltage and extra-fast transient overvoltage cases are based on simulation calculations. In [5], a lightning surge simulation was carried out by establishing an equivalent circuit parameter model for the proportional model of the excitation winding of an autotransformer, but the inter-turn voltage percentage was not calculated and could not provide a reference for transformer engineering estimation. In [6], an equivalent circuit model was developed for every two turns along the radial direction, and the interlayer voltage distribution under lightning surge overvoltage was solved, but the safety margin of the interlayer insulation was not analysed and could not be used as a reference for engineering applications. The literature [7] proposes a hybrid circuit model for the need of extra-fast transient overvoltage simulation of EHV power transformers, using an MTL circuit to simulate the head of the winding and a collective parametric RLC circuit to simulate the rest of the winding. This approach can increase the size of the coils that can be modelled, but the difference in the number of equivalent circuits between the winding head and the rest of the winding is large and may result in coefficient matrix singularities. A centralised parametric model of a transformer winding consisting of a new Π-type equivalent circuit per turn was developed in the literature [8], which nearly doubles the applicable frequency for inter-turn voltage calculations of transformer windings of different voltage levels under extra-fast transient overvoltages, but ignores the coupling between adjacent coils. Most of these studies are based on wave process calculations with a single winding equivalent circuit model. As the accuracy of the parameters is related to the actual coil coupling of the coils, further studies are needed to analyse the transformer coil transient voltage distribution based on an accurate transformer coil high-frequency model.

In this paper, for the same type of 35 kV dry-type transformer with several faults during operation in different wind farms, an MTL model is first established considering the interaction of the primary and secondary windings. A two-dimensional axisymmetric model is also established based on the actual transformer winding dimensions. The finite element method is applied to calculate the distribution parameters of the transformer windings at high frequencies. The transformer winding is then subjected to overvoltage calculations based on the MTL model taking into account the influence of the secondary winding. A standard lightning surge overvoltage is loaded and an exceptionally fast transient overvoltage waveform generated by the opening of a vacuum circuit breaker in a wind farm is considered. The inter-pie and inter-turn voltage distributions in the transformer winding are analysed and calculated for the two cases. The waveforms of the winding voltage as a function of time, the maximum overvoltage value between pies and the maximum overvoltage value between turns are obtained. The winding structure is optimised in combination with the simulation analysis to reduce the maximum turn-to-turn voltage. Finally, the accuracy of the winding structure optimisation simulation study is verified by testing the transformer surge voltage before and after optimisation.

2. Simulation Modelling of a Transformer Considering the Primary and Secondary Windings

2.1. MTL Model Considering Primary and Secondary Windings

The MTL model can consider the distribution parameters of all windings and simulate the winding voltage distribution under transient overvoltage more accurately. Therefore, the MTL model is considered in this paper to analyse the influence of the secondary winding on the voltage distribution of the primary winding under transient voltages in wind farms. For the special structure of the transformer coil, the winding is split along the axial plane, ignoring the effect of bending of the turns, and the turns are spread into straight lines, with each turn becoming a "transmission line" [9]. These multiple "transmission lines" are connected end to end according to the coil winding relationship; the entire transformer winding constitutes a multi-conductor transmission line model, as shown in Figure 1. The boundary conditions for the wire turns are that the voltage and current at the end of the i-th wire are equal to the voltage and current at the beginning of the i-th + 1st wire ($I = 1, 2, \ldots, n-1$); the first wire is connected to a voltage source at the beginning and the n-th wire is grounded at the end, then the corresponding multi-conductor transmission line model can be regarded as a $2n$-port network with a total of $2n$ boundary conditions. The telegraph equation for the MTL in its time domain can be expressed as

$$\frac{\partial[U(x,t)]}{\partial z} = -\left([R][I(x,t)] + [L]\frac{\partial[I(x,t)]}{\partial t}\right) \quad (1)$$

$$\frac{\partial[I(x,t)]}{\partial z} = -\left([G][I(x,t)] + [C]\frac{\partial[U(x,t)]}{\partial t}\right) \quad (2)$$

where $U(z, t)$, $I(z, t)$ are the column vectors of voltage and current, respectively, and $[R]$, $[L]$, $[G]$, $[C]$ are the matrices of resistance, inductance, conductance and capacitance parameters of the multi-conductor system, respectively.

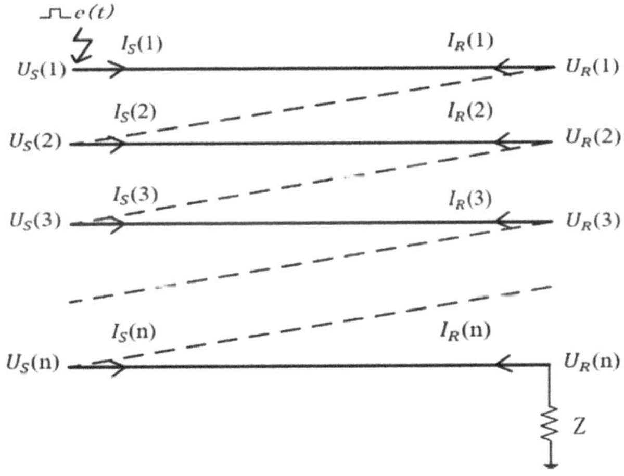

Figure 1. MTL model of transformer.

When considering the effect of the secondary winding, the secondary winding is grounded at both ends and the boundary conditions change accordingly; the specific derivation process can be found in [10].

2.2. Distribution Parameter Model for Transformers

The total number of turns in the high-voltage winding of a 35 kV dry-type transformer is 959, and the winding is divided into 14 pies; the medium voltage winding is 96 turns and the winding is divided into 5 pies. The resulting 2D axisymmetric simulation model of the transformer is shown in Figure 2.

In the internal structure of the 35 kV dry-type transformer, the core is made of silicon steel. The low-voltage and medium-voltage coils are foil-type and the high-voltage coils are epoxy-cylinder-type. The transformer model material parameters are shown in Table 1.

Table 1. Material parameters.

	Electrical Conductivity (20 °C) (S/m)	Relative Permeability	Relative Dielectric Constant	Materials
Conductors	3.45×10^7	1	1	Aluminium
Insulation	0	1	3.6	Epoxy resin
Iron cores	2×10^6	4×10^3	1	Silicon steel
Air	0	1	1	-

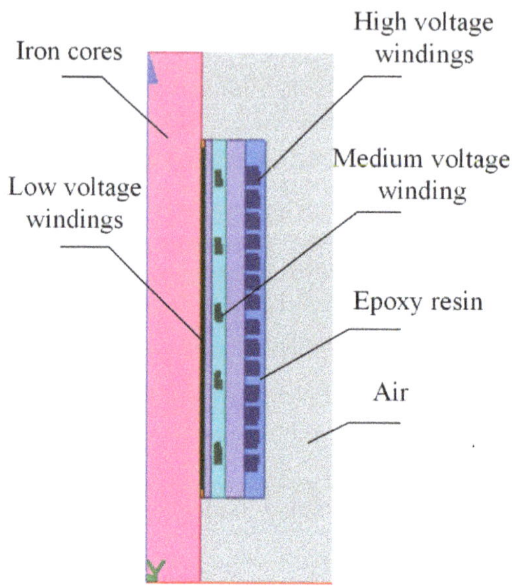

Figure 2. Two-dimensional axisymmetric simulation diagram.

2.3. Calculation of Distribution Parameters

2.3.1. Calculation of Capacitance Parameters

The formulae for calculating the capacitance parameters refer to the literature [11]. Taking the 1st pie coil as an example, the results of the calculation of its capacitance parameters are analysed as follows. The sequence structure of the 1st pie coil is shown in Figure 3, and the resulting self-capacitance of each turn is shown in Figure 4. The self-capacitance of each turn is highly dependent on the position of the turns. The outermost and innermost coils are further away from ground on the outside and have a smaller self-capacitance. The last two turns of the 2nd layer coil (wire turns 11 and 12) are the same as the 1st layer, with the outer side further away from ground and with a smaller self-capacitance. The middle part of the coil is surrounded by other coils and has a higher self-capacitance. The mutual capacitance parameters of coil 1 and the other coils are shown in Figure 5. As can be seen from Figure 5, the mutual capacitance of the coils adjacent to coil 1 is higher. Coil 1 and coils 2 and 10 are adjacent to each other, where the area directly opposite the coils in the vertical direction is smaller than the area directly opposite the coils in the horizontal direction, thus the inter-turn capacitance of coils 1 and 2 is much smaller than the inter-turn capacitance of coils 1 and 10. The inter-turn capacitance of coils not adjacent to coil 1 is approximately zero.

2.3.2. Calculation of Inductance and Resistance Parameters

The inductance and resistance parameters of the winding are related to the frequency value. With surge voltages, the frequency components are very high and can reach MHz and above. In this case, a strong skin effect is generated in the core, making it difficult for the magnetic flux to pass through, and the corresponding inductance values are reduced. Therefore, an analysis of the time-harmonic magnetic field is required. In this paper, frequencies of MHz and above are taken and the inductance–resistance matrix is calculated and analysed using an infinitely long core model with no iron yoke and constant permeability.

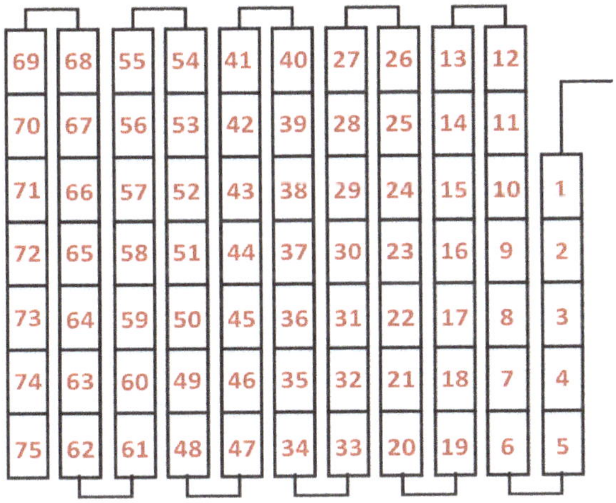

Figure 3. The sequential structure of the first cake.

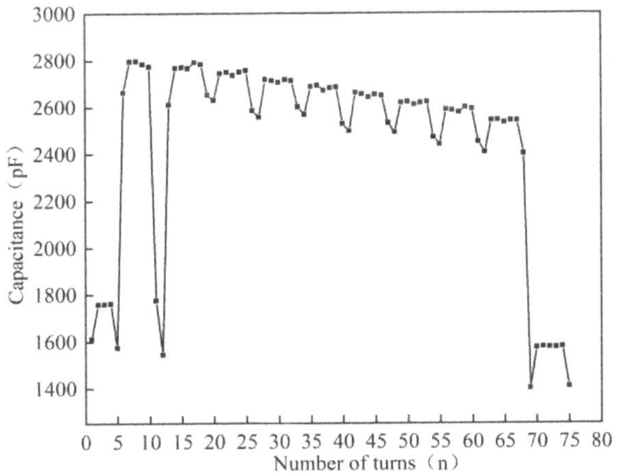

Figure 4. The self-capacitance of each turn in cake 1.

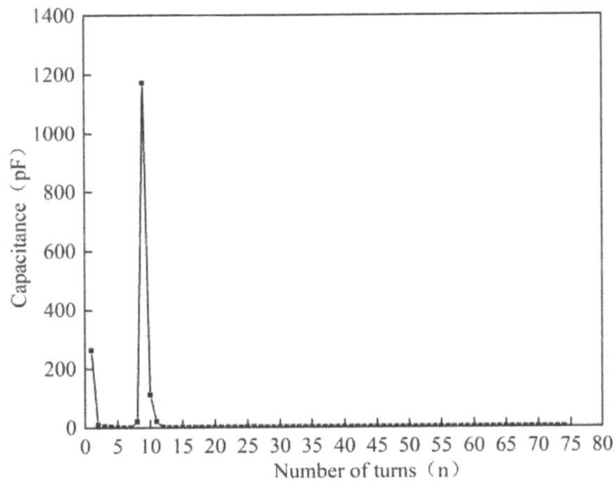

Figure 5. The mutual capacitance between turn 1 and each turn in cake 1.

The self-inductance of the 1st cake coil, for example, is shown in Figure 6. As can be seen from Figure 6, the self-inductance varies from turn to turn in each layer due to the skinning effect at high frequencies. In this case, the self-inductance of the coils in the same layer decreases gradually from the end to the middle. The mutual inductance parameters of coil 1 and the other coils are shown in Figure 7. It can be seen that the difference between the mutual inductance parameters of the adjacent coils and their self-inductance parameters is very small. The mutual inductance tends to decrease with the coils further apart, but still by the same order of magnitude.

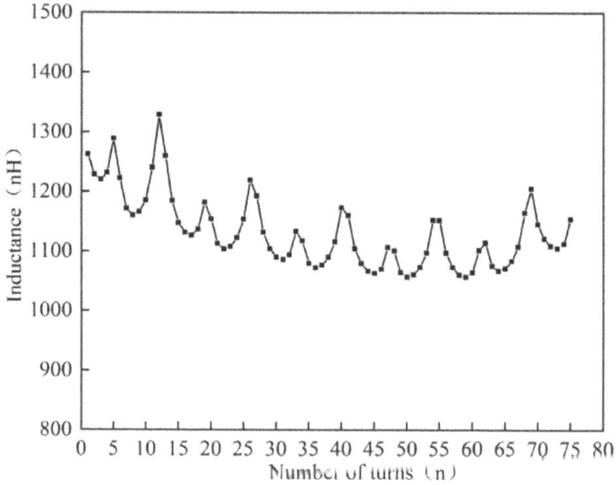

Figure 6. Self-inductance of each turn in cake 1.

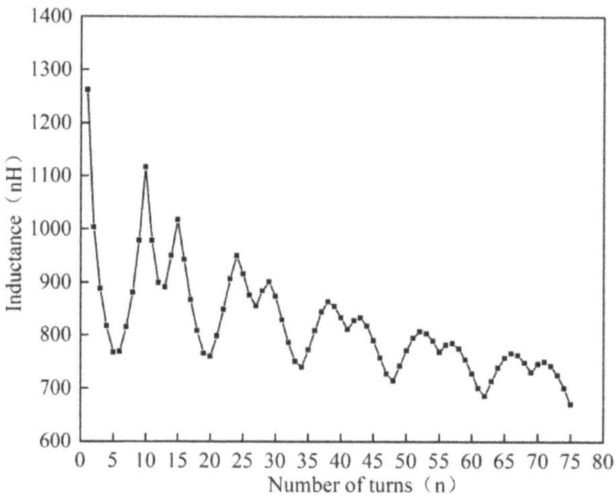

Figure 7. The mutual inductance of coil 1 and other coils in cake 1.

The resistance parameters of the first cake winding are shown in Figure 8. As can be seen from Figure 8, the resistance per turn varies due to skinning effects and differences in the radii of the different winding layers, but the values are small and have little effect on the distribution of the inrush voltage.

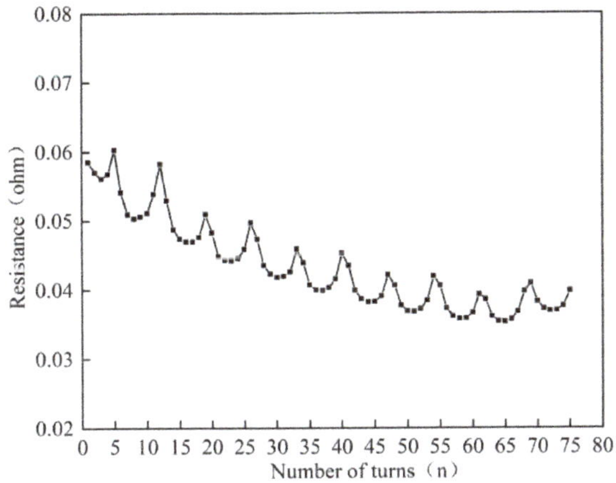

Figure 8. Self-resistance of each turn in cake 1.

3. Transient Voltage Distribution of the Transformer Considering the Effect of the Secondary Side

3.1. Windings Voltage Distribution under Lightning Overvoltage

In order to study the winding inter-turn voltage distribution under lightning shock, a standard lightning shock voltage with a transient overvoltage amplitude of 111 kV and a wavefront time/wave tail time of 1.2/50 μs was loaded at the transformer inlet. The resulting winding inter-turn voltage distribution is shown in Figure 9. As can be seen from Figure 9, the maximum voltage difference between adjacent line cakes is between line cake 1 and line cake 2, with an amplitude of approximately 20.43 kV; the minimum voltage difference is between line cake 9 and line cake 10, with an amplitude of approximately 6.129 kV.

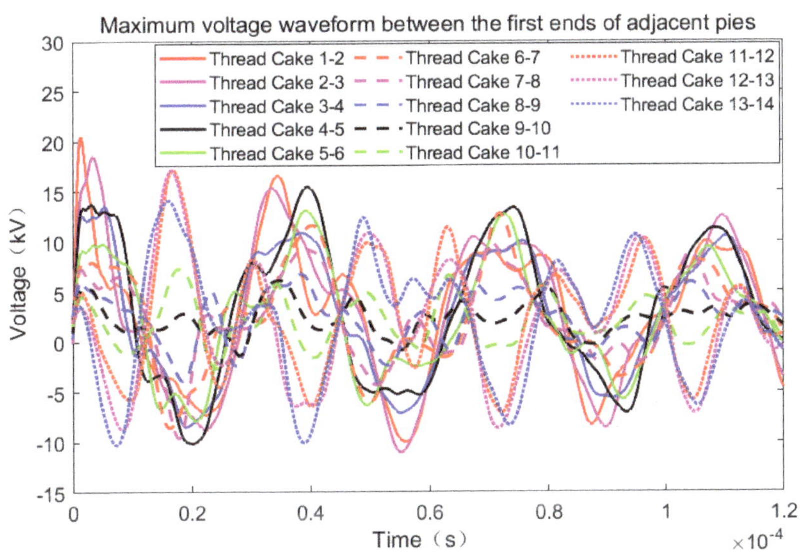

Figure 9. The voltage difference between winding cakes.

Inter-turn voltage difference in the vertical direction was small; considering the effect of capacitance to ground, the maximum inter-turn voltage difference was at the farthest interval between adjacent layers (e.g., D2 and E3), the node number marker diagram is shown in Figure 10. The voltage difference between the layers is shown in Figures 11 and 12. The voltage peak between turns D10-E11 in the horizontal direction of line cake 1 is the largest between the layers, reaching 5.282 kV; the voltage peak between turns D2-E3 is the smallest between the layers of the cake, approximately 4.431 kV. Peak voltage between turns D13-E14 in the horizontal direction of line cake 2 is the maximum between the layers of this cake and amounts to 4.339 kV; peak voltage between turns D14-E15 is the minimum between the layers of this cake and amounts to approximately 4.281 kV.

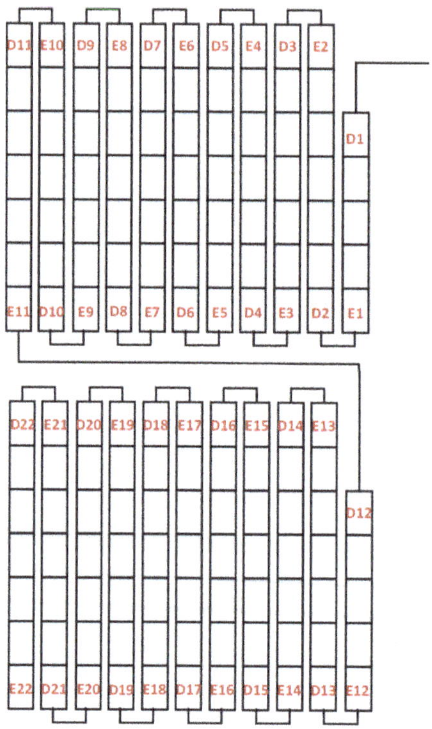

Figure 10. Label diagram of node numbers.

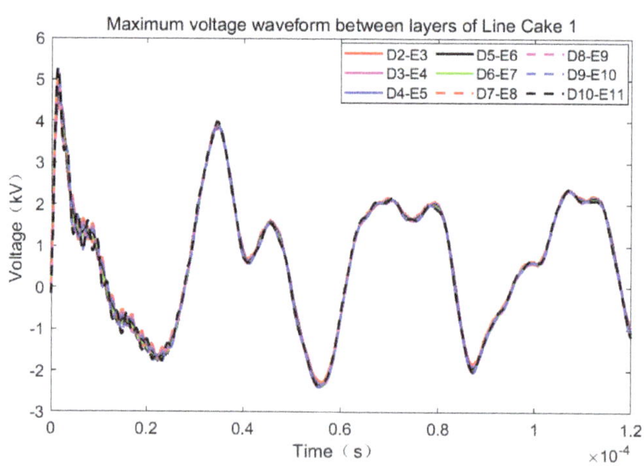

Figure 11. Maximum voltage difference between layers of cake 1.

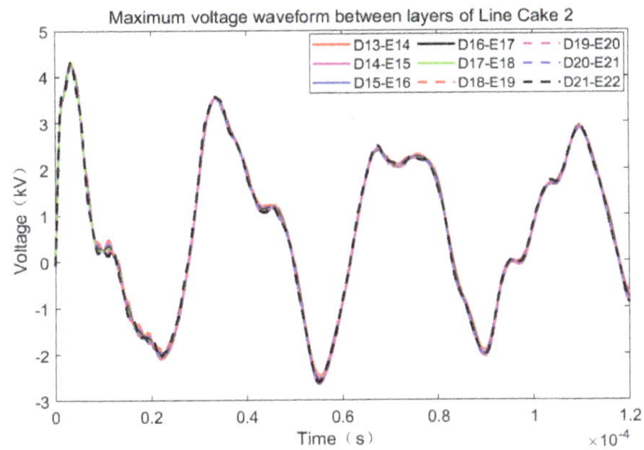

Figure 12. Maximum voltage difference between layers of cake 2.

Considering the effects of electrothermal ageing in the long-term operation of dry-type transformers, the breakdown field strength of the epoxy resin drops to about 20 kV/mm [12]. The insulation thickness between turns is 0.64 mm in both horizontal and vertical directions, and the required voltage for turn-to-turn breakdown is approximately 12.8 kV. The maximum overvoltage between layers 1 and 2 of the line cake is 5.282 kV when a lightning surge overvoltage occurs, which is much less than 12.8 kV, making it more difficult for insulation breakdown to occur between turns of the transformer under a lightning surge.

3.2. Windings Voltage Distribution under Extra-Fast Transient Overvoltage

In this paper, the overvoltage simulation results of the wind farm circuit breaker breakdown model at the transformer line ends from the literature [13] were used and the section of the waveform with the greatest wavefront steepness was selected as the excitation for the transformer windings. This waveform has a voltage amplitude of 145 kV, a rising edge of 0.2 μs and a wavefront steepness of 700 kV/μs, with frequencies mainly concentrated in the range 1–20 MHz.

The transient voltage distribution at the first end of each bobbin is shown in Figure 13. The maximum values of the first four cakes were 114 kV, 89 kV, 68.3 kV and 51.5 kV. The voltage differences between the winding cakes are shown in Figure 14. The maximum voltage difference between adjacent cakes was between cake 1 and cake 2, with an amplitude of approximately 32.06 kV; the minimum voltage difference was between cake 11 and cake 12, with an amplitude of approximately 2.463 kV. As can be seen from Figures 13 and 14, the potential distribution between the wire cakes in the area of the high-voltage winding near the first end was extremely uneven, with the maximum value occurring between wire cake 1 and wire cake 2. Therefore, the interlayer potential difference between the 1st and 2nd cakes of the high-voltage winding was analysed to determine whether the maximum interlayer overvoltage would cause damage or failure of the insulation material, thus providing further theoretical support for optimising the coil insulation design.

The analysis of the voltage difference waveforms between the layers is shown in Figures 15 and 16. The voltage peak between turns D2-E3 in the horizontal direction of line cake 1 was the largest between layers, reaching 11.6 kV; the voltage peak between turns D10-E11 was the smallest between layers of the cake, about 3.983 kV. The voltage peak between turns D13-E14 in the horizontal direction of line cake 2 was the largest between layers of the cake, reaching 7.512 kV; the voltage peak between turns D19-E20 was the smallest between layers of the cake, about 3.432 kV. The maximum inter-turn voltage difference is shown in Figure 17 (example of the first 2 pies). It can be seen that the maximum voltage difference was between C5 and C6, with an amplitude of approximately 6.653 kV.

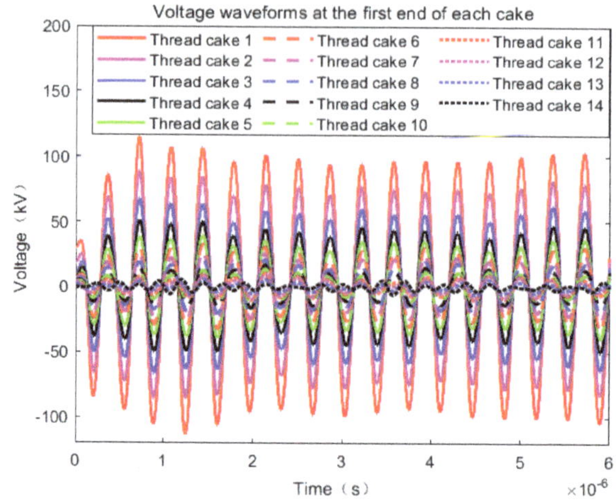

Figure 13. The voltage waveforms at the first end of each cake.

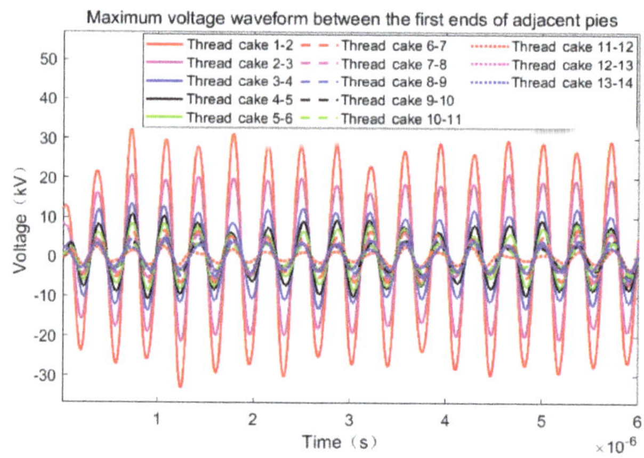

Figure 14. The voltage difference between winding cakes.

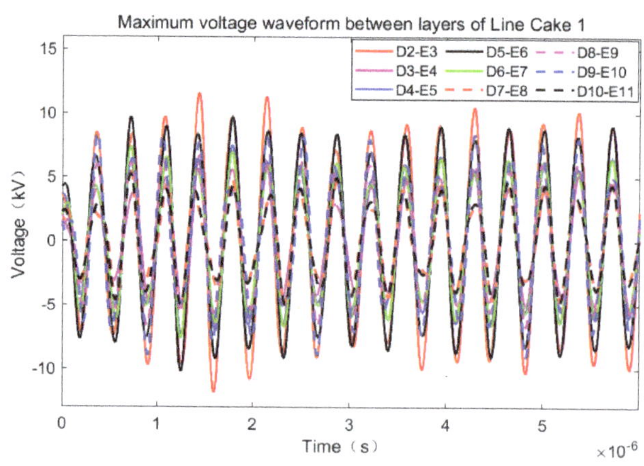

Figure 15. Maximum voltage difference between layers of cake 1.

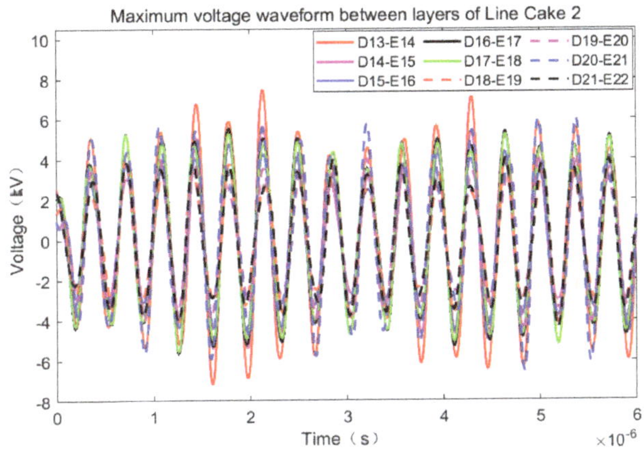

Figure 16. Maximum voltage difference between layers of cake 2.

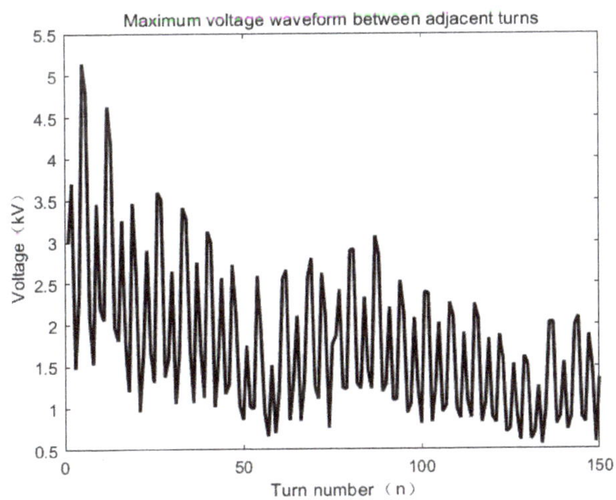

Figure 17. The maximum voltage difference between turns of the first two cakes.

Taking into account the effects of electrothermal ageing under long-term operation of dry-type transformers, the voltage required for turn-to-turn insulation breakdown of the winding is approximately 12.8 kV. Since the generation of operational transient overvoltages of very high amplitude and frequency is limited, the maximum overvoltage between layers 1 and 2 of the line cake when wind farm extra-fast transient overvoltages occur is 11.6 kV, which is close to 12.8 kV. Thus, in this case, turn-to-turn insulation breakdown may occur, but the number of occurrences is limited. By optimising the winding structure of the transformer, the insulation margins can be made larger, thus reducing the likelihood of breakdown.

4. Transformer Winding Structure Optimisation Analysis

4.1. Windings Voltage Distribution after Structural Optimisation

The above analysis shows that the maximum interlayer voltage of 11.6 kV is close to the horizontal interturn breakdown voltage of 12.8 kV for a transformer winding with 75 turns in the first pie, and therefore measures need to be taken to reduce the winding overvoltage. To address the shortcomings of various overvoltage protection methods [14,15] and transformer insulation designs [16,17] and consider the fact that in actual transformer production, either the installation of capacitor rings and capacitor turns or the insertion of shielding wires [18,19] increases the size and production costs of the transformer, this paper uses a change in the number of turns of the first cake for winding optimisation. Due to the uneven voltage distribution between the line cakes in the area of the high-voltage winding near the first end, the maximum interlayer voltage difference is influenced by the capacitance to ground. Consider reducing the number of turns in each layer of the transformer winding in the first end area to improve the voltage distribution pattern.

In this section, the first and second pancake coils are changed to 43 and 65 turns, respectively, and the voltage distribution between adjacent pancakes and between layers is analysed for the optimised winding under extra-fast transient overvoltage. The transformer high-voltage winding is divided into 14 pancakes and the voltage waveforms at the first end of each pancake are shown in Figure 18. The maximum values of the first four cakes were 111.4 kV, 98.48 kV, 88.67 kV and 70.42 kV, respectively. The voltage difference between the winding pancakes is shown in Figure 19. The maximum voltage difference between adjacent cakes was between cake 1 and cake 2, with an amplitude of approximately 28.3 kV; the minimum voltage difference was between cake 10 and cake 11, with an amplitude of approximately 2.293 kV.

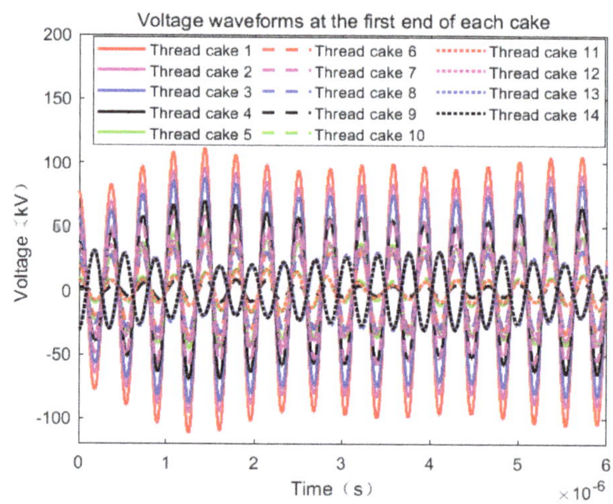

Figure 18. The voltage waveforms at the head end of each cake.

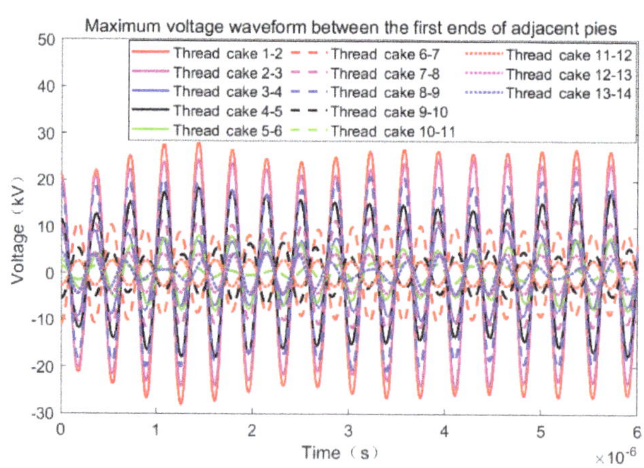

Figure 19. The voltage difference between winding cakes.

The node number marker diagram is shown in Figure 20. The analysis of the voltage difference waveform between each layer is shown in Figures 21 and 22. As the number of turns per winding layer decreases, the ground capacitance also decreases and the winding voltage distribution is relatively more uniform. The voltage peak between turns D2-E3 in the horizontal direction of wire cake 1 is the largest between layers, reaching 9.104 kV; the voltage peak between turns D8-E9 is the smallest between layers of the cake, at approximately 1.352 kV. The voltage peak between turns D20-E21 in the horizontal direction of wire cake 2 is the largest between layers of the cake, reaching 5.379 kV; the voltage peak between turns D18-E19 is the smallest between layers of the cake, at approximately 1.043 kV. The maximum inter-turn voltage difference is shown in Figure 23 (for the first 3 cakes). It can be seen that the maximum voltage difference is between C3 and C4, with an amplitude of approximately 3.389 kV.

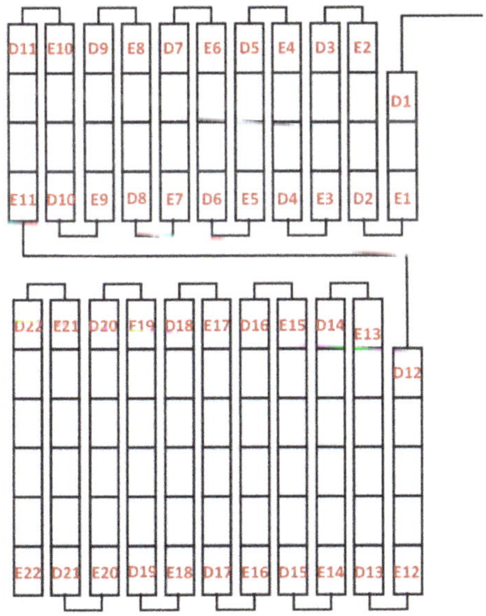

Figure 20. Label diagram of node numbers.

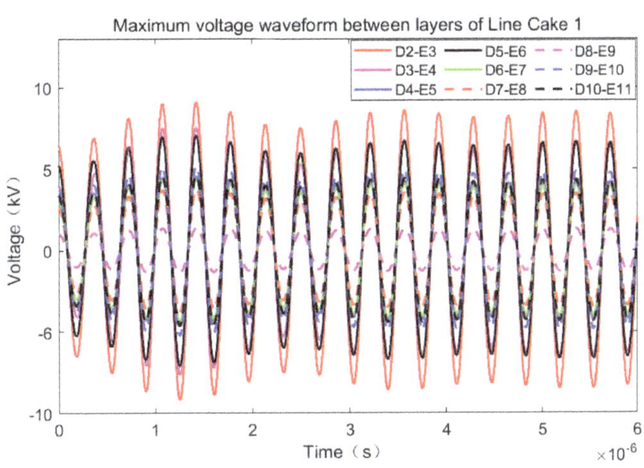

Figure 21. Maximum voltage difference between layers of cake 1.

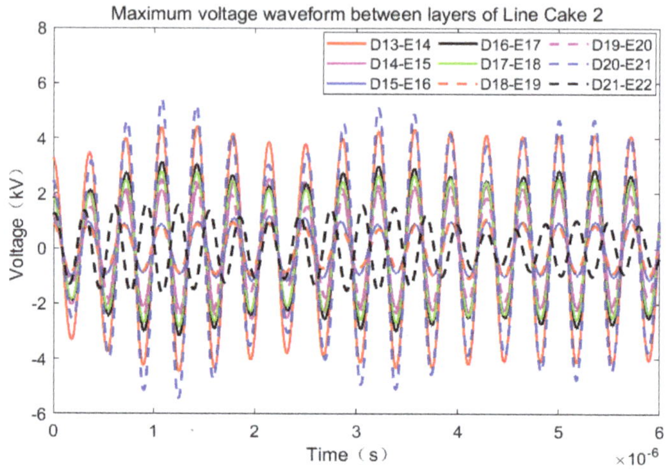

Figure 22. Maximum voltage difference between layers of cake 2.

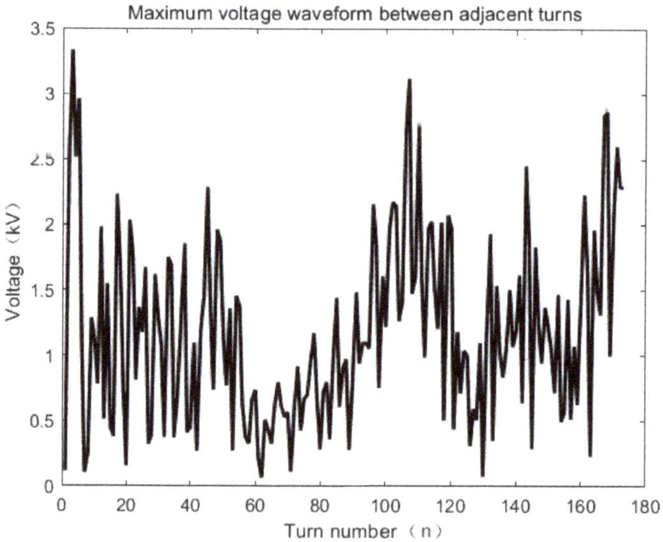

Figure 23. The maximum voltage difference between turns of the first three cakes.

A comparison of the maximum voltage difference and position before and after winding structure optimisation is shown in Table 2. When the winding structure was changed to 43 turns in the first pie, the maximum overvoltage between the first and second pie was reduced by 11.73%; the maximum voltage difference between turns was reduced by 49.1%; the maximum interlayer overvoltage was changed from 12 kV to 9.104 kV, the overvoltage amplitude was reduced by 24.1% and the insulation breakdown was also reduced.

4.2. Test Verification

Since the access of the measurement leads changes the winding structure and distribution parameters, affecting the original transformer structure, this makes the measurement results between each cake and turn not match the real situation. Therefore, this paper experimentally investigates the actual effect of the surge voltage on the transformer before and after optimisation and compares the test results with the simulation conclusions for verification. The test transformer parameter table is shown in Table 3. The actual layout diagram of the transformer withstand voltage test is shown in Figure 24. The schematic diagram of the actual winding of the transformer coil is shown in Figure 25. Using a surge voltage generator to simulate a very fast transient overvoltage with a voltage amplitude of 145 kV, a wavefront time of 0.2 μs and a wavefront steepness of 700 kV/μs, three withstand voltage tests were conducted on each of the three transformers before optimisation of the winding structure. The three transformers passed the first two tests, but under the action of the third shock, two of the transformers suffered breakdown near the first end. The breakdown diagram is shown in Figure 26. The three transformer models with optimised high-voltage winding structure remained intact under the same overvoltage five times each. The test results verify the accuracy of the winding structure optimisation simulation study and provide a practical basis for the safe and stable operation of the optimised transformers.

Table 2. Parameter table for test transformers.

Parameter	Data
Frequency (Hz)	50
HV rated power (kVA) (AF)	4778
HV rated voltage (V)	33,000
HV rated current (A)	83
HV BIL (KV)	170
HV AC (KV) (1500 m)	70
MV rated power (kVA) (AF)	4157
MV rated voltage (V)	6000
MV rated current (A)	400
MV BIL (KV)	60
MV AC (KV) (1500m)	20
LV rated power (kVA) (AF)	838
LV rated voltage (V)	690
LV rated current (A)	701
LV BIL (KV)	-
LV AC (KV) (1500 m)	3
HV-MV impedance (based on 4779 kVA) (%)	8.5 ± 10
HV-LV impedance (based on 4779 kVA) (%)	14.6 ± 10
MV-LV impedance (based on 4779 kVA) (%)	3~6
Sound power level	<95 dBA
Efficiency (PEI)(%)	≥99.354
Insulation class/temperature rise	F/75K

Table 3. Comparison of maximum voltage difference and position before and after optimisation of winding structure.

Winding Construction	When the First Pie Is 75 Turns	When the First Pie Is 43 Turns
Inter-segment (max. voltage)	32.06 kV	28.3 kV
Position	Between the first and second pie	Between the first and second pie
Interlayer (max. voltage)	11.6 kV	9.104 kV
Position	Between D2 and E3	Between D2 and E3
Inter-turn (max. voltage)	6.653 kV	3.389 kV
Position	Between turns 5 and 6	Between turns 3 and 4

Figure 24. Actual arrangement of transformer voltage test.

Figure 25. Schematic diagram of actual winding of transformer coils.

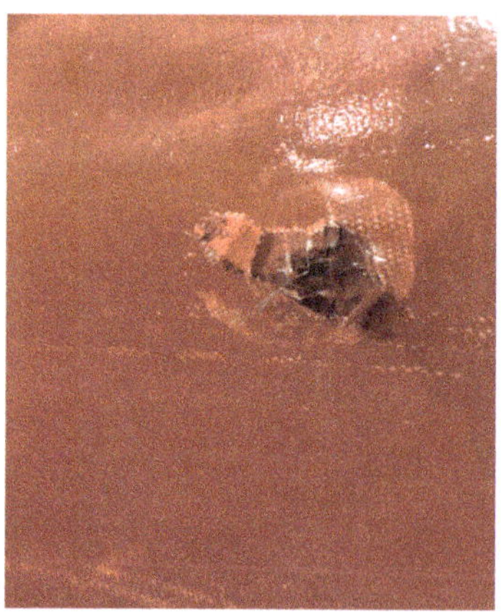

Figure 26. Winding breakdown after voltage test of transformer with unoptimised winding structure.

5. Conclusions

The analysis of the inter-turn voltage distribution and winding structure optimisation of transformer windings in wind farms under the action of overvoltage leads to the conclusion that:

(1) The maximum intercake voltage under lightning strikes is located between line cakes 1 and 2 with an amplitude of 20.43 kV; the maximum interlayer voltage is located at D10-E11 of line cake 1 with an amplitude of 5.282 kV, and the insulation margin in this case reaches 12.8 kV and is not prone to breakdown.
(2) The maximum inter-pancake voltage under extra-fast transient overvoltage is located between pancake 1 and 2 with an amplitude of 32.06 kV; the maximum inter-layer voltage is located at D2-E3 of pancake 1 with an amplitude of 11.6 kV, which meets the insulation design requirements. However, as the insulation margin of the transformer decreases under long-term operation, the maximum interlayer voltage is close to the insulation breakdown voltage. The winding structure therefore needs to be optimised to make the insulation margin larger.
(3) By optimising the winding structure, the maximum interlayer overvoltage was changed from 11.6 kV to 9.104 kV and its insulation breakdown was reduced by 24.1%.
(4) The transformer was subjected to a shock voltage test, which resulted in insulation breakdown at the first end of the high-voltage winding before structural optimisation, while the high-voltage winding after structural optimisation was left intact. This experiment demonstrates the accuracy of the simulation results.

Author Contributions: This article is the result of the collaboration of all co-authors. Writing—review and editing, Z.P.; writing—original draft preparation, X.Y.; resources, Y.W.; software, H.L.; validation, Z.F. All authors have read and agreed to the published version of the manuscript.

Funding: This research received no external funding.

Institutional Review Board Statement: Not applicable

Informed Consent Statement: Not applicable.

Data Availability Statement: Not applicable.

Conflicts of Interest: The authors declare no conflict of interest.

References

1. Theocharis, A.; Popov, M.; Seibold, R.; Voss, S.; Eiselt, M. Analysis of Switching Effects of Vacuum Circuit Breaker on Dry-Type Foil-Winding Transformers Validated by Experiments. *IEEE Trans. Power Deliv.* **2014**, *30*, 351–359. [CrossRef]
2. Zhou, Q.; Cheng, Y.; Bian, X.; Liu, F.; Zhao, Y. Analysis of Restrike Overvoltage of Circuit Breakers in Offshore Wind Farms. *IEEE Trans. Appl. Supercond.* **2016**, *26*, 1–5. [CrossRef]
3. Sarajčev, P.; Goić, R. A Review of Current Issues in State-of-Art of Wind Farm Overvoltage Protection. *Energies* **2011**, *4*, 644–668. [CrossRef]
4. Nie, H.; Liu, X.; Wang, Y.; Yao, Y.; Gu, Z.; Zhang, C. Breaking Overvoltage of Dry-Type Air-Core Shunt Reactors and its Cumulative Effect on the Interturn Insulation. *IEEE Access* **2019**, *7*, 55707–55720. [CrossRef]
5. Zang, Y. Research on Wave Process and Winding Model of Extra-High Voltage Power Transformer. Master's Thesis, Shandong University, Jinan, China, 2017.
6. Chen, L. Hollow Reactor Wave Process Calculation and Result Analysis. Master's Thesis, Harbin Institute of Technology, Harbin, China, 2020.
7. Yang, Y.; Wang, Z. Frequency domain segmentation modeling of large power transformer coils for ultra-fast transient simulation. *Chin. J. Electr. Eng.* **2010**, *30*, 66–71.
8. Chen, Z.; Zhang, K.; Ou, F. Modeling of transformer winding concentration parameters under the action of VFTO. *Transformer* **2013**, *50*, 10–15.
9. Chen, M. Optimum Design of Multi-band Transformer with Multi-section for Two Arbitrary Complex Frequency-dependent Impedances. *Chin. J. Electron.* **2012**, *21*, 160–164.
10. Xie, Q.; Li, J.; Wang, T. Study on the wave process of 800 kV converter transformer scaling model. *Transformer* **2016**, *53*, 12–16.
11. Zupan, T.; Trkulja, B.; Obrist, R.; Franz, T.; Cranganu-Cretu, B.; Smajic, J. Transformer Windings' RLC Parameters Calculation and Lightning Impulse Voltage Distribution Simulation. *IEEE Trans. Magn.* **2015**, *52*, 1–4. [CrossRef]
12. Mi, Y.; Liu, L.; Deng, S.; Gui, L.; Ouyang, W. Electrothermal aging characteristics of epoxy resin under bipolar exponential decay pulse voltage and its insulation life evaluation based on Cole-Cole model. *IEEE Trans. Dielectr. Electr. Insul.* **2019**, *26*, 784–791. [CrossRef]

13. Xin, Y.; Zhao, B.; Liang, Q.; Zhou, J.; Qian, T.; Yu, Z.; Tang, W. Development of Improved Suppression Measures Against Reignition Overvoltages Caused by Vacuum Circuit Breakers in Offshore Wind Farms. *IEEE Trans. Power Deliv.* **2021**, *37*, 517–527. [CrossRef]
14. Awad, E.A.; Badran, E.A.; Youssef, F.M. Mitigation of temporary overvoltages in weak grids connected to DFIG-based wind farms. *J. Electr. Syst.* **2014**, *10*, 431–444.
15. Smugala, D.; Piasecki, W.; Ostrogorska, M.; Florkowski, M.; Fulczyk, M.; Granhaug, O. Wind Turbine Transformers Protection Method Against High-Frequency Transients. *IEEE Trans. Power Deliv.* **2015**, *30*, 853–860. [CrossRef]
16. Gabrić, P.; Mikulecky, A.; Ilić, D. A concept for experimental testing of oil-barrier insulation system. *Teh. Vjesn.-Tech. Gaz.* **2017**, *24*, 355–362. [CrossRef]
17. Wang, W.; Liu, Y.; He, J.; Ma, D.; Hu, L.; Yu, S.; Li, S.; Liu, J. An Improved Design Procedure for a 10 kHz, 10 kW Medium-Frequency Transformer Considering Insulation Breakdown Strength and Structure Optimization. *IEEE J. Emerg. Sel. Top. Power Electron.* **2022**, *10*, 3525–3540. [CrossRef]
18. Li, P. Study on the Distribution of Lightning Shock and VFTO on Transformer Windings. Master's Thesis, Taiyuan University of Technology, Taiyuan, China, 2017.
19. Wu, M. Power Transformer Winding Wave Process and Optimization of Winding Longitudinal Insulation Structure. Master's Thesis, Southeast University, Nanjing, China, 2016.

energies

MDPI

Article

Simulation and Protection of Reignition Overvoltage in Wind Farm Considering Microscopic Dielectric Recovery Process of Vacuum Circuit Breaker

Ziheng Pu [1], Hao Liu [1,*], Yaoqiang Wang [2], Xinyun Yu [1] and Tian Wu [1]

1 College of Electrical Engineering and New Energy, China Three Gorges University, Yichang 443002, China
2 Hainan Jinpan Electric Research Institute Co., Ltd., Wuhan 430074, China
* Correspondence: only6hao@163.com

Abstract: The high amplitude and steep overvoltage generated by the breaking of the vacuum circuit breaker in the wind farm damages the inter-turn insulation of the transformer. There is a certain difference between the simulation results of the traditional reignition model and the measured overvoltage. It is necessary to improve the simulation model to simulate the overvoltage condition of the transformer more accurately and then select appropriate overvoltage protection measures. In this paper, based on the physical process of dielectric recovery during the opening process of the vacuum circuit breaker, a model of dielectric strength recovery is built to simulate the arc reignition of the vacuum circuit breaker. The model was applied to compare the overvoltage protection effects of RC snubbers, surge arresters, and choke coils. The simulation results show that the overvoltage amplitude and reignition times calculated by the model proposed in this paper are closer to the measured values. Compared with the traditional linear curve reignition model, the accuracy was increased by 24% and 51.2%, respectively. The parameter value of RC snubbers, the connection mode of surge arresters, and the combination mode of choke coil have an influence on overvoltage suppression. Finally, a suitable suppression scheme is proposed by installing a combined arrester on the high-voltage side of the transformer and connecting a choke coil in series, which can limit the phase-to-ground voltage and the phase-to-phase voltage to 2.43 p.u and 3.24 p.u, respectively, and reduce the steepness from 157.2 kV/μs to 22.3 kV/μs.

Keywords: wind farm; vacuum circuit breaker; microscopic dielectric; reignition; overvoltage protection

Citation: Pu, Z.; Liu, H.; Wang, Y.; Yu, X.; Wu, T. Simulation and Protection of Reignition Overvoltage in Wind Farm Considering Microscopic Dielectric Recovery Process of Vacuum Circuit Breaker. *Energies* **2023**, *16*, 2070. https://doi.org/10.3390/en16042070

Academic Editor: Tomasz Norbert Kołtunowicz

Received: 25 January 2023
Revised: 9 February 2023
Accepted: 11 February 2023
Published: 20 February 2023

1. Introduction

Multiple reignitions of the vacuum circuit breaker opening in wind farms generate transient overvoltages of very high amplitude and steepness, which cause shocks to the equipment and threaten the safe and stable operation of the power system [1,2]. The overvoltage of wind farms simulated using the traditional vacuum circuit breaker model is somewhat different from the test results, and the analysis of the operating state of the equipment and risk judgment is not accurate enough; there are still accidents such as turn-to-turn and phase-to-phase insulation breakdown of the step-up transformer at the end of the wind turbine [3,4]. For this reason, it is necessary to improve the vacuum circuit breaker simulation model to more accurately simulate the overvoltage condition of transformers in wind farms and provide a reference for overvoltage protection.

Existing studies show that whether reignition occurs after a vacuum circuit breaker is opened depends on the dielectric recovery process after arc extinguishing [5]. The Helmer reignition model ignores the complex diffusion process of residual ions, electrons, and neutral particles after the arc; it considers that the gap breakdown depends only on the static withstand voltage strength, and the dielectric insulation strength increases linearly with time. It also considers the interception and the high-frequency arc extinguishing capability to realize the simulation of vacuum circuit breaker reignition [6]. Ref. [7] found that the linear fit of the dielectric recovery rate is too fast, and the simulation error is significantly based on the reignition test; further considering the mechanical properties of the contacts, the deduced dielectric insulation strength rises quadratically and nonlinearly with time. In Ref. [8], the dispersion of the gap breakdown voltage was found to be significant, so the random nature of the breakdown voltage was considered, the breakdown voltage was considered to occur in the region between two dielectric insulation strength curves, and the nonlinear contact motion was considered. The results show that considering the random breakdown voltage in a vacuum with the contact motion characteristics makes the overvoltage assessment more accurate compared to the conventional method. Ref. [9] considered that this dispersion is related to the diffusion of microscopic particles and the decrease in dielectric recovery strength caused by free metal particles. Based on multiple sets of measured breakdown voltage distribution statistics, the dielectric insulation strength curve was fitted in segments. The accuracy of the simulation was improved after considering the sudden drop in breakdown voltage caused by microscopic particles. Combined with the above studies, the simulation of wind farm reignition takes less account of the influence of the microscopic dielectric. This paper improves the dynamic insulation strength of the dielectric in the traditional reignition model from this aspect.

The residual ions, electrons, and neutral particles left in the gap after the arc extinguishment of the vacuum circuit breaker affect the breakdown voltage of the fracture gap. The recovery process of the microscopic dielectric has been studied extensively by scholars. In Ref. [10], the dielectric recovery process after arc extinguishment was divided into three stages: sheath growth, metal vapor attenuation, and static withstand voltage stage by vacuum switch dielectric recovery strength test. Most of the analytical models are used for their numerical simulation, such as the continuous transition model (CTM) describing the sheath growth, the Farrar formula for metal vapor attenuation, and the average vapor attenuation formula [11]. Based on the above theory, Ref. [12] established, for the first time, a reignition model for microscopic dielectric recovery of vacuum circuit breakers; it simulates and calculates the operating overvoltage when opening a no-load transformer and compares it with the average recovery rate, and the results show that the established three-stage recovery strength calculation model is closer to the actually measured recovery characteristics. In Ref. [13], a dielectric recovery microscopic reignition model was established to study the influence law of load circuit parameters on the reignition of small inductive currents in circuit breaker opening. Ref. [14] found that in high-frequency arc opening, the arc burning time is short, the rate of change in current is large, the cathode spot can not be fully diffused, the cathode spot distribution diameter is smaller than the diameter of the contact, the traditional analytical model is improved, the plasma is calculated, metal vapor density is increased by order of magnitude, the dielectric recovery rate becomes slower, and the probability of reignition increases. The above reignition model takes into account the physical process of microscopic dielectric recovery, which is closer to the actual vacuum circuit breaker breaking characteristics. At present, the model is less used in the study of wind farm overvoltage. This paper intends to use this model to analyze the transient overvoltage endured by wind farm transformers more accurately.

For switching overvoltages caused by vacuum circuit breakers, the main suppression measures are the installation of surge arresters and RC snubbers. Surge arresters can suppress overvoltage amplitude but not steepness; the use of RC snubbers leads to a low probability of reignition occurrence [15–17]. Recent studies have shown that the use of choke coils can reduce the steepness of overvoltages, with insignificant suppression of voltage magnitude, and the installation of both surge arresters and choke coils is recommended [18,19]; however, the wiring method and the number of surge arresters have not been studied and require further discussion. The high-voltage side of transformers in wind farms is usually triangularly wired, which is subjected to higher voltage stresses compared to star wiring; therefore, the effect of suppression measures on the phase-to-phase voltage also needs to be analyzed.

In this paper, firstly, an improved vacuum breaker reignition model is built in the electromagnetic transient software based on the physical process of post-arc dielectric recovery of a vacuum breaker. The model takes into account the three stages of post-arc dielectric recovery, the dynamic changes in cathode spot distribution diameter, and the actual breaking characteristics of the contacts. The proposed model is compared with the measured values in the related literature to verify the accuracy of the model, and based on the model, the reignition overvoltage of wind farms is simulated and analyzed. Finally, a systematic simulation analyses the impact of different parameter values; wiring methods; and combinations of RC snubbers, surge arresters, and choke coils on the overvoltage suppression effect. Additionally, the best solution for overvoltage suppression is proposed.

2. Vacuum Circuit Breaker Micro-Dielectric Recovery Reignition Model

A vacuum circuit breaker opening between the contacts produces an arc; the arc current occurs before reaching the natural over-zero point. If the current value is small, the arc becomes unstable and is forced to extinguish; it then enters the post-arc dielectric recovery process. The contact ends bear the transient recovery voltage (TRV), and dielectric recovery strength competes with each other. The three stages bearing the TRV size are different. Reignition and open failure may occur, and static can withstand the voltage stage if the TRV peak can be withstood; it then opens successfully. The vacuum circuit breaker opening flow chart is shown in Figure 1.

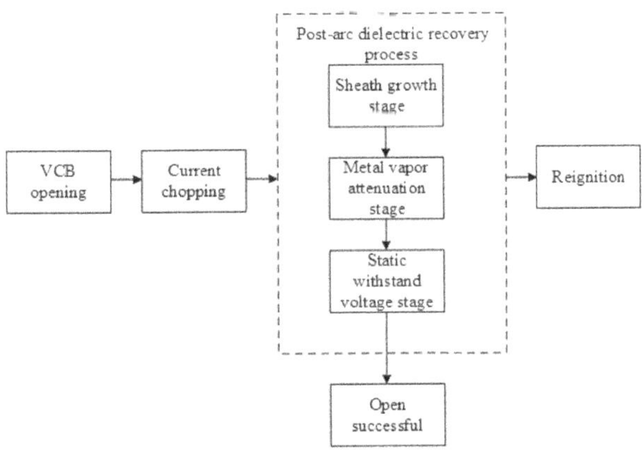

Figure 1. Vacuum circuit breaker opening flow chart.

It is crucial to simulate the dielectric recovery process accurately. In order to solve the problem in the traditional reignition model where the description of the dielectric recovery process after the arc is too simple, this paper considers the physical process of sheath development and metal vapor decay, establishes a mathematical model for each stage of dielectric recovery after the arc, and considers the vacuum breaker interception and high-frequency arc extinguishing characteristics to realize the simulation of arc reignition.

2.1. Mathematical Model of the Post-Arc Dielectric Recovery Process

2.1.1. Sheath Growth

The sheath growths were analyzed using the continuous transition model (CTM), which is modeled as follows [14]:

$$l^2 = \frac{4\varepsilon_0 u_0}{9eZN_i}\left[\left(1+\frac{u_{TRV}}{u_0}\right)^{\frac{3}{2}} + 3\frac{u_{TRV}}{u_0} - 1\right] \quad (1)$$

$$u_0 = \frac{M_i}{2e}\left(v_i - \frac{dl}{dt}\right)^2 \quad (2)$$

$$N_i = N_{i0} \exp(-\frac{t-t_0}{\tau})(\delta_{AMP}\frac{l^2}{d^2}+1) \tag{3}$$

$$N_{i0} = \frac{4I_0}{v_i \pi D^2 Ze} \tag{4}$$

$$I_0 = \frac{dI}{dt} t_{vod} \tag{5}$$

The physical meaning and values of each parameter are shown in Table 1.

Table 1. Physical meaning and values of each parameter.

Parameter	Physical Meaning	Value or Unit
l	sheath length	mm
ε_0	vacuum dielectric constant	8.85×10^{-12} F/m
e	electronic charge	1.602×10^{-19} C
Z	average charge carried by ions	1.3~1.5
u_0	sheath potential	V
δ_{AMP}	inter-polar ion space charge distribution coefficient	5
I_0	Initial value of post-arc current	A
M_i	metal ion mass	1.062×10^{-25} kg
v_i	copper ion movement velocity	5000 m/s
D	cathode spot distribution diameter	mm
N_i	plasma density	m^{-3}
τ	ion diffusion decay time parameter	0.5~10 μs
d	gap distance between contacts	Max. 20 mm
t_{vod}	the time from the moment of current zero crossing to the beginning of the sheath growth	100 ns

Sheath stage reignition criterion:

$$E = 2\sqrt{\frac{eZN_i}{\varepsilon_0}}(\sqrt{u_{TRV}u_0 + u_0^2} - u_0) \tag{6}$$

Heavy breakdown occurs when the electric field strength E at the cathode surface is greater than the critical electric field strength $E_c(5 \times 10^9$ V/m) [13]. The metal vapor attenuation phase starts when the sheath length l is the same as the gap length d.

2.1.2. Metal Vapor Attenuation

The cathode spot of the vacuum arc is the main source of metal vapor. Under diffusive arc mode conditions, the metal vapor produced by the electrode at a sinusoidal current $i = Im\sin\omega t$ can be written as follows [20]:

$$n(t) = \frac{S_m \omega}{\omega^2 + \beta^2}\left[\frac{\beta}{\omega}\sin\omega t - \cos\omega t + e^{-\beta t}\right]e^{-\beta t} \tag{7}$$

$$\omega = 2\pi f \tag{8}$$

$$\beta = \frac{1}{2R}\sqrt{\frac{8KT}{\pi M}} \tag{9}$$

$$S_m = \frac{K_e E I_m}{M \pi R^2 d} \tag{10}$$

where $n(t)$ is the average density of metal vapor; Im is the amplitude of sinusoidal current; β is the decay coefficient of metal vapor; T is the temperature of the metal vapor in the gap after the current crosses zero, taken as 2000 K; K is Boltzmann's constant, and its value is 1.38×10^{-23} J/K; E is the corrosion rate, taken as 61 μg/C; K_e is the evaporation coefficient of the electrode, taken as 4; M is the metal vapor atomic mass, where copper is 1.062×10^{-25} kg; and R is the cathode spot distribution radius.

The limit metal vapor density of the vacuum circuit breaker is in accordance with the Paschen curve law. When the product of metal vapor density and gap opening distance is about 3×10^{19} m^{-2}, the gap is considered to be broken [5].

When the average free travel of electrons is the same as the gap length, the dielectric recovery process after the vacuum circuit breaker arc enters the static pressure resistance stage. The relationship between the average free travel of electrons λ_e and the critical density of metal vapor n_c is given by the following equation [21].

$$n_c = \frac{1}{\sqrt{2}\lambda_e \pi r_{Cu}^2} \tag{11}$$

where the copper ion radius r_{Cu} is taken as 1.278×10^{-10} m.

2.1.3. Static Withstand Voltage Stage

At this stage, the vacuum circuit breaker gap returns to the vacuum state; the breakdown at this time is a vacuum breakdown, and the relationship between the impact withstand voltage of the 40.5 kV vacuum interrupter, and the contact gap length is as follows [22]:

$$u_d = 70.12d^{0.56} \tag{12}$$

The contact gap length d is obtained from the measured breaking speed characteristic curve [23]. The magnitude of the transient recovery voltage (u_{TRV}) is compared with the static shock voltage (u_d) of the circuit breaker, and the vacuum gap is broken when u_{TRV} is greater than u_d.

2.1.4. Cathode Spot Distribution Diameter

The calculation of the initial plasma density and metal vapor density requires the cathode spot distribution diameter D. When the vacuum circuit breaker reignites, the contact gap flows through the high-frequency current; the arc burning time is short; the cathode spot is difficult to spread to the edge of the contact; the contact utilization rate is reduced; D takes the diameter of the contact, which makes the calculated density value small; there are deviations from the actual; and the model needs to be introduced for the acquisition of D. Assuming that the cathode spots are distributed circularly, and the velocity v of the cathode spots in all directions is the same, D can be written as follows [20]:

$$D = \begin{cases} 2vt(2vt < D_c) \\ D_c(2vt \geq D_c) \end{cases} \tag{13}$$

where v is the diffusion rate of the cathode spot, proportional to $(di/dt)^{1/2}$ [14], and t is the arc-burning time. When the arc-burning time is long enough, D equals the diameter of the contact Dc.

2.2. *High-Frequency Arc Extinguishing Capability*

Reignition leads to the gap flow through the high-frequency current, and its superposition on the frequency current appears as a large number of over-zero points; due to the initial arc current change rate being large, the vacuum circuit breaker can not cut off the high-frequency current. With the decay of the current, the current change rate gradually decreases after a few cycles. When the current is over zero, and the current change rate is small enough, the high-frequency current is cut off, so the critical value of the current change rate can be used on behalf of the vacuum. Therefore, the critical value of the current change rate can be used to represent the high-frequency current extinguishing ability of the vacuum circuit breaker. The range of the critical value is 100~600 A/μs [3]. In this paper, the critical current rate of change in simulation is set to 300 A/μs.

2.3. Vacuum Circuit Breaker Opening and Closing Logic

The built vacuum circuit breaker reignition model is shown in Figure 2. The model consists of an ideal switch, logic control module, and stray components. The ideal switch is set to infinity resistance (100 MΩ) when it is broken, and the arc resistance is replaced by a very small resistance (0.01 Ω) when it is closed. The logic control module controls the opening and closing state of the switch; the capacitance of the stray component is set to 200 pF, the resistance is set to 50 Ω, and the inductance is set to 50 nH.

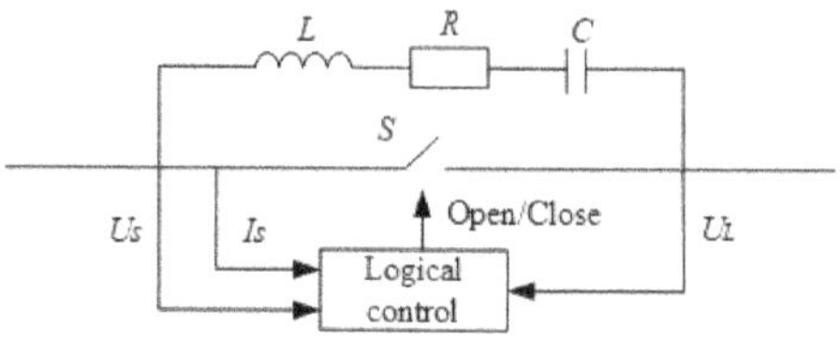

Figure 2. Simulation model of vacuum circuit breaker.

The simulation frequency cut-off current value I_b is taken as 3 A, and the simulation step is taken as 0.02 μs. The specific simulation flow is shown in Figure 3.

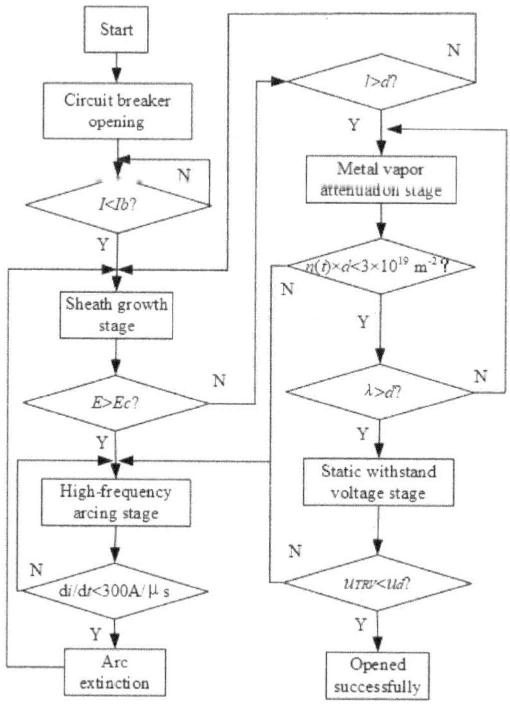

Figure 3. Simulation flowchart of the reignition process.

3. Wind Farm Vacuum Circuit Breaker Reignition Overvoltage Model Validation

3.1. Overall Wind Farm System Model Building

The typical electrical layout of the wind farm is shown in Figure 4, and its collector line wiring structure is chain-shaped, with the entire wind farm consisting of four feeders in parallel and eight wind turbines connected to a single feeder with a 700 m spacing between neighboring turbines on the feeder. The cable length from the bottom of the turbine to the nacelle step-up transformer is about 80 m [24]. The voltage at the wind turbine end is 690 V, which is raised to 35 kV through the booster transformer, and the high voltage side of each booster transformer is connected to each other using collector lines to form a joint unit, which is finally connected to the booster station for voltage boosting.

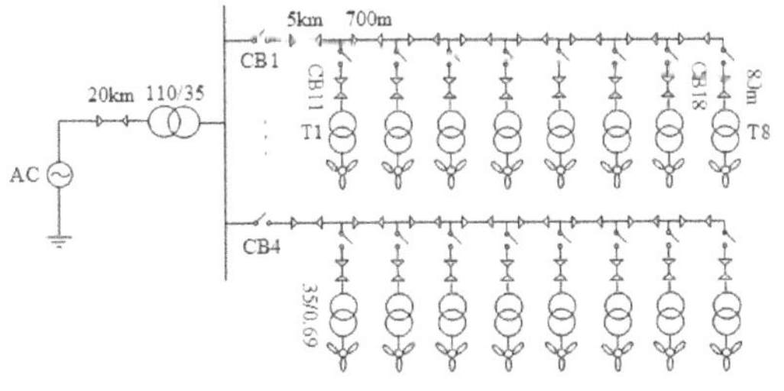

Figure 4. Typical distribution diagram of wind farm cable system.

In the simulation model, the step-up transformer models its high-frequency characteristics by connecting stray capacitors in parallel, with the high-voltage side capacitor C_H = 1 nF, the low-voltage side capacitor C_L = 3.1 nF, and the inter-high-voltage capacitor C_{HL} = 3 nF [3]. The switching transients (about 3–5 ms) generated during the vacuum circuit breaker shutdown are much larger than the doubly fed fan control circuit response time (about tens to hundreds of milliseconds); therefore, for transient studies, the DFIG is mostly modeled as a high-frequency induction motor impedance, which is connected in parallel with its control circuit impedance as well as in series with a harmonic filter. See Figure 5 for the stator-side equivalent resistance Rs = 0.00047 Ω and inductance Ls = 0.0283 mH, rotor-side equivalent resistance Rr = 0.00053 Ω and inductance Lr = 0.0305 mH, motor excitation inductance L_m = 1.21 mH, stator and grid-side filter inductance L_{fr} = 1 mH and L_{fg} = 0.73 mH, and filter capacitance C_f = 100 μF [25]. The frequency-dependent (phase) model provided by the transient electromagnetic software was chosen for the three-core cable, and the metal shield was grounded via a 0.1 Ω resistance [26].

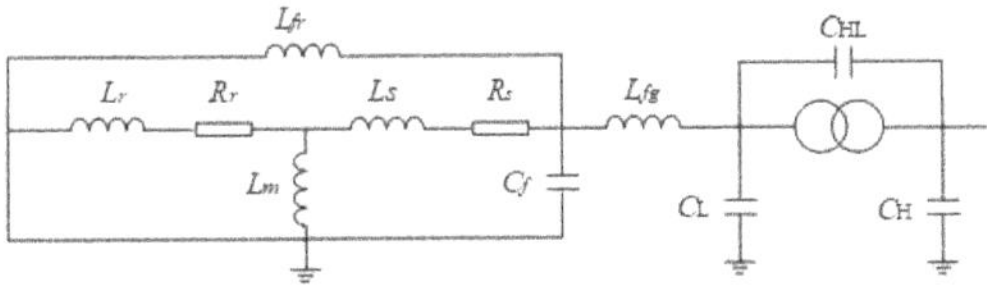

Figure 5. High-frequency model of transformer and DFIG.

3.2. Vacuum Breaker Model Validation

In order to verify the effectiveness of the vacuum breaker model built in this paper, the same simulation model as the wind farm testbed arrangement in the paper [27] was built in the transient electromagnetic software (see Figure 6). The transformer, cable, and vacuum breaker were used in the high-frequency model mentioned above, and the wind turbine model was replaced by the reactor equivalent as in the paper [27] test. The simulation results are shown in Figure 7, which includes the comparison of the effect of traditional calculation methods (linear and quadratic fitting of dielectric dynamic insulation strength curve). The peak voltage, voltage steepness, number of reignitions, and duration of reignition were selected as the comparison indexes [9], and the comparison results are shown in Table 2.

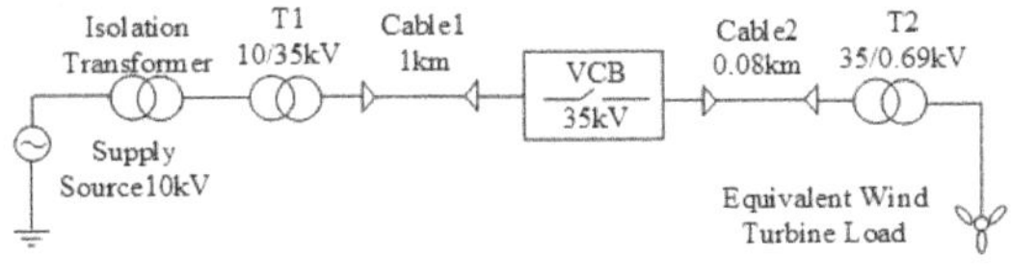

Figure 6. Layout of wind farm experimental platform [27].

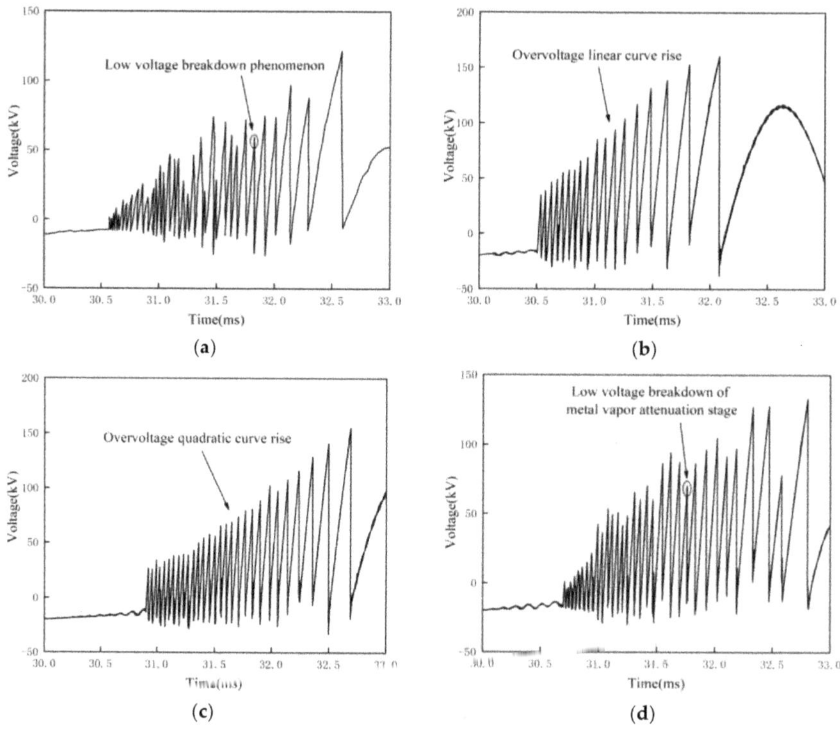

Figure 7. Overvoltage at transformer T2 simulated by different methods: (**a**) real measurement [27] (peak voltage: 123.01 kV, steepness: 148.31 kV·μs^{-1}, number of reignition: 43, reignition duration: 2.1 ms); (**b**) linear dielectric dynamic insulation strength curve (peak voltage: 156.18 kV, steepness: 130.64 kV·μs^{-1}, number of reignition: 18, reignition duration: 1.6 ms); (**c**) quadratic dielectric dynamic insulation strength curve (peak voltage: 148.67 kV, steepness: 126.32 kV·μs^{-1}, number of reignition: 30, reignition duration: 1.9 ms); (**d**) dielectric recovery (peak voltage: 126.61 kV, steepness: 157.89 kV·μs^{-1}, number of reignition: 40, reignition duration: 2.2 ms).

Table 2. Comparison of characteristic parameters of reignition overvoltage.

Method	Voltage (kV)	Steepness (kV·μs^{-1})	Number of Reignition	Reignition Duration (ms)
Real measurement	123.01	148.31	43	2.1
Linear	156.18	130.64	18	1.6
Quadratic	148.67	126.32	30	1.9
Dielectric recovery	126.61	157.89	40	2.2

Among the above three models, the values of critical parameters for the linear and quadratic methods were taken from the empirical values given in the papers [4,7,9]. When combining the data in Figure 7 and Table 2, it can be seen that the peak overvoltage of the conventional model simulation is larger than the measured value, and the errors of the linear and quadratic methods are 26.9% and 20.8%, respectively. The steepness, number of reignitions, and duration of reignition are smaller than the measured values, with errors of −11.9%, −58.1%, and −23.8% and −14.8%, −30.2%, and −9.5%, respectively. Considering the vacuum circuit breaker microscopic dielectric recovery simulation, the voltage peak, steepness, number of reignitions, and reignition duration are close to the measured values with errors of 2.9%, 6.5%, −6.9%, and 4.8%, respectively. It was also found that the overvoltage of the conventional method rises smoothly according to the fitted linear and quadratic curves, while the overvoltage simulated by the proposed method rises to fluctuate, the waveform is more accurate to the measured value, and the reignition process is more precise.

The analysis shows that the traditional linear method assumes that the contact gap is linearly proportional to time (constant opening speed), and the withstand voltage rises linearly with time, which is too fast to recover compared with the actual withstand voltage characteristics and ignores the low-voltage breakdown situation. Additionally, the gap withstand voltage value calculated at the later stage is large, resulting in a higher breakdown voltage when reignition occurs, a reduced number of reignitions, and a shorter reignition duration. The secondary method assumes constant acceleration of contacts, linear increase in opening speed, and nonlinear increase in withstand voltage with time. The simulation has a good fit in the early stage and the number of reignitions increases. However, because the low voltage breakdown is also ignored, the overall situation is lower than the actual measurement, and the breakdown voltage is large in the later stage.

By using the reignition model in this paper, the simulation found that the arc reignition mainly occurs in the metal vapor attenuation and static withstand voltage stage. The TRV of metal vapor attenuation stage is smaller than that of static withstand voltage stage, so when reignition occurs in metal vapor attenuation stage, the over-voltage after the reignition is significantly lower than the over-voltage of the last reignition. To achieve the low voltage breakdown in the measured waveform, the number of simulated reignition is higher than the traditional model. The static withstand voltage stage determines the overvoltage of the final reignition, and its withstand voltage value is obtained according to the measured breaking speed, so the peak overvoltage is closer to the actual measurement. Compared with the traditional linear curve reignition model, the reignition model built in the paper significantly improved the accuracy of overvoltage peak, reignition number, and reignition duration by 24%, 51.2%, and 19%, respectively. Compared with the traditional secondary curve reignition model, the accuracy of the overvoltage peak and reignition number improved significantly by 17.9% and 23.3%, respectively. It can be seen that the establishment of a detailed model of the physical process of post-arc dielectric recovery further improves the simulation accuracy of overvoltage.

4. Simulation of Reignition Overvoltage Protection Measures

The overvoltage protection measures used in this section are capacitance absorbers, surge arresters, and choke coils. The capacitance absorber consists of resistance and capacitance in a series. Surge arresters are essentially nonlinear resistors, which use the V-I characteristic curve of residual voltage 73 kV at a nominal discharge current of 1 mA DC. The equivalent circuit of the choke coil is composed of inductance and resistance in parallel. The per-unit voltage of the 35 kV power system is as follows:

$$U_{p.u.} = 40.5 \times \frac{\sqrt{2}}{\sqrt{3}} = 33.1 \text{ kV} \tag{14}$$

Reignition overvoltage caused by the opening operation of the vacuum circuit breaker at the bottom of the wind farm tower is the most serious [4]. In this paper, a simulation model was established according to the typical wind farm layout in Figure 3, in which the vacuum circuit breaker, transformer, cable, and wind turbine models adopt the proposed method. Without suppression measures, the overvoltage simulation waveform of the high-voltage side of the terminal transformer T1 is shown in Figures 8 and 9 of the vacuum circuit breaker CB11 at the bottom of the opening tower, and the amplitude and steepness of the relative ground voltage are 153.7 kV (4.6 p.u.) and 157.2 kV/μs, respectively. The interphase overvoltage amplitude is 262.8 kV (7.9 p.u.), and the reignition duration is 3.7 ms. The relative operating overvoltage in a 35 kV system shall not exceed 132 kV (4.0 p.u.), and the relative overvoltage shall not exceed 1.4 times the relative overvoltage between phases, i.e., 185 kV (5.6 p.u.) [28]. It can be seen that the opening operation is easy to damage the insulation of the terminal transformer of the wind farm, and effective protection is required.

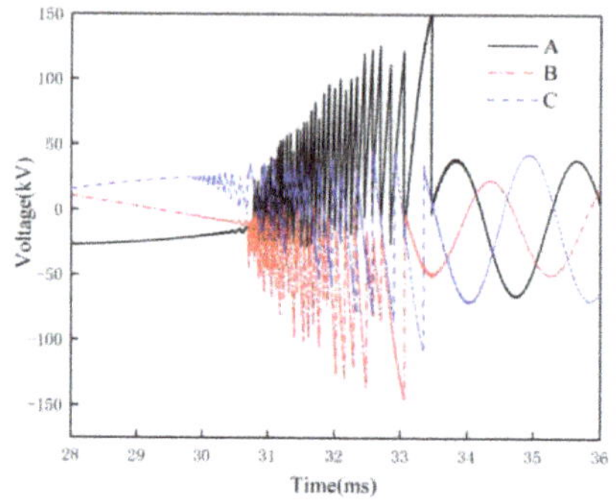

Figure 8. Simulation waveform of transformer T1 phase-to-ground voltage without suppression measures.

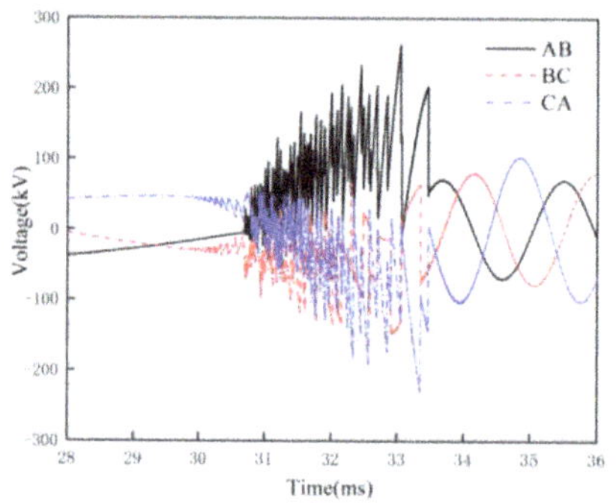

Figure 9. Simulation waveform of transformer T1 phase-to-phase voltage without suppression measures.

4.1. Effect of RC Snubbers on Overvoltage

RC snubbers are connected in parallel to the protection equipment and can effectively reduce the overvoltage amplitude and frequency. Since there is no national standard for RC snubbers, the capacitance parameters are based on many years of actual operating experience at home and abroad. In order to effectively suppress the reignition overvoltage in wind farms, this paper combines different capacitance parameters for simulation, with resistance selected from 100 to 500 Ω, and capacitance is taken as 0.01 µF, 0.05 µF, and 0.1 µF. The overvoltage situation under different capacitance parameters is shown in Figure 10.

As can be seen from Figure 10, (1) resistance and capacitance affect the suppression effect. Resistance remains constant; the larger the capacitance value, the more obvious the overvoltage suppression. Additionally, capacitance remains constant; the resistance increases; the overvoltage amplitude decreases; the capacitance value is 0.05 µF; the resistance value is 400 Ω; and the voltage amplitude appears as a significant decline, with a drop of 58.3%. The increase in resistance value reduces the oscillation frequency, prolongs the moment of high-frequency current over zero, reduces the chance of reignition, and thus suppresses the overvoltage amplitude. In order to reduce the loss and heat, the resistance value should not be selected too large, generally about 400 Ω. (2) Regarding 400 Ω/0.05 µF is compared with the commonly used 100 Ω/0.1 µF RC snubbers protection effect; the latter eliminated the occurrence of reignition, and the former occurred with three reignitions, but the relative ground and phase voltage did not exceed the specified standard value. The simulation results are shown in Table 3 and Figures 11 and 12, and the two protection effect is very close. The former is able to overcome the problem of excessive capacitance current and resistance burnout of the latter in specific cases, and the capacitance value is reduced

Abbreviations

CTM	continuous transition model
VCB	vacuum circuit breaker
TRV	transient recovery voltage
DFIG	double-fed induction generator
BIL	basic insulation level

References

1. Ghafourian, S.M.; Arana, I.; Holboll, J.; Sorensen, T.; Popov, M.; Terzija, V. General Analysis of Vacuum Circuit Breaker Switching Overvoltages in Offshore Wind Farms. *IEEE Trans. Power Deliv.* **2016**, *31*, 2351–2359. [CrossRef]
2. Kafshgari, N.A.; Ramezani, N.; Nouri, H. Effects of high frequency modeling & grounding system parameters on transient recovery voltage across vacuum circuit breakers for capacitor switching in wind power plants. *Int. J. Electr. Power Energy Syst.* **2019**, *104*, 159–168.
3. Abubakar, U.; Mekhilef, S.; Mokhlis, H.; Seyedmahmoudian, M.; Horan, B.; Stojcevski, A.; Bassi, H.; Rawa, M.J.H. Transient Faults in Wind Energy Conversion Systems: Analysis, Modelling Methodologies and Remedies. *Energies* **2018**, *11*, 2249. [CrossRef]
4. Guo, Y.; Jiang, X.; Chen, Y.; Zheng, M.; Liu, G.; Li, X.; Tang, W. Reignition overvoltages induced by vacuum circuit breakers and its suppression in offshore wind farms. *Int. J. Electr. Power Energy Syst.* **2020**, *122*, 106227. [CrossRef]
5. Wang, J.; Geng, Y.; Liu, Z. *The Theory and Technology of Single-Break Vacuum Circuit Breaker for Transmission Level*; Machinery Industry Press: Beijing, China, 2017.
6. Helmer, J.; Lindmayer, M. Mathematical modeling of the high frequency behavior of vacuum interrupters and comparison with measured transients in power systems. In Proceedings of the 17th International Symposium on Discharges and Electrical Insulation in Vacuum, Berkeley, CA, USA, 21–26 July 1996; IEEE: New York, NY, USA, 1996; Volume 1, pp. 323–331.
7. Abdulahovic, T.; Thiringer, T.; Reza, M.; Breder, H. Vacuum circuit-breaker parameter calculation and modelling for power system transient studies. *IEEE Trans. Power Deliv.* **2014**, *32*, 1165–1172. [CrossRef]
8. Razi-Kazemi, A.A.; Fallah, M.R.; Rostami, M.; Malekipour, F. A new realistic transient model for restrike/prestrike phenomena in vacuum circuit breaker. *Int. J. Electr. Power Energy Syst.* **2020**, *117*, 105636. [CrossRef]
9. Badrzadeh, B.; Hogdahl, M.; Isabegovic, E. Transients in Wind Power Plants-Part I: Modeling Methodology and Validation. *IEEE Trans. Ind. Appl.* **2012**, *48*, 794–807. [CrossRef]
10. Farrall, G.A. Recovery of dielectric strength after current interruption in vacuum. *IEEE Trans. Plasma Sci.* **1978**, *6*, 360–369. [CrossRef]
11. Mo, Y.; Shi, Z.; Jia, S. Study of the influences of different factors on the charged particles absorbed by the post-arc anode during the post-arc sheath expansion process in vacuum circuit breakers. *AIP Adv.* **2021**, *11*, 015317. [CrossRef]
12. An, C.; Sima, W.; Liao, R.; Li, S. Research on 35 kV circuitry vacuum circuit breaker interrupting no load transformer overvoltage. *Proc. CSEE* **2002**, *08*, 100–104.
13. Wang, F.; Wang, Z.; Hao, M. Dielectric recovery process of multiple re-ignition phenomena after small inductive current interruption in vacuum. *High Volt. Eng.* **2018**, *44*, 2019–2026.
14. Liu, L.; Yuan, Z.; Liu, S.; Chen, L.; Wu, K.; Pan, Y. Research on the Postarc Sheath Growth Process Considering the Plasma Motion and Distribution. *IEEE Trans. Plasma Sci.* **2020**, *48*, 4289–4297. [CrossRef]
15. Liljestrand, L.; Lindell, E. Efficiency of surge arresters as protective devices against circuit-breaker-induced overvoltages. *IEEE Trans. Power Deliv.* **2015**, *31*, 1562–1570. [CrossRef]
16. Nilges, M.; Schegner, P. Influence of RC snubbers on the steepness of switching transients caused by vacuum circuit breakers. In Proceedings of the 2018 IEEE Power & Energy Society General Meeting (PESGM), Portland, OR, USA, 5–10 August 2018; IEEE: New York, NY, USA, 2018; pp. 1–5.
17. Zhou, Q.; Cheng, Y.; Bian, X.; Liu, F.; Zhao, Y. Analysis of restrike overvoltage of circuit breakers in offshore wind farms. *IEEE Trans. Appl. Supercond.* **2016**, *26*, 1–5. [CrossRef]
18. Smugala, D.; Piasecki, W.; Ostrogorska, M.; Florkowski, M.; Fulczyk, M.; Granhaug, O. Wind Turbine Transformers Protection Method against High-Frequency Transients. *IEEE Trans. Power Deliv.* **2015**, *30*, 853–860. [CrossRef]
19. Ghasemi, S.; Allahbakhshi, M.; Behdani, B.; Tajdinian, M.; Popov, M. Probabilistic analysis of switching transients due to vacuum circuit breaker operation on wind turbine step-up transformers. *Electr. Power Syst. Res.* **2020**, *182*, 106204. [CrossRef]
20. Liu, L.; Yuan, Z.; Liu, S.; Chen, L.; He, J.; Pan, Y. Experiment research of post-arc current and cathode spots distribution in medium-high frequency vacuum arc. *Phys. Plasmas* **2020**, *27*, 063507. [CrossRef]
21. Liu, Z.Y.; Wang, J.M.; Yuan, S. Physical process of interrupting ac current in vacuum interrupter and mathematical evaluation of the interrupting capacity. In Proceedings of the International Symposium on Discharges and Electrical Insulation in Vacuum, ISDEIV, Xi'an, China, 18–22 September 2000; Volume 2, pp. 431–434.
22. Ding, J.; Liu, X.; Yao, X.; Liu, Z.; Geng, Y.; Wang, J. Prebreakdown Current Characteristics of Vacuum Gaps with Various Gap Distances. In Proceedings of the 2020 29th International Symposium on Discharges and Electrical Insulation in Vacuum (ISDEIV), Padova, Italy, 26–30 September 2021; IEEE: New York, NY, USA, 2021; pp. 15–18.

23. Rong, M.; Wang, X.; Yang, W.; Jia, S. Theoretical and experimental analyses of the mechanical characteristics of a medium-voltage circuit breaker. *IET Sci. Meas. Technol.* **2005**, *152*, 45–49. [CrossRef]
24. Glasdam, J.; Bak, C.L.; Hjerrild, J. Transient Studies in Large Offshore Wind Farms Employing Detailed Circuit Breaker Representation. *Energies* **2012**, *5*, 2214–2231. [CrossRef]
25. Xin, Y.; Zhao, B.; Liang, Q.; Zhou, J.; Qian, T.; Yu, Z.; Tang, W. Development of Improved Suppression Measures against Reignition Overvoltages Caused by Vacuum Circuit Breakers in Offshore Wind Farms. *IEEE Trans. Power Deliv.* **2021**, *37*, 517–527. [CrossRef]
26. Ghafourian, S.M. *Switching Transients in Large Offshore Wind Farms*; The University of Manchester: Manchester, UK, 2015.
27. Xin, Y.L.; Tang, W.H.; Zhou, J.J.; Yang, Y.; Liu, G. Sensitivity analysis of reignition overvoltage for vacuum circuit breaker in offshore wind farm using experiment-based modeling. *Electr. Power Syst. Res.* **2019**, *172*, 86–95. [CrossRef]
28. GB 311.1—1997; Insulation Coordination for High Voltage Transmission and Distribution Equipment. Xi'an High Voltage Apparatus Research Institute: Xi'an, China, 1997.

energies MDPI

Article

Partial Discharge-Originated Deterioration of Insulating Material Investigated by Surface-Resistance and Potential Mapping

Marek Florkowski * and Maciej Kuniewski

Department of Electrical and Power Engineering, Faculty of Electrical Engineering, Automatics, Computer Science, and Biomedical Engineering, AGH University of Krakow, al. Mickiewicza 30, 30-059 Krakow, Poland; maciej.kuniewski@agh.edu.pl
* Correspondence: marek.florkowski@agh.edu.pl

Abstract: The endurance of medium- and high-voltage electrical insulation is a key reliability element in a broad spectrum of applications that cover transmission and distribution levels, the transportation segment, the industrial environment, and power electronics-based energy-conversion systems. The high electric-field stress and high-frequency switching phenomena as well as the impact of environmental conditions lead to the occurrence of partial discharges (PD) and the subsequent deterioration of electrical insulation. Partial discharges usually occur inside solid insulation materials in tiny voids that may either be located adjacent to the electrodes or in the bulk of dielectric material. This effect refers to both AC and DC systems; however, AC voltage is usually much more intensive as compared to DC voltage. This paper describes a novel combined approach based on surface-resistance and potential mapping to reveal the effects of internal processes and the deterioration of insulating material due to the actions of partial discharges. To realize the research objective, the following two-step approach was proposed. Multi-point resistance mapping enables us to identify the spots of discharge channels, manifesting a-few-orders-of-magnitude-lower surface resistance as compared to untreated areas. In addition, surface-potential mapping that was stimulated by corona-charge deposition reflects quasi-equipotential clusters and the related polarity-dependent dynamics of charge decay. A high spatial and temporal resolution allows for the precise mapping and tracing of decay patterns. Experiments were carried out on polyethylene (PE) and Nomex specimens that contained embedded voids. During PD events, the effective discharge areas are identified along with the memory effects that originate from the accumulation of surface charges. Long-term aging processes may drive the formation of channels that are initiated from the deteriorated micro clusters, in turn, penetrating the bulk isolation. The presented methodology and experimental results extend the insight into PD mechanisms and internal surface processes.

Keywords: partial discharges; surface-resistance mapping; surface-potential decay; electrical insulation degradation; aging

Citation: Florkowski, M.; Kuniewski, M. Partial Discharge-Originated Deterioration of Insulating Material Investigated by Surface-Resistance and Potential Mapping. *Energies* **2023**, *16*, 5973. https://doi.org/10.3390/en16165973

Academic Editor: Tomasz Norbert Kołtunowicz

Received: 10 June 2023
Revised: 17 July 2023
Accepted: 2 August 2023
Published: 14 August 2023

1. Introduction

The endurance of medium- and high-voltage (HV) electrical insulation is a key reliability element in a broad spectrum of applications that cover transmission and distributions levels, the transportation segment, the industrial environment, and power electronics-based energy-conversion systems [1–6]. The high electric field stress and high-frequency switching phenomena as well as environmental impacts such as temperature, humidity, or mechanical stresses lead to the occurrence of partial discharges (PD) and the subsequent deterioration of electrical insulation. This stimulates ongoing research on various characteristics of insulating materials, such as enhanced electrical endurance, high thermal conductivity, high energy storage density, extreme environmental resistance, and environmental friendliness [1,6]. Partial discharges usually occur inside solid insulation materials in tiny voids (gaseous inclusions) that may either be located adjacent to the electrodes or in the bulk of the dielectric material. This effect refers to both AC and DC systems; however, AC voltage is usually much more intensive as compared to DC voltage. The long-term penetration of PDs may lead to local surface degradation or treeing propagation, whose origins often occur at sharp edges where electric field amplification occurs. The deterioration of a void's surface may result in radical surface-resistivity impairment—often by a few orders of magnitude; this creates local clusters that can even act as equipotential sheets. Such a conductivity non-homogeneity may thus impose a space-charge accumulation. This effect has an important role in HVDC (High Voltage Direct Current) insulation systems—especially related to power cables (considering their long-term endurance) [2]. A locally accumulated space charge may elevate operational stresses in HVDC polymeric cable systems—especially during polarity inversion [7]. The actions of partial discharges in gaseous inclusions result in changes in void surface conditions (for example, their conductivity, morphology, or roughness), which impacts the charge accumulation and transport effects [8–13]. Special attention is paid to investigations related to effective discharge and memory effects [10,14–16]. New insight was provided by mapping the discharge channels in the void creating an effective partial discharge area [17–19]. The modified surface conductivity of void walls impacts the decay time of deposited charges during consecutive PDs (thus, inception conditions in the long run) [20–23].

In this paper, partial discharge-originated deterioration was investigated by surface-resistance and surface potential-decay mapping. In this new approach, the effects of PD dynamics can be evaluated—especially revealing discharge channel clusters. The long-term exposure to partial discharges leads to variations in the discharge areas inside void walls, accelerating the clustering effect. In such a way, the evolution of so-called "effective discharge areas" may be observed [14,15,19,20], influencing the creation of discharge spots (thus, contributing to the stochastic character of the magnitude of partial discharges [21–32], which can be revealed during the acquisition by phase-resolved PD-pattern modulation [14–19]). These phenomena often contribute to memory effects [27,33–35]. The localization of surface clusters may even be impacted by the magnetic field, which influences the discharge trajectory [36,37]. The presented experiment was conducted on polyethylene (PE) and Nomex specimens that were subjected to accelerated deterioration stresses at 1 kHz. The samples took the form of a flat round gaseous inclusion with a diameter that was much larger than the thickness of an inclusion. The presented approach allowed for the localization of changes in a dielectric structure that was subjected to multi-channel PDs; these were identified through the detection of multi-point clusters with reduced surface resistances. Allocation to the clusters with different resistances by surface mapping reflected the energetic impact that is caused by partial discharges. The formation of the semiconducting surface inside voids may cause the creation of channels that penetrate toward the bulk in the next stage (e.g., the early stages of treeing). Such investigations aimed to better understand discharge mechanisms on the one hand and lead to the proper selection of the appropriate materials for HV insulation systems on the other. The novelty aspect of the paper refers also to the introduced measurement approach, which may be applied to the diagnostics of certain electrical power components, such as insulators, spacers, or power transformer insulating components during revisions. The compact version of the measurement rig will allow easy transportation to the inspection site, enabling easy diagnostics of the deterioration of the insulation materials.

The rest of this paper is organized as follows. The experimental methodology is presented in Section 2, followed by the specimen and measurement setup description in Section 3. The experimental results and discussion are in Sections 4 and 5, respectively. Finally, the conclusion of this paper is drawn in Section 6.

2. Experimental Methodology

The intention of the performed experiments was to get inside the degradation processes that take place in sources of discharges that are embedded in dielectric insulation. The experimental methodology comprises five distinct phases, which are illustrated in Figure 1. The scope comprised the two-step approach proposed in this paper, related to surface-resistance and potential mapping. In the preparatory step, each specimen was cleaned and discharged while checking its condition with a measurement of the surface potential. The flat sample was formed on sheet layers of insulating material that contained an embedded round void. The void walls were explored during the experiments. A detailed description of the specimen's geometry is presented in the following section. The investigations started with the accelerated degradation of the specimen after being subjected to a sinusoidal voltage of 18 kV and a frequency of 1 kHz for a predefined time period of 100 min. In this way, intensive partial discharges penetrated the void volume. In the second step, the specimens were opened in a clean environment, and both the upper and bottom walls were subjected to surface-resistance mapping that was performed automatically on a defined region. The investigated surface comprised the zone of the gaseous inclusion as well as an untreated external part for reference.

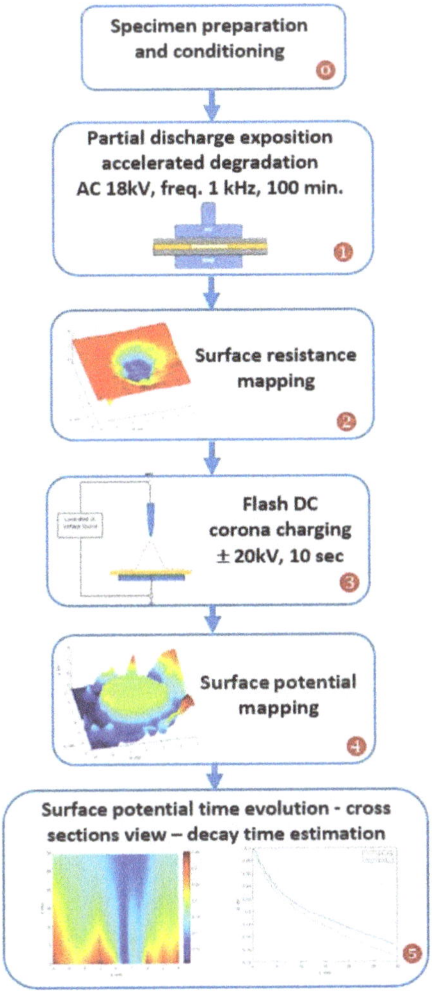

Figure 1. Sequence of experimental methodology steps.

The third step was related to the DC corona charging of the void walls in the setup that is described in the next section. The polarity of the deposited charges was defined upfront. This phase was necessary to perform in the following fourth step the surface-potential mapping, including the time-decay observation along the marked cross-sections (fifth step). In this way, the partial discharge-deteriorated insulating material inside the void was analyzed by means of both surface-resistance testing and potential mapping.

3. Specimen and Measurement Setup

In this section, detailed descriptions of the specimens, setup, and measurement instrumentation are provided.

3.1. Specimens

The presented experiment was conducted on polyethylene (PE) and Nomex® (DuPont, Wilmington, NC, USA, Type 418 flexible paper with a rated temperature range of 220 °C and a sheet thickness of 120 μm) specimens that were subjected to an accelerated deterioration stress at 1 kHz. Both materials are extensively used in high-voltage applications; polyethylene is a basic material for the main insulation of polymeric high-voltage cables, while Nomex is applied in the transformer layer and ground insulation as well as the insulation of motor coil wraps and conductors (the latter's calendared blend of mica and aramid provides increased voltage corona endurance). The investigated Nomex 418 contained a 50% share of mica lamellas. The aramid paper eliminated the main drawbacks of cellulose, i.e., reduced moisture absorption as well as increased resistance to high temperatures. The typical electrical properties for PE and Nomex 418 are presented in Table 1.

Table 1. Typical properties of Nomex Type 418 and PE specimens [18].

	Nomex 418	PE
Dielectric Strength, AC [kV/mm]	30	40
Dielectric constant, dry	2.3	2.2
Dissipation factor	0.006	0.0009
Volume resistivity [Ω·m]	10^{14}	6×10^{15}
Surface resistivity [Ω]	4×10^{13}	5×10^{13}
Layer thickness [mm]	$0.96 = 8 \times 0.12$	1

Specimens from both materials (format 100 × 100 mm) were designed according to the geometry that is shown in Figure 2; these contained flat round gaseous inclusions with a diameter of d = 38 mm, which is much larger than the thickness of an inclusion (a = 1 mm—$d >> a$). The main specimen insulation was composed of three layers (each 1 mm thick) with a punched void in the middle one. Two glass plates (1 mm thick each) were pressed on both sides of the samples to safeguard their mechanical stability. The investigated surfaces (the upper and bottom void walls) are marked in red in Figure 2. The surface-resistance and potential mapping was carried out on these layers after opening the specimen.

Both high-voltage and ground electrodes made of stainless steel (diameter—50 mm) ensured a uniform electric field distribution. The measurements were performed in insulating oil in order to prevent surface discharges on the specimen. The following environmental conditions were present during the experiments: a temperature of 24 °C, a humidity level of 31%, and an atmospheric pressure of 999 hPa.

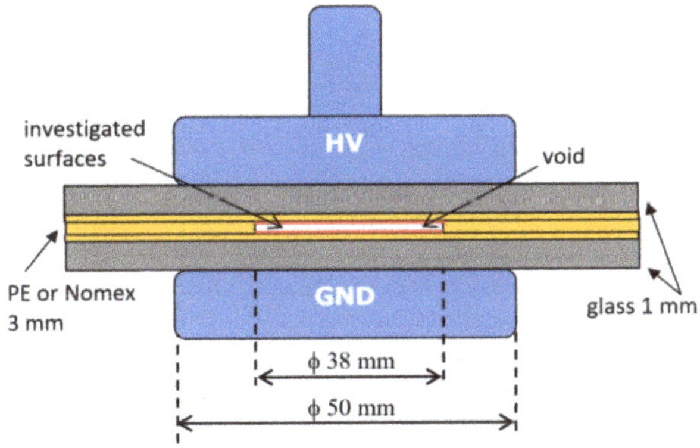

Figure 2. Geometry of test specimen with embedded void.

3.2. Measurement Setup

The measurement setup elements that are described in this section refer to the consecutive stages in the experimental framework; i.e., accelerated deterioration of specimens with AC voltage, DC charge sprinkling in corona mode, and both surface-resistance and potential mapping.

3.2.1. Accelerated Specimen Degradation

The simultaneous accelerated degradation of both the PE and Nomex specimens was conducted with a sinusoidal voltage of 18 kV and a frequency of 1 kHz for a predefined time period of 100 min. The electrical exposure setup is shown in Figure 3. The high voltage was delivered from a Trek 20/20B amplifier (Advanced Energy, Denver, CO, USA), which was controlled by a SIGLENT function generator (Siglent, Shenzhen, China). The HV source was protected with a Z = 2 MΩ resistor. The level of the high voltage was measured by means of a Tektronix Probe (Tektronix, Beaverton, OR, USA, Model P6015A—depicted in Figure 3) represented by an R_1 and R_2 resistive divider, with an input impedance equal to 100 MΩ, providing an attenuation factor of 1:1000.

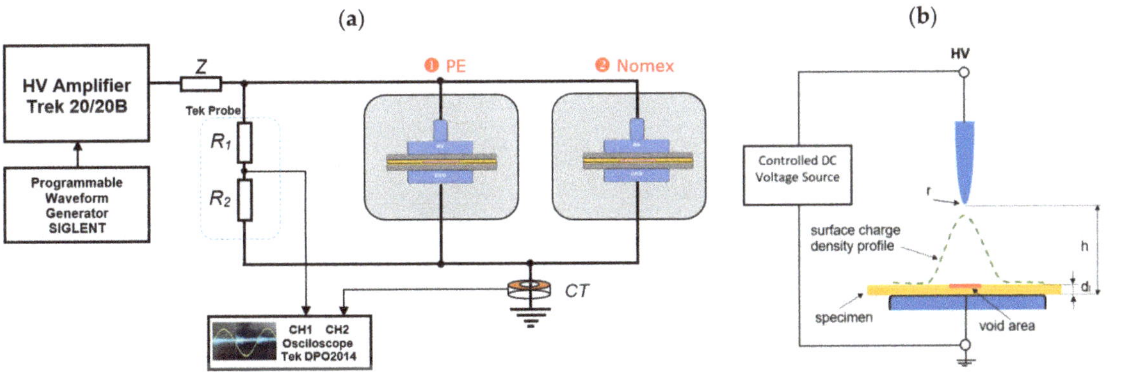

Figure 3. Electrical exposure setup: (**a**) AC accelerated high-frequency deterioration; (**b**) DC charge sprinkling.

The presence of partial discharges was detected by a CT wideband current transformer that was terminated at 50 Ω and placed on the ground side and observed on an oscilloscope. The aging conditions were set based on the performed tests. The voltage was adjusted to the level above the partial discharge inception voltage. The higher frequency was ensuring the acceleration of the specimen deterioration. The duration of the test was set according to the visible aging effects.

3.2.2. DC Charge Sprinkling

The charge-sprinkling setup (shown in Figure 3b) was used to accumulate the charge of the defined polarity on the specimen surface. In corona mode, DC voltage that was equal to ±18 kV was applied to an HV needle electrode with a radius of 700 µm that was positioned at a distance *h* of 100 mm above the sample surface (with thickness d_l located on the ground planar electrode). The relatively great distance between the needle tip and the specimen surface resulted in a quite-uniform charge deposition. The void region is marked in red in Figure 3b. The duration of the flash-sparkling process was 10 s. The voltage potential related to the charge density was measured by using the system that is described in the next section.

3.2.3. Surface-Resistance Mapping

The measurements of the surface-resistance mapping on the void walls were performed in the setup that is presented in Figure 4. The multipoint method that was described in [18] was applied to visualize the potential distribution in the frame of the investigated region. The scanning was executed fully automatically along a programmed line-by-line pattern. The interelectrode distance of the probe was equal to 5 mm. The mobile micro two-electrode probe was automatically positioned along the *x*- and *y*-axes above the defined zone of investigation. The control along the *z*-axis ensured the proper pressure and contact between the measuring electrodes and the specimen surface. The probe was driven by three USB-controlled motors (M_X, M_Y, and M_Z), which positioned the resistance probe in three dimensions at defined coordinates in space. The measurement of the surface resistance was performed by an electrometer through a current measurement at an applied voltage of 1 kV. The results were sent to the computer via a USB interface. The mapping sequence was executed along the defined trajectory at defined grid points. The surface-resistance mapping was carried out in a matrix of 25 × 25 = 625 points, i.e., covering a region of 60 × 60 mm with a raster of 2.5 mm. Both the upper and lower void walls were investigated.

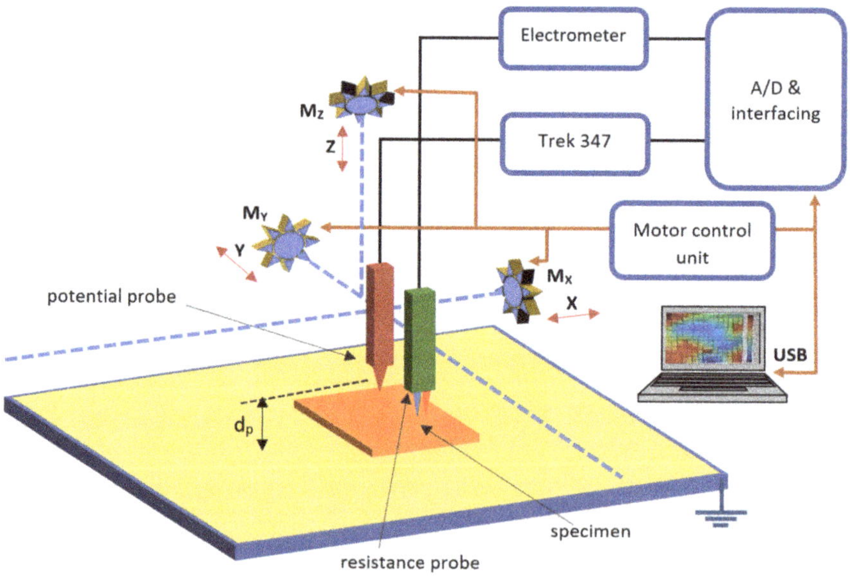

Figure 4. Instrumentation for automatic surface-resistance and potential mapping.

3.2.4. Surface-Potential Mapping

The measurements of the surface-potential mapping and decay-time estimation were performed in the setup that is presented in Figure 4. Based on this approach, the surface-charge density could also be calculated [38–41]. One side of the void wall specimen material was put on the grounded electrode. The mobile-readout probe was automatically positioned along the *x*- and *y*-axes above the selected region of investigation. The control along the *z*-axis allowed for the precise positioning of the probe above the specimen. The head that fixed the probe was driven by three USB-controlled motors (M_x, M_y, and M_z), which positioned the probe in three dimensions. The readout of the surface potential was carried out by means of an electrostatic voltmeter (Trek Model 347), which was directly connected to a 16-bit A/D converter and a computer-interfacing USB module. The potential measurement was based on a non-contacting field-nulling approach [38]. In the surface-potential mapping mode, the probe was positioned at a distance of d_p = 1 mm above the specimen. The surface-potential mapping was carried out in a matrix of 25 × 25 = 625 points, i.e., covering a region of 60 × 60 mm with a raster of 2.5 mm. The probe-to-plane distance was 1 mm. The upper and lower sheets of the void specimen were investigated. For visualization in both the surface-resistance and surface-potential mapping cases, the interpolation of the surface color distribution was applied using the measurement values from the grid nodes. For the interpolation, a cubic convolution from Matlab (ver. R2022a) was used.

4. Results

The measurements were executed according to the methodology that was described in the previous section (distinguishing several stages). The experiments started with an accelerated deterioration of specimens at an AC voltage of 18 kV and a frequency of 1 kHz for 100 min. The experimental setup of the PD exposition place was presented in Figure 3. The intensive action of the partial discharge inside the embedded void resulted in internal surface degradation. The individual channels on both the upper and lower inclusions' voids were created. After a longer stressing time, these local spots formed bigger clusters that were characterized by lower surface resistance. The identification of these effects in both specimens is the subject of this paper and will be described in the following parts.

4.1. Surface-Resistance Mapping

One of the degradation-assessment methodologies that are presented in this paper is surface-resistance mapping. The applied instrumentation allowed for the automatic execution of the mapping of a defined region. The surface resistances of the untreated non-aged materials were higher than $R_s > 10^{14}$ Ω in the case of PE and higher than $R_s > 10^{12}$ Ω for Nomex. The surface-resistance mapping of the bottom side of the PE and Nomex specimens that was carried out on region (x, y) = (−30, 30; −30, 30) [mm] is shown in Figure 5. The void wall could be very well-distinguished from the untreated external zone. The deteriorated regions manifested two-orders-of-magnitude-lower resistance as compared to the non-aged parts; i.e., for PE, the value dropped from 3.1×10^{13} to 10^{11} Ω, and in the case of Nomex, it dropped from 1.6×10^{12} to 1.6×10^{10} Ω. The region of the accelerated deterioration is well-marked in the 3D (Figure 5c,d) and cross-section views (Figure 5e,f). The resistance mapping reveals the creation of a quasi-equipotential zone (corresponding to the void diameter) on one side with a significantly increased level of surface conductivity. On the other hand, one can even observe the clusters and zones in that region that corresponded to the more- and less-intensive discharge channels.

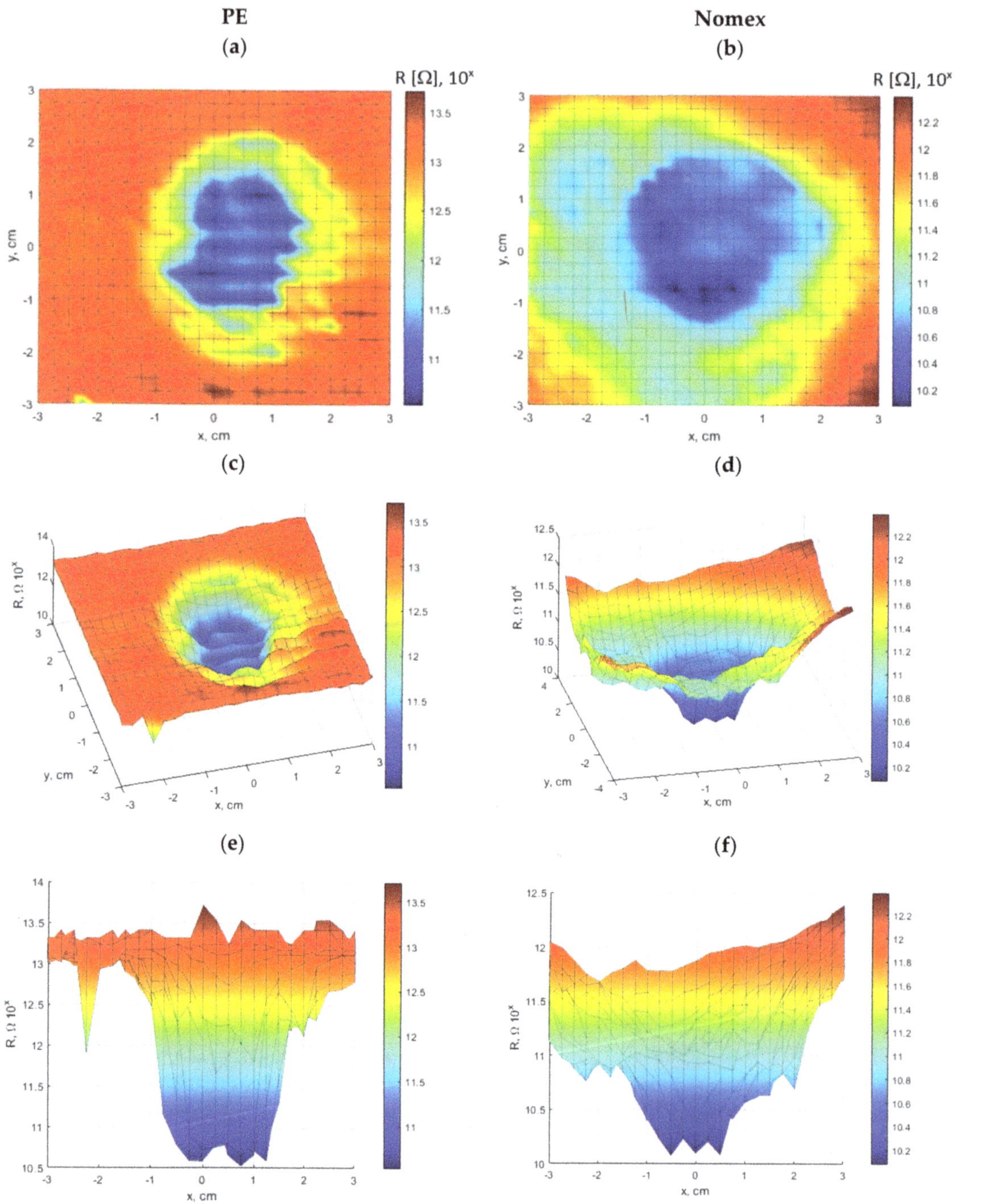

Figure 5. Surface-resistance mapping inside voids on PE (left column) and Nomex (right column) specimens (bottom sides): (**a**,**b**) map view; (**c**,**d**) 3D view; (**e**,**f**) cross-section view.

4.2. Surface-Potential Mapping

In order to reveal the deteriorated clusters, the void walls were exposed to a charge deposition with a defined polarity in the corona mode in the setup that is shown in Figure 3b. Charge sprinkling at ±20 kV DC was conducted during a flash-exposition time of 10 s. Immediately after the charge deposition, the electrostatic-potential surface mapping was performed by using the instrumentation that is described in Section 3.2.4.

The surface-potential mapping on the void walls of the PE specimens (bottom sides) is shown in Figure 6a,b (3D view) for the negative polarity and Figure 6c,d (3D view) for the positive polarity of the accumulated charge. An analogic sequence of the surface-potential mapping on the void walls of the Nomex specimens is shown in Figure 7a for the negative and Figure 7b for the positive sprinkling. In both cases, the void regions were well-defined; the contours of the voids (with a diameter of 38 mm) are clearly visible in the patterns that are shown in Figures 6 and 7. The course of the surface potential within the void wall zone in fact reflects the equipotential cluster, which was identified by the surface-resistance mapping as a region with significantly reduced surface resistance.

The potential measurements indicated the PD accumulated-charge zones, which were concentrated quasi-radially. Due to the intensive accelerated deterioration test (high-frequency and high-electric field stress), the spatial distribution of the PD filamentary channels became more and more uniform in their shapes. The mechanism of these effects also referred to the memory mechanism that was related to the individual PD events [12,21,27,35].

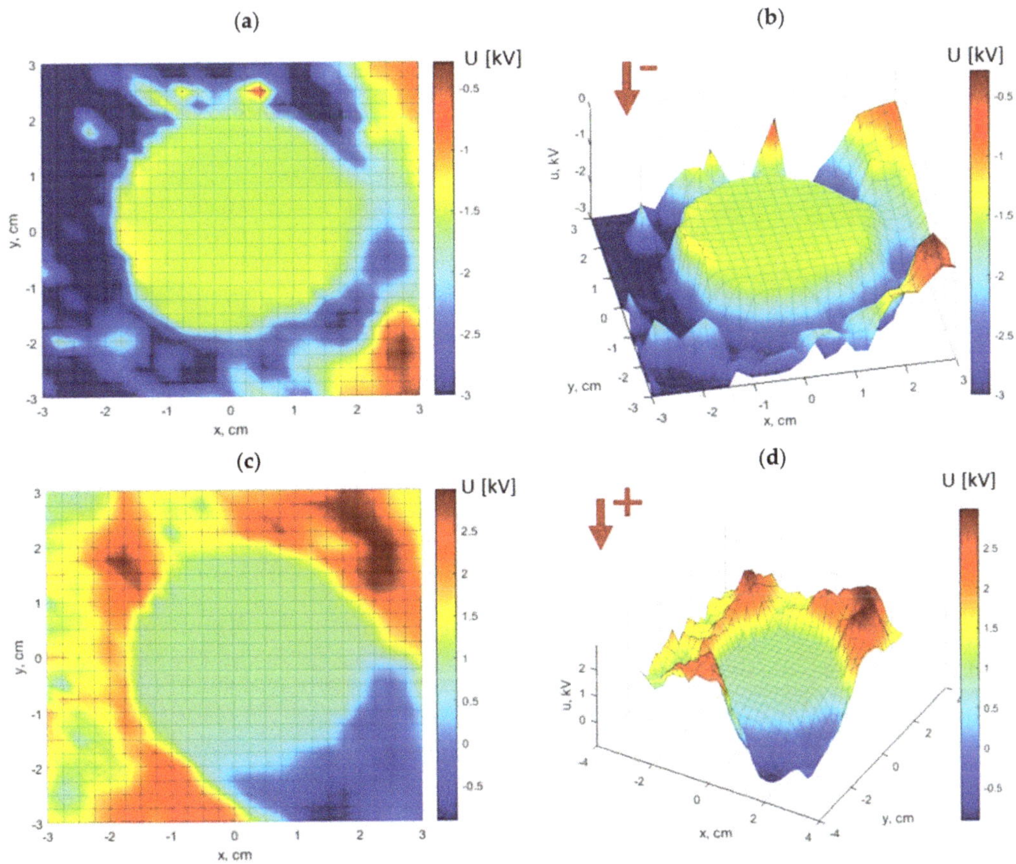

Figure 6. Surface-potential mapping on void walls of PE specimens (bottom side): (**a**,**c**) map views; (**b**,**d**) 3D views; (**a**,**b**) negative polarity of corona charging; (**c**,**d**) positive polarity of corona charging.

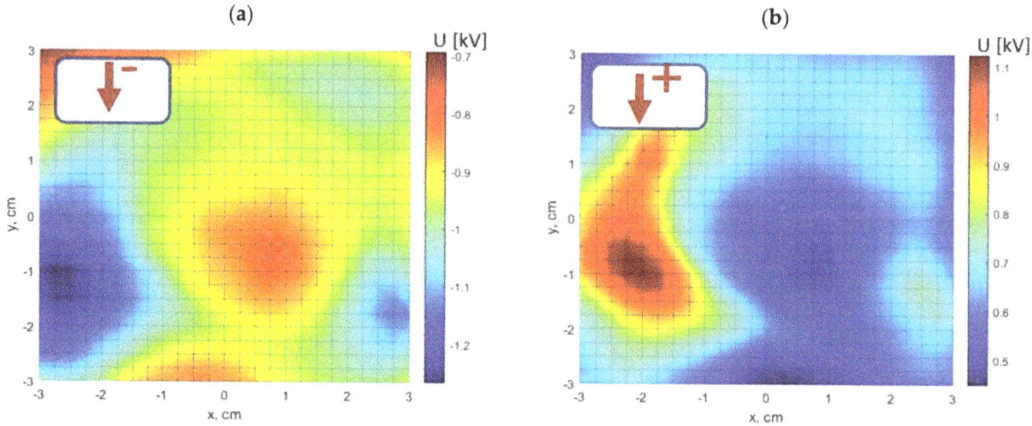

Figure 7. Surface-potential mapping on void walls of Nomex specimens (bottom sides)—map view: (**a**) negative polarity of corona charging; (**b**) positive polarity of corona charging.

5. Discussion

The analysis of the partial discharge-originated degradation of insulating material that is presented in this paper was carried out jointly by two techniques: namely, surface-resistance mapping and surface-potential mapping. The former was performed directly after accelerated deterioration tests at 1 kH, while the latter was executed in the second step immediately after the DC charge deposition with a known polarity. There are mainly three physical mechanisms that are associated with surface-charge decay: namely, gas neutralization due to environmental exposition (incl. the impacts of humidity and temperature), surface conduction, and bulk transport [27,33,34]. All of these mechanisms may have been present in the experiment; however, they would have manifested different dynamics. Surface conduction may have especially been present in the deteriorated regions on the gaseous inclusion walls. In this way, even separate spots that exhibited elevated levels of surface conductivity may have formed joint clusters over time. Also, the void boundary that represented the strong electric field tangential component due to the charge non-homogeneity may have supported this effect. In turn, the bulk transport could have been characterized by charge injection (dependent on the mobility and density of the charge carriers), polarization, and intrinsic conduction [34]. In the case of the flat specimen geometry and neglecting the existence of a space charge in the material bulk, the surface potential could have been assumed to be proportional to the surface-charge density. The applied combination of the surface-resistance and potential mapping revealed the effects of partial discharge action on a dielectric surface in two domains in a quantitative way; one reflecting the degradation level, and the second revealing the dynamics of the charge neutralization and transport. The graphical visualization of the mapping effects on the surface of the deteriorated insulating material is illustrated in Figure 8. The shown symbolic grid layout density depends on the applied measurement conditions, and the resolution can be separately defined for the resistance and potential measurements. There is a tradeoff between the raster spatial resolution and the total scanning time. The boundary of the void is marked in the graph, indicating the upper and bottom walls as well as distinguishing the untreated and deteriorated areas.

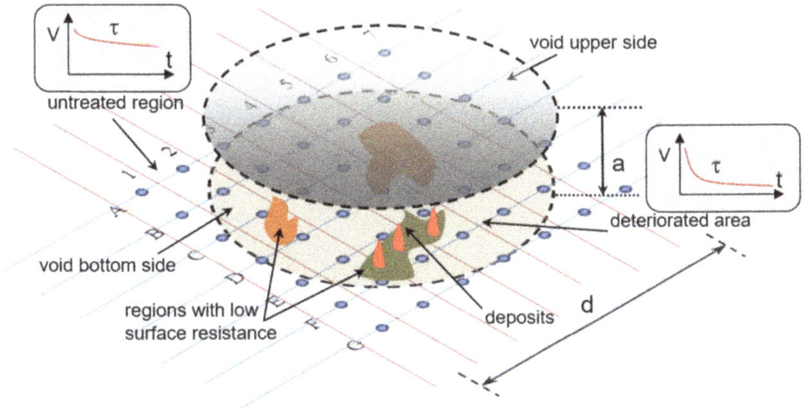

Figure 8. Visualization of mapping effects on surface of deteriorated insulating material.

The surface resistance-mapping results that were presented in the previous section (Figure 5) exhibit the PD channel-formation spots on the void walls with high resolution. The inclusion contour can be well-distinguished as well as the clusters inside the void that corresponded to the intensities of the PD at different spots. After extraction, the exact void areas for both specimens are shown in Figure 9.

(a) **(b)**

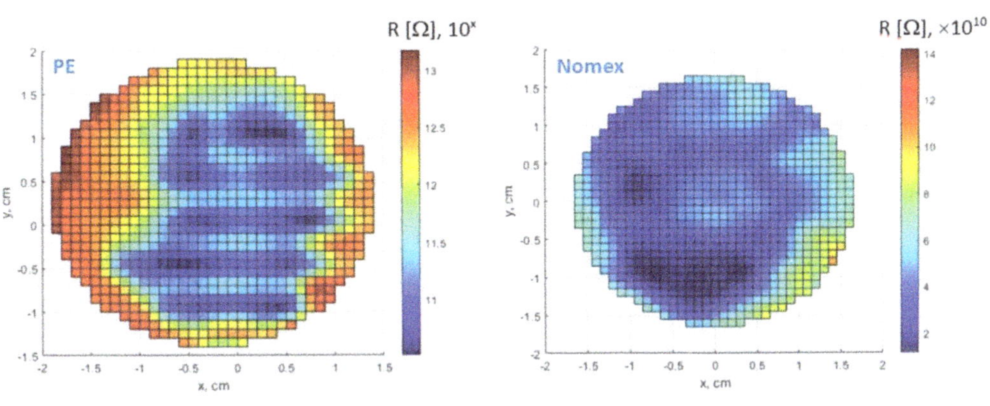

Figure 9. Extracted void areas deteriorated by partial discharges in high-frequency accelerated test (1 kHz, 18 kV) based on surface-resistance mapping: (**a**) PE; (**b**) Nomex.

In order to assess the degree of surface degradation, the following classification procedure was applied [19]. Threshold levels on the surface resistance were defined, which allowed us to reveal the S_i regions, which dropped one, two, or more orders of magnitude from the reference untreated level. The corresponding areas were calculated according to the following formula:

$$S_i = \iint_{x,y} S_\rho(x,y)dxdy \qquad \text{for } \mathrm{Th}_{\mathrm{low\_i}} < \rho(x,y) < \mathrm{Th}_{\mathrm{high\_i}}, \tag{1}$$

where $\mathrm{Th}_{\mathrm{low\_i}}$, $\mathrm{Th}_{\mathrm{high\_i}}$ define the thresholds that specified the sub-regions. In particular, $\mathrm{Th}_{\mathrm{low\_i}}$ can be set to zero; in this way, the area with a resistance below the $\mathrm{Th}_{\mathrm{high\_i}}$ value will be visualized. The percentage share P_i of this region with respect to the whole area of the void with diameter d yields the following:

$$P_i = \frac{S_i}{\pi d^2/4} \quad . \tag{2}$$

According to the above procedure, the degradation degrees (expressed by P_i) for both the PE and Nomex specimens are shown in Table 2. The threshold levels in the case of PE showed a drop in surface resistance R by three orders of magnitude, whereas Nomex showed a drop of two orders. The surface-resistivity mapping for the PE specimen with the applied thresholding (going down order by order of magnitude) is presented in Figure 10.

Table 2. Threshold levels and percentage shares (P_i) for PE and Nomex specimens.

Thresholds Th_{high_i} [Ω] Bottom Void Wall	PE P_i [%]	Thresholds Th_{high_i} [Ω] Bottom Void Wall	Nomex P_i [%]
$10^{10} < R \leq 1 \times 10^{11}$	17.0	$10^9 < R \leq 2.5 \times 10^{10}$	14.4
$1 \times 10^{11} < R \leq 1 \times 10^{12}$	39.3	$2.5 \times 10^{10} < R \leq 5 \times 10^{10}$	55.4
$1 \times 10^{12} < R \leq 1 \times 10^{13}$	42.6	$5 \times 10^{10} < R \leq 7.5 \times 10^{10}$	23.0
$1 \times 10^{13} < R \leq 1 \times 10^{14}$	1.1	$7.5 \times 10^{10} < R \leq 10^{11}$	6.0

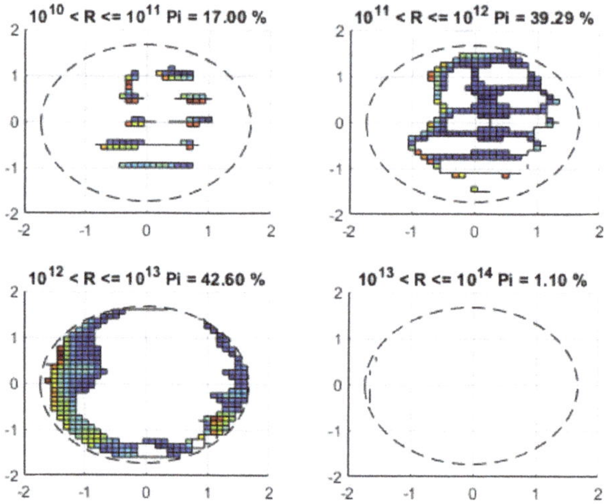

Figure 10. Surface-resistivity mapping for PE specimen with applied thresholding (Table 2): thresholds are marked on top of each plot, and the degradation degree is expressed by P_i factor.

The highest conductance ($R < 10^{11}$) was attributed to 17% of the void area, while the majority was occupied by a cluster where $R < 10^{12}$ (56.3%) and $R < 10^{13}$ (98.9%). However, radically different patterns were revealed in these two cases. The centric and radial focus in the first case that reflected the high-intensity PDs can be noticed, whereas the latter one represents the discharges that occurred around the boundary of the outer gas inclusion. In the cases of the regions with the lowest resistance ($R < 10^{11}$), the tiny hot spots are well-visible; these can be attributed to those places where the PDs had the greatest impacts.

In the case of Nomex (Figure 11), individual thresholding was applied since the span was smaller.

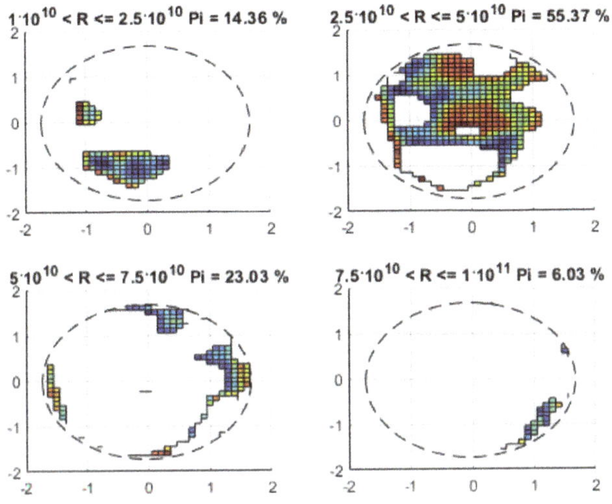

Figure 11. Surface-resistivity mapping for Nomex specimen with applied thresholding (Table 2): thresholds are marked on top of each plot, and the degradation degree is expressed by P_i factor.

The highest conductance (R < 2.5 × 10^{10}) could be attributed to 14.4% of the void area, whereas the majority was occupied by clusters for those resistances that were within a range of 2.5 × 10^{10} < R < 5 × 10^{10} (P_i = 55.4%). This is represented by a rather strongly clustered pattern that was centrally located in the void. The spot was relatively big in this case, unlike the individual dots that could be observed in the PE case above (Figure 10). The strongest PD action could be observed when applying the threshold (R < 2.5 × 10^{10}).

After the positive and negative charge depositions (shown in Figures 6 and 7 for the PE and Nomex specimens, respectively), the surface potential-mapping results exhibited the boundaries of the equipotential areas. From this perspective, the void interior represented a quite uniform and clustered pattern, even though the individual sub-regions that were distinguished by the resistance mapping may have varied in the order of the magnitude of their surface-resistance values. To analyze the dynamics of the surface-potential decay, the measurements were performed along a cross-section of the inclusion (along the x-axis at y = 0), passing both the untreated region and alongside the void diameter. In all of the cases, the charge was deposited in corona mode at 20 kV over a short time period (10 s). The time evolutions that were acquired over 35 min for the Nomex sample (bottom side) are shown in Figure 12a,b (negative polarity) and Figure 12c,d (positive polarity). To reveal the decay-time constants, the cuts that are marked by the white dashed lines in Figure 12a,c are shown in the plot in Figure 12b,d. For the deposition of the negative charge, the blue line in the plot (Figure 12b) traveled through the middle of the void and had a decay time of τ = 8.6 min, whereas the brown line (representing the untreated region) yielded τ = 16.5 min.

In the case of the positively polarized Nomex specimen's surface, a cross-section is shown in Figure 12c, and the corresponding cuts along the decay time are shown in Figure 12d. The blue line that is marked through the middle of the void had a decay time of τ = 9.1 min, whereas the brown line (representing the untreated region) yielded τ = 30.9 min. The polarity-dependent decay times were radically different for the non-aged region (and were much longer for the positive charging), whereas the potential-decay times inside the deteriorated void were quite similar for both polarities.

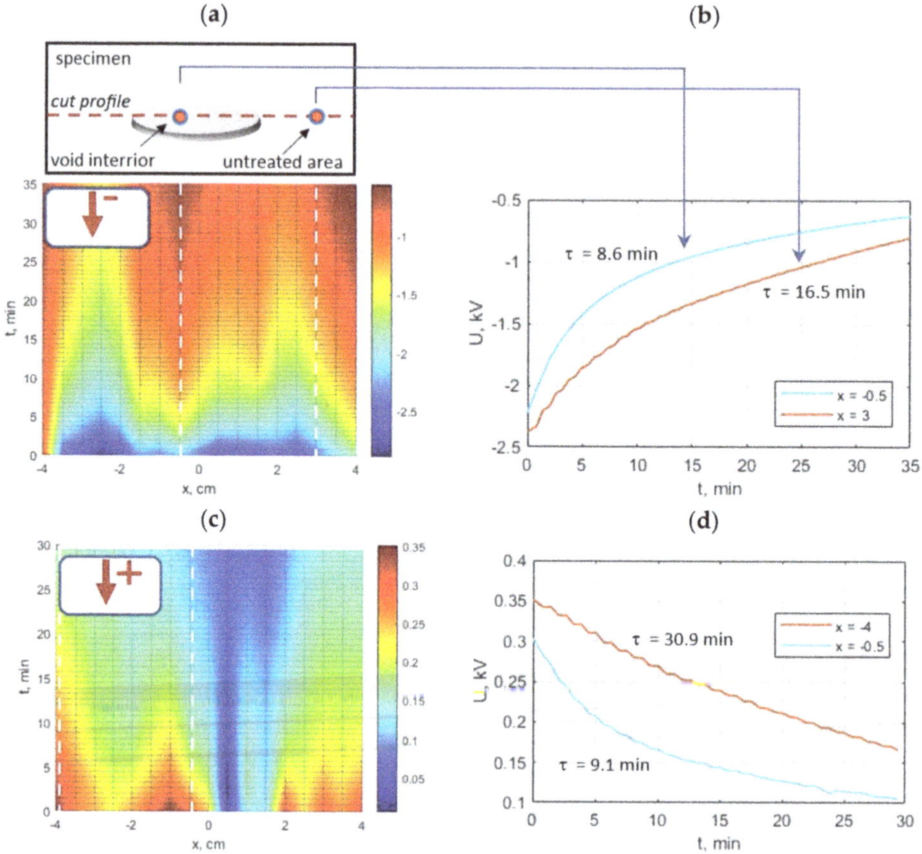

Figure 12. Surface-potential mapping-time evolution for Nomex—cross-section alongside x-axis at $y = 0$: (**a**) planar mapping—negative charging; (**b**) potential decay at selected cross-sections; (**c**) planar mapping—positive charging; (**d**) potential decay at selected cross-sections (charge deposition over 10 s: (**a**,**b**) DC −20 kV; (**c**,**d**) +20 kV—white dashed lines represent time evolution cutting profiles).

The presented approach revealed the non-homogenous deterioration of the void walls, which reflected the exact spots of strong PD penetration. In both cases (PE and Nomex), more-intensive PD action could be observed in the central and radial directions than at the inclusion's external round and boundary zones. The long-term partial discharge exposure resulted in the transition of the effective discharge area. The presented approach may also be utilized in research on the various charge-decay mechanisms.

6. Conclusions

The goal of this paper was to describe a novel combined approach that was based on surface-resistance and potential mapping to reveal the effects of the internal processes and deterioration of insulating materials due to the actions of partial discharges. A dedicated measurement methodology was elaborated in order to properly address these measurements. The experiments were performed on flat PE and Nomex specimens that contained embedded voids. Multi-point resistance mapping allowed us to identify the spots of the discharge channels, which manifested a few orders of magnitude lower surface resistance as compared to the untreated areas. In addition, surface-potential mapping that was stimulated by corona-charge deposition reflected the quasi-equipotential clusters and related dynamics of the charge decay. The derived time evolution of the surface-potential mapping revealed quite similar decay times for the deteriorated void interior (negative charging of Nomex—τ = 8.6 min; positive charging—τ = 9.1 min); these reflected the corresponding drops in the surface resistance. The high spatial and temporal resolution allowed for the precise mapping and tracing of the decay patterns. During the PD events, the effective discharge areas were identified with the memory effects that originated from the accumulation of the surface charges. The experimental results showed the formation of clusters in the deteriorated area, which revealed a few orders of magnitude reduced surface resistance; the developed methodology allowed for the precise localization of these regions. The surface-potential mapping indicated those spots that represented the equipotential behavior and quantitatively showed the potential decay processes. The two measurement approaches that were presented reflected a quantitative correlation and correspondence between the untreated and aged regions (pinpointing subclusters in the latter case). It could be observed that the PD action resulted in a few orders of magnitude surface resistance drop in the deteriorated void areas. Furthermore, the potential mapping revealed the creation of equipotential clusters. The patterns of the clusters and the individual spots reflected the local PD intensities. It can be stated that the void walls were not homogenously stressed nor deteriorated, leading to the formation and transition of the effective discharge area within the aging time. The long-term aging processes may have driven the formation of channels that were initiated from the deteriorated micro clusters, in turn, penetrating the bulk isolation. The introduced measurement approach may be applied to the diagnostics of certain electrical power components, such as high-voltage insulators, GIS/GIL spacers, or power transformer insulating component during revisions. The compact version of the measurement rig will allow easy transportation to the inspection site. The follow up work will be two-fold, on one site will refer to the extensions of the measurement setup capabilities in terms of the further improvement of the sensitivity, probe design and measurement speed of the overall procedure, on the other side will focus on the theoretical foundations and interpretation of the obtained results. The presented methodology and experimental results extend the insight into PD mechanisms and internal surface processes.

Author Contributions: Conceptualization & methodology, M.F. and M.K.; formal analysis, M.F. and M.K. investigation and resources, M.F. and M.K.; data curation, M.F. and M.K.; writing—original draft preparation, M.F. and M.K.; visualization, M.F. and M.K.; supervision and project administration, M.F. and M.K. All authors have read and agreed to the published version of the manuscript.

Funding: This research received no external funding.

Data Availability Statement: Not applicable.

Conflicts of Interest: The authors declare no conflict of interest.

References

1. Li, C.; Yang, Y.; Xu, G.; Zhou, Y.; Jia, M.; Zhong, S.; Gao, Y.; Park, C.; Liu, Q.; Wang, Y.; et al. Insulating materials for realising carbon neutrality: Opportunities, remaining issues and challenges. *High Volt.* **2022**, *7*, 610–632. [CrossRef]
2. Mazzanti, G. Issues and Challenges for HVDC Extruded Cable Systems. *Energies* **2021**, *14*, 4504. [CrossRef]
3. Rajashekara, K. Power Conversion Technologies for Automotive and Aircraft Systems. *IEEE Electrif. Mag.* **2014**, *6*, 50–60. [CrossRef]
4. Rumi, A.; Lusuardi, L.; Cavallini, A.; Pastura, M.; Barater, D.; Nuzzo, S. Partial Discharges in Electrical Machines for the More Electrical Aircraft. Part III: Preventing Partial Discharges. *IEEE Access* **2021**, *9*, 30113–30123. [CrossRef]
5. Fabiani, D.; Montanari, G.C.; Cavallini, A.; Mazzanti, G. Relation between Space Charge Accumulation and Partial Discharge Activity in Enameled Wires under PWM-like Voltage Waveforms. *IEEE Trans. Dielectr. Electr. Insul.* **2004**, *11*, 393–405. [CrossRef]
6. Lebey, T.; Rumi, A.; Cavallini, A. Challenges for Electrical Insulation Systems in High Voltage Aviation Applications. *IEEE Electr. Insul. Mag.* **2022**, *38*, 5–11. [CrossRef]
7. Mazzanti, G.; Chen, G.; Fothergill, J.; Fu, M.; Hozumi, N.; Lee, J.H.; Li, J.; Marzinotto, M.; Mauseth, F.; Morshuis, P.; et al. The insulation of HVDC extruded cable system joints. Part 1: Review of materials, design and testing procedures. *IEEE Trans. Dielectr. Electr. Insul.* **2019**, *26*, 964–972. [CrossRef]
8. Tanaka, T. Internal partial discharge and material degradation. *IEEE Trans. Dielectr. Electr. Insul.* **1986**, *EI-21*, 899–905. [CrossRef]
9. McAllister, I.W. Decay of charge deposited on the wall of gaseous void. *IEEE Trans. Electr. Insul.* **1992**, *27*, 1202–1207. [CrossRef]

10. Tanaka, T.; Uchiumi, M. Two kinds of decay time constants for interfacial space charge in polyethylene-laminated dielectrics. In Proceedings of the 1999 Annual Report Conference on Electrical Insulation and Dielectric Phenomena, Austin, TX, USA, 17–20 October 1999; pp. 472–475.
11. Niemeyer, L. A generalized approach to partial discharge modeling. *IEEE Trans. Dielectr. Electr. Insul.* **1995**, *2*, 4. [CrossRef]
12. Florkowski, M. Influence of Insulating Material Properties on Partial Discharges at DC Voltage. *Energies* **2020**, *13*, 4305. [CrossRef]
13. Gutfleisch, F.; Niemeyer, L. Measurement and simulation of PD in epoxy voids. *IEEE Trans. Dielectr. Electr. Insul.* **1995**, *2*, 729–743. [CrossRef]
14. Wu, K.; Suzuoki, Y. Effects of discharge area on PD patterns in voids. In Proceedings of the 1999 Annual Report Conference on Electrical Insulation and Dielectric Phenomena, Austin, TX, USA, 17–20 October 1999; pp. 227–230.
15. Florkowska, B. Partial discharge measurements with computer aided system in polyethyleneterephthalate and polypropylene films. In Proceedings of the 8th International Symposium on High Voltage Engineering, Yokohama, Japan, 23–27 August 1993; pp. 41–44.
16. Florkowski, M.; Florkowska, B.; Włodek, R. Investigations on post partial discharge charge decay in void using chopped sequence. *IEEE Trans. Dielectr. Electr. Insul.* **2017**, *24*, 3831–3838. [CrossRef]
17. Wu, K.; Ijichi, T.; Kato, T.; Suzuoki, Y.; Komori, F.; Okamoto, T. Contribution of surface conductivity to the current forms of partial discharges in voids. *IEEE Trans. Dielectr. Electr. Insul.* **2005**, *12*, 1116–1124.
18. Florkowski, M.; Florkowska, B.; Kuniewski, M.; Zydron, P. Mapping of discharge channels in void creating effective partial discharge area. *IEEE Trans. Dielectr. Electr. Insul.* **2018**, *25*, 2220–2228. [CrossRef]
19. Florkowski, M.; Kuniewski, M. Correspondence Between Charge Accumulated in Voids by Partial Discharges and Mapping of Surface Charge with PEA Detection. *IEEE Trans. Dielectr. Electr. Insul.* **2022**, *29*, 2199–2208. [CrossRef]
20. Rachmawati; Kojima, H.; Kato, K.; Zebouchi, N.; Hayakawa, N. Electric Field Grading and Discharge Inception Voltage Improvement on HVDC GIS/GIL Spacer With Permittivity and Conductivity Graded Materials (ε/σ-FGM). *IEEE Trans. Dielectr. Electr. Insul.* **2022**, *29*, 1811–1817. [CrossRef]
21. Florkowski, M. *Partial Discharges in High-Voltage Insulating Systems—Mechanisms, Processing, and Analytics*; AGH Press: Kraków, Poland, 2020; ISBN 978-83-66364-75-2.
22. Li, C.; Shahsavarian, T.; Baferani, M.A.; Caoa, Y. Tailoring insulation surface conductivity for surface partial discharge mitigation. *Appl. Phys. Lett.* **2021**, *119*, 032903. [CrossRef]
23. Kozioł, M.; Nagi, Ł.; Boczar, T.; Nadolny, Z. Quantitative Analysis of Surface Partial Discharges through Radio Frequency and Ultraviolet Signal Measurements. *Energies* **2023**, *16*, 3641. [CrossRef]
24. Wu, K.; Suzuoki, Y.; Dissado, L.A. Improved simulation model for PD pattern in void considering effects of discharge area. In Proceedings of the 2003 Annual Report Conference on Electrical Insulation and Dielectric Phenomena, Albuquerque, NM, USA, 19–22 October 2003; pp. 32–35.
25. Illias, H.; Chen, G.; Lewin, P. Partial discharge within a spherical cavity in a dielectric material as a function of cavity size and material temperature. *IET Sci. Meas. Technol.* **2012**, *6*, 52–62. [CrossRef]
26. Wu, K.; Pan, C.; Meng, Y.; Cheng, Y. Effect of void area on PD magnitude uniformity. In Proceedings of the 2013 Annual Report Conference on Electrical Insulation and Dielectric Phenomena, Shenzhen, China, 20–23 October 2013; pp. 1165–1168.
27. Pan, C.; Wu, K.; Chen, G.; Gao, Y.; Florkowski, M.; Lv, Z.; Tang, J. Understanding Partial Discharge Behavior from the Memory Effect Induced by Residual Charges: A Review. *IEEE Trans. Dielectr. Electr. Insul.* **2020**, *27*, 1951–1965.
28. Florkowski, M. Imaging and simulations of positive surface and airborne streamers adjacent to dielectric material. *Measurement* **2021**, *186*, 110170. [CrossRef]
29. Callender, G.; Lewin, P.L. Modeling partial discharge phenomena. *IEEE Electr. Insul. Mag.* **2020**, *36*, 29–36. [CrossRef]
30. Lemke, E.; Muhr, M.; Hauschild, W. Modeling of Cavity Discharges under AC and DC Voltage—Part II: Opportunities of the Dipole-based PD Model. *IEEE Trans. Dielectr. Electr. Insul.* **2020**, *27*, 1853–1860. [CrossRef]
31. Imburgia, A.; Romano, P.; Rizzo, G.; Viola, F.; Ala, G.; Chen, G. Reliability of PEA Measurement in Presence of an Air Void Defect. *Energies* **2020**, *13*, 5652. [CrossRef]
32. Wu, K.; Okamoto, T.; Suzuoki, Y. Effects of discharge area and surface conductivity on partial discharge behavior in voids under square voltages. *IEEE Trans. Dielectr. Electr. Insul.* **2007**, *14*, 461–470. [CrossRef]
33. Pan, C.; Wu, K.; Du, Y.; Meng, Y.; Cheng, Y.; Tang, J. The effect of surface charge decay on the variation of partial discharge location. *IEEE Trans. Dielectr. Electr. Insul.* **2016**, *23*, 2241–2249. [CrossRef]
34. Kumara, J.R.S.S.; Serdyuk, Y.V.; Gubanski, S.M. Surface potential decay on LDPE and LDPE/Al_2O_3 nano-composites: Measurements and modeling. *IEEE Trans. Dielectr. Electr. Insul.* **2016**, *23*, 3466–3475. [CrossRef]
35. Pan, C.; Wu, K.; Meng, Y.; Cheng, Y.; Tang, J. The effect of discharge area variation on stochastic characters of PD magnitude. *IEEE Trans. Dielectr. Electr. Insul.* **2017**, *24*, 217–226. [CrossRef]
36. Florkowski, M. Magnetic field modulated dynamics of partial discharges in defects of high voltage insulating materials. *Nat. Sci. Rep.* **2022**, *12*, 22048.
37. Wang, M.; Du, B.X.; Kong, X.; Liang, H.; Ma, Y. High Magnetic Field Affecting Dielectric Polarization and Partial Discharge Behavior of Epoxy Resin. *IEEE Trans. Dielectr. Electr. Insul.* **2022**, *29*, 1267–1274.
38. Noras, M.A.; Pandey, A. Surface charge density measurements. *IEEE Ind. Appl. Mag.* **2010**, *7–8*, 41–47.

39. Takamura, T.; Yashima, M.; Kawamoto, T. Principle of surface charge measurement for thick insulating specimens. *IEEE Trans. Dielectr. Electr. Insul.* **1998**, *5*, 497–504.
40. Nikonov, V.; Bartnikas, R. The Influence of Dielectric Surface Charge Distribution Upon the Partial Discharge Behavior in Short Air Gaps. *IEEE Trans. Plasma Sci.* **2001**, *29*, 866–874.
41. Kacprzyk, R.; Pelesz, A. Problems of potential distribution measurements on objects with a constant charge. *Prz. Elektrotech.* **2018**, *94*, 172–175.

energies

MDPI

Article

Advanced Diagnostic Approach for High-Voltage Insulators: Analyzing Partial Discharges through Zero-Crossing Rate and Fundamental Frequency Estimation of Acoustic Raw Data

Kaynan Maresch [1,*], Luiz F. Freitas-Gutierres [1], Aécio L. Oliveira [1], Aquiles S. Borin [1], Ghendy Cardoso, Jr. [1], Juliano S. Damiani [1], André M. Morais [2], Cristian H. Correa [3] and Erick F. Martins [3]

[1] Laboratory for Analysis and Protection of Electrical Systems, Technology Center, Federal University of Santa Maria, Santa Maria 97105-900, RS, Brazil
[2] High and Extra High Voltage Laboratory, Institute of Technology, Federal University of Pará, Pará 66075-110, PA, Brazil
[3] Engineering Board, CPFL Transmission, Porto Alegre 90230-181, RS, Brazil
[*] Correspondence: msc.maresch@gmail.com

Abstract: Acoustic inspection is a valuable technique that can detect early stage defects in equipment, thereby facilitating predictive maintenance. In recent times, ultrasonic sensors have made detecting partial discharges through acoustic sensing increasingly feasible. However, interpreting the acoustic signals can pose challenges, as it requires extensive expertise and knowledge of equipment configuration. To address this issue, a technique based on zero-crossing rate and fundamental frequency estimation has been proposed to standardize insulator diagnosis. In an experiment involving a database of 72 raw acoustic signals with frequencies ranging from 0 to 128 kHz, various types of pollution and defects were introduced to a chain of insulators. By employing the proposed technique, the occurrence of partial discharges can be detected and classified according to type, such as corona or surface discharges. This advanced approach to diagnosis simplifies the process while providing valuable insights into the severity of observed phenomena in the field.

Keywords: classification; high-voltage; insulator; partial discharge; acoustic inspection

Citation: Maresch, K.; Freitas-Gutierres, L.F.; Oliveira, A.L.; Borin, A.S.; Cardoso, G., Jr.; Damiani, J.S.; Morais, A.M.; Correa, C.H.; Martins, E.F. Advanced Diagnostic Approach for High-Voltage Insulators: Analyzing Partial Discharges through Zero-Crossing Rate and Fundamental Frequency Estimation of Acoustic Raw Data. *Energies* **2023**, *16*, 6033. https://doi.org/10.3390/en16166033

Academic Editor: Tomasz Norbert Kołtunowicz

Received: 14 June 2023
Revised: 3 July 2023
Accepted: 6 July 2023
Published: 17 August 2023

1. Introduction

Partial discharges (PDs) are spatially limited and incomplete dielectric breakdowns, in which the voltage gradient is only strong enough to sustain the electric discharge in the vicinity of the point of inception [1]. According to [2], there are four types of PDs: corona, internal, surface and electrical treeing discharges.

As shown in Figure 1A, corona discharges occur if the electric field is sufficiently non-uniform in the surrounding air. This type of PD is established, commonly, near sharp points or where conductors are curved on transmission lines. Corona discharges may be accompanied by a bluish luminescence and a hissing noise, containing ultrasound and audible acoustic signals. Moreover, corona discharges could have a transient or steady state behavior, and they may lead to insulation deterioration.

Figure 1B illustrates an internal discharge that may occur in voids or cavities inside the insulator layer, where there is low dielectric strength. The continuous occurrence of internal discharges in a solid dielectric may lead to major defects in insulation and the formation of discharge channels, thus resulting in electrical trees, see Figure 1C. Furthermore, surface discharges, shown in Figure 1D, occur when there is a relatively high parallel component of the electric field to a dielectric surface [2].

PDs are typically accompanied by the emission of sound, light, heat and chemical reactions [1]. Therefore, the most common techniques to measure PDs are based on acoustic, electrical, optical, thermal and chemical methods of detection. Electrical methods are premised on the coupling of a measuring instrument (MI) in the system that has equipment subjected to PDs. Through this MI, voltage and current signals are taken and PDs are evaluated. The measurement system for high-voltage (HV) tests is presented in [3].

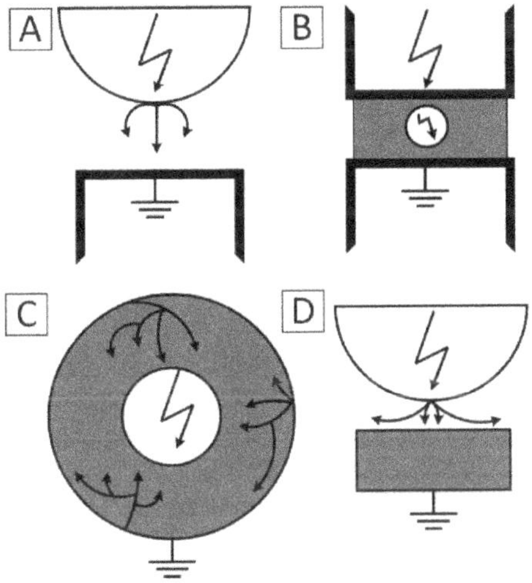

Figure 1. Types of partial discharges: (**A**) corona; (**B**) internal; (**C**) treeing channel; (**D**) surface discharge.

Visual or optical detection is based on the emission of light from PDs. It may require a dark room or, alternatively, a photographic record could be used throughout a long exposure time [2]. The detection of PDs made from chemical reactions is more commonly used in systems with oil or gas insulators. It consists of the analysis of decomposition products dissolved in the insulating media, e.g., humidity and gas [1].

Acoustic methods are based on the measurement of the sound produced by electrical discharges. In this regard, audible sound is, typically, generated by severe PDs and electric arcs. By comparison, the satisfactory detection of incipient PDs depends on ultrasound-based techniques [2]. The use of ultrasound signals to evaluate the occurrence of PDs is a well-established technique and has been used in previous studies, such as [4,5]. In these articles, the authors discuss theoretical and practical foundations for the online (or otherwise stated, energized) PD evaluation in gas-insulated substations [4] and power transformers [5].

More accurate ultrasonic sensors have been introduced on the market, resulting in an emergence of sophisticated methods for evaluating PDs. A feature that turns acoustic detection into a powerful technique is the immunity to electromagnetic interference (EMI), making it a suitable method for evaluating PDs in energized equipment [6]. In this regard, ultrasound inspection is used in different types of equipment, including generators, transformers, cables and insulators. However, the acoustic measurement could be contaminated with noise, which makes the identification of PDs difficult and complex [6].

To take account of the external noise in acoustic measurement, new approaches developed in [7] combine the use of distinct sensors with signal processing to eliminate noise and generate diagnostics of the state of the equipment.

The wavelet transform (WT) is widely used in signal noise reduction, e.g., the techniques in [8–11]. In [8], the authors reported the use of wavelet energy coefficient and principal component analysis (PCA) for glass and polymer insulators; wavelet packet transform (WPT) for glass insulators [9]; and WT, filtering and noise reduction for polymeric insulators [10].

Figure 2 shows the main types of sensors, signal processing techniques and diagnostic methods applied to acoustic inspection.

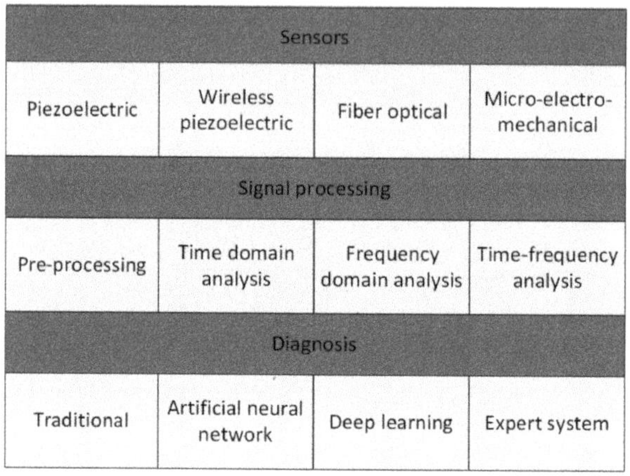

Figure 2. An overview of acoustic inspection. Adapted from [6].

In [11], a method is proposed for the evaluation of surface discharges in HV glass insulators. This method is based on WT for signal processing and on radial basis function neural networks for identifying PDs. A similar technique is also proposed in [12]. In this approach, a hybrid model that combines particle swarm optimization (PSO), WT and neural networks is used to classify PDs in a gas-insulated switchgear (GIS).

An ultrasonic diagnostic method is proposed in [13] for monitoring PDs in medium-voltage pin-type insulators. This work takes advantage of a normalized frequency, obtained from measurements in the insulators, to identify the magnitude of PDs. In [14], the state echo network is used for the classification of insulators based on ultrasound signals. This study, [14], demonstrates that the proposed neural network presented an 87.36% accuracy for multiclassification. The insulators are evaluated in the laboratory under controlled conditions at a voltage of 13.8 kV (phase-to-phase). Furthermore, this work has shown that the echo state network performed better than multilayer and support vector-machines.

This paper presents a novel methodology for detecting and categorizing PDs in HV glass insulators using raw (unprocessed) acoustic signals. The acoustic signals are captured within a frequency range of 0 to 128 kHz without undergoing any prior processing, such as heterodyning (a widely employed technique in acoustic inspection equipment). The conducted experiments involve insulators contaminated with various types of artificial pollutants to simulate real-world conditions that can give rise to PDs. For the diagnosis of the raw acoustic signals, two techniques are employed: zero-crossing rate and fundamental frequency estimation. These techniques contribute to the identification of anomalies and the categorization of PD severity. The proposed method standardizes the diagnosis of insulators and categorizes PDs into corona and superficial types. This, in turn, helps to predict failures and reduce unwarranted shutdowns by facilitating scheduled maintenance.

The paper's primary contributions are as follows:

- Firstly, it introduces a method that does not rely on extensive training datasets for neural networks or artificial intelligence models. This departure from traditional machine learning approaches eliminates the need for large-scale data collection and time-consuming training processes.
- Furthermore, the proposed technique offers distinct advantages for acoustic inspections, particularly in the context of energized substations. By enabling online diagnosis directly in the field, it enhances the efficiency and effectiveness of HV glass insulator assessments. This capacity empowers maintenance teams to make real-time decisions and take proactive measures to ensure the reliable operation of equipment.
- Categorizing PDs according to type (using the proposed methodology) and understanding the spatial distribution of acquired acoustic emissions (via the measuring device) enables the assessment of the severity of these phenomena and the associated

risks to electrical equipment. The potential offered by the proposed technique can be further enhanced when integrated with regular acoustic inspections and the historical monitoring of the equipment.

- Finally, the technique greatly mitigates challenges associated with interpreting acoustic data in the electrical sector. Presently, acoustic signal interpretation heavily relies on the experience of inspectors and, often, the subjective act of listening to signals using headphones or auscultation devices. This reliance on human expertise inherently introduces limitations and inconsistencies. The proposed method offers a standardized approach, reducing reliance on individual experience and establishing a more reliable and objective framework for acoustic signal analysis.

The structure of this paper is outlined as follows. Section 2 presents the mathematical foundation underlying the proposed method. The experiments and the characteristics of the insulators tested in a laboratory are described in Section 3. Subsequently, Section 4 reports the evaluation and analysis of the measurements with the proposed technique. Section 5 substantiates the validation of the diagnostic approach, while Section 6 presents the concluding remarks and findings derived from this study.

2. Mathematical Basis of the Proposed Method

Two techniques underlie the evaluation of PDs in this paper: zero-crossing rate (ZCR) and fundamental frequency estimation. As will be discussed in the following sections, ZCR is a technique used to detect anomalies in audio frames. In this sense, anomalies are defined in this paper as the presence of recurrent and accentuated peaks in the ultrasound signal, which lead to changes in the ZCR result. When changes occur, the fundamental frequency is calculated and utilized to classify ultrasound signals as normal or abnormal. In the event of identifying anomalies, they can be categorized as corona or surface PDs, or may indicate the presence of an inconclusive phenomenon. The theoretical aspects of ZCR and fundamental frequency estimation are discussed below.

2.1. Zero-Crossing Rate (ZCR)

The zero-crossing rate (ZCR) of an audio measurement represents the rate of changes of the signal during a frame. In other words, it is equal to the number of times the signal modifies from a positive to a negative value, divided by the length of the frame [15], in the following manner:

$$\mathrm{ZCR}(i) = \frac{1}{2W_L}\sum_{n=1}^{W_L}|sgn[x_i(n)] - sgn[x_i(n-1)]| \tag{1}$$

where sgn is the function (2), WL is the frame length, x_i is the instantaneous value and n is the number of the sample. In this case, the ZCR result is dimensionless.

$$sgn[x_i(n)] = \begin{cases} 1, x_i(n) \geq 0 \\ -1, x_i(n) < 0 \end{cases} \tag{2}$$

Moreover, ZCR may also reflect, in a coarse manner, the spectral characteristics of a signal [16].

2.2. Fundamental Frequency Estimation

Frequency estimation is performed through the power spectral density (PSD). The PSD technique describes how the power content of a signal distributes as a function of the frequency. The power spectrum is obtained from the Fourier transform of the autocorrelation function of the acoustic signal. The autocorrelation function is a powerful mathematical tool to evidence repeating patterns, such as an unknown periodic signal hidden by noise. Additionally, it can identify the fundamental frequency of a signal with harmonics through the linear relationship between delayed values in a time-series data [17,18].

For random signals, the periodogram power spectral density (S_x) is given by (3);

$$S_x(e^{j\omega}) = \lim_{N\to\infty} E\left\{ \frac{1}{2N+1} \left| \sum_{n=-N}^{N} x(n)e^{-j\omega n} \right|^2 \right\} \quad (3)$$

where E is the evaluation, that is, the mathematical expectation.

Considering a finite amount of data with lengths equal to $N-1$, (3) can be rewritten as (4):

$$S_x(e^{j\omega}) = \frac{1}{N} \left| \sum_{n=0}^{N-1} x(n)e^{-j\omega n} \right|^2 = \frac{1}{N} \left| X(e^{j\omega}) \right|^2 \quad (4)$$

where $X(e^{j\omega})$ is the discrete-time Fourier transform (DTFT) of the windowed signal, which in turn is given by (5). The rectangular window has the same length as the input signal and the frequency sample is equivalent to 256 kHz.

$$\omega = \frac{2\pi k}{N}, k = 0, 1, 2, \ldots N-1 \quad (5)$$

Through PSD, the fundamental frequency and harmonics characteristics can be calculated, as performed in Section 4.

3. Laboratory Experiments

The purpose of laboratory tests is to produce different types and intensities of PDs in a chain of HV glass insulators. PDs are generated during an increase in the applied voltage. Artificial pollution was applied to the insulators to replicate real operational conditions that may influence the occurrence of PDs. For each voltage step, ultrasound measurements were made at a fixed distance.

The measurement methodology, the description of the artificial pollution and the ultrasound measurement process are described below.

3.1. Measurement Methodology

The methodology of the laboratory experiments is shown in Figure 3.

The equipment illustrated in Figure 3 consists of:

1. Resonant controller;
2. Power transformer;
3. Resonant source;
4. Device under test—insulator chain (IC);
5. Capacitive divider;
6. Digital Digital ultrasound testing device, positioned at a distance of 12.5 m from the device under test;
7. Oscilloscope;
8. Acquisition system of the capacitive divider;
9. Computer.

In the tests, a chain of HV glass insulators with three elements was used, in which different types of artificial pollution were applied. The technical sheet of the insulator is presented in [19].

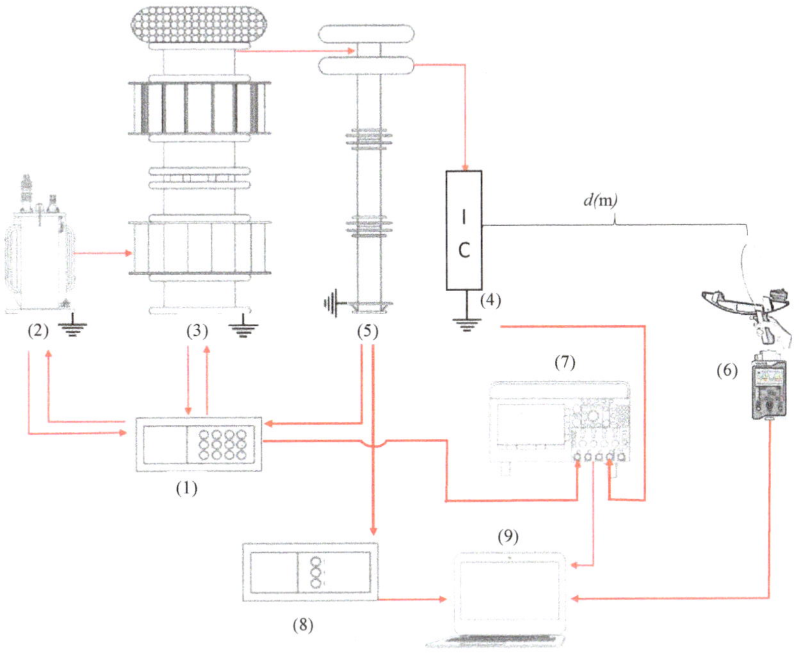

Figure 3. Experimental method organization.

3.2. Description of Artificial Pollution

The tests are intended to reproduce real conditions, to which an insulator chain, as presented in Figure 4a, may be subjected. For example, the formation of a conductive microfilm deposited by rain, wind, or other environmental conditions; the presence of mud, which can be formed by dust and/or bird droppings; and partial damage and degradation.

The first case is illustrated in Figure 4b and it is characterized by a conductive microfilm produced with saline pollution, distributed partially or totally along the superficies of the insulators. For the second case, mud pollution (a mixture of soil and water) was applied as shown in Figure 4c. Finally, partial damage was performed on the surface of the insulator. In Figure 4d, small cracks are perceptible in the insulator.

Table 1 shows the condition of the insulators and the voltage range used in the laboratory experiments. The insulator chain was subjected to voltage levels ranging from 10 kV to the highest level supported by the laboratory's resonant source, until the overcurrent protection system intervened.

Table 1. Test cases.

Case	Condition	Voltage Range (kV)
1	Clean and undamaged insulator chain.	10 to 100
2	Partial pollution with saline solution. Location: Upper element of the insulator chain.	10 to 80
3	Total pollution with saline solution. Location: All elements of the insulator chain.	10 to 70
4	Partial pollution with mud. Location: Upper element of the insulator chain.	10 to 120
5	Partial pollution with mud. Location: Upper and intermediate elements of the insulator chain.	10 to 120
6	Total pollution with mud. Location: All elements of the insulator chain.	10 to 100
7	Upper element of the insulator chain with partial and surface damages.	10 to 130

Figure 4. (**a**) Clean and undamaged insulator chain; (**b**) total pollution with saline solution; (**c**) partial and surface damages; (**d**) partial pollution with mud.

Importantly, the maximum voltage levels depend on the conditions of the insulator (pollution levels) and on environmental conditions, since humidity and temperature influence the occurrence of PDs, owing to either favorable or unfavorable ionization conditions in the air. Therefore, laboratory tests carried out on days with distinct environmental conditions with the same chain of insulators may lead to different maximum voltage levels required for partial dielectric breakdown.

3.3. Ultrasound Measurement Process

Ultrasound measurements were made at a fixed distance of 12.5 m with a sampling rate of 256 kHz during a period of 10 s. The measurements are composed of raw audio signal sized in microvolts and frequency spectrum, both time-varying. Figures 5–11 show the amplitude and frequency spectrogram for each test case under different voltage intensities.

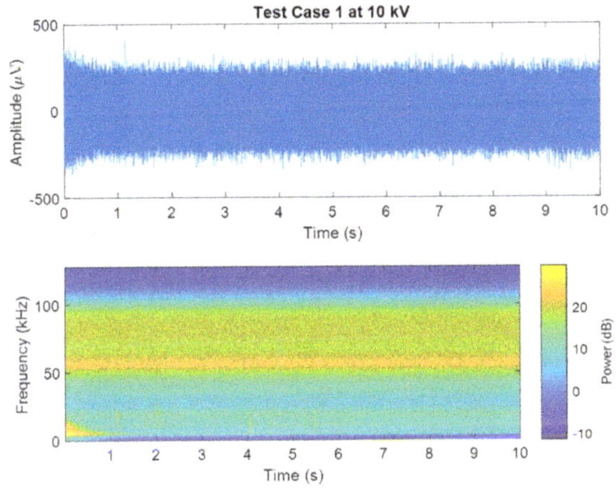

Figure 5. Amplitude and frequency spectrogram of the ultrasound signal for Test Case 1 at 10 kV.

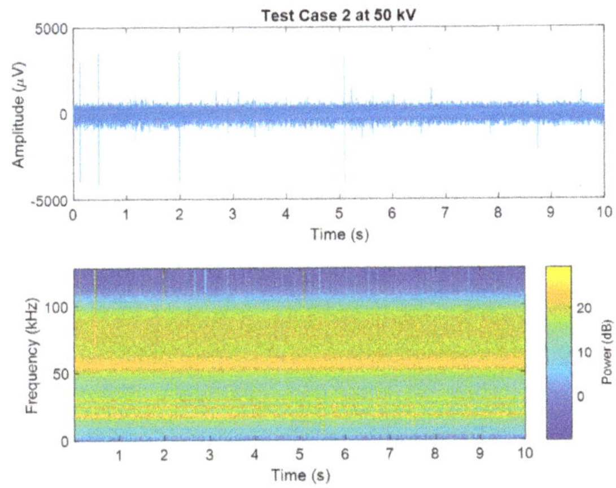

Figure 6. Amplitude and frequency spectrogram of the ultrasound signal for Test Case 2 at 50 kV.

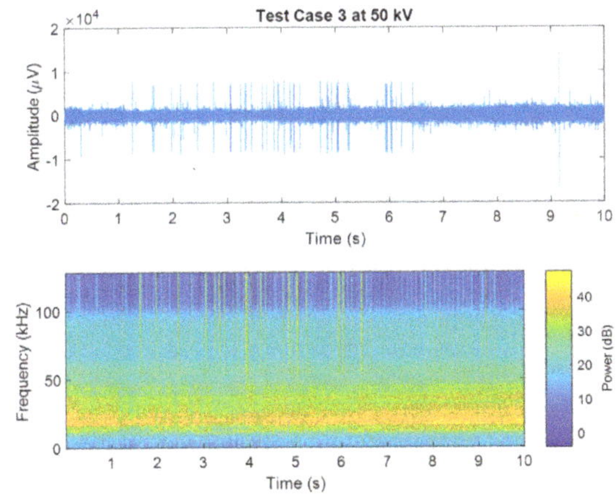

Figure 7. Amplitude and frequency spectrogram of the ultrasound signal for Test Case 3 at 50 kV.

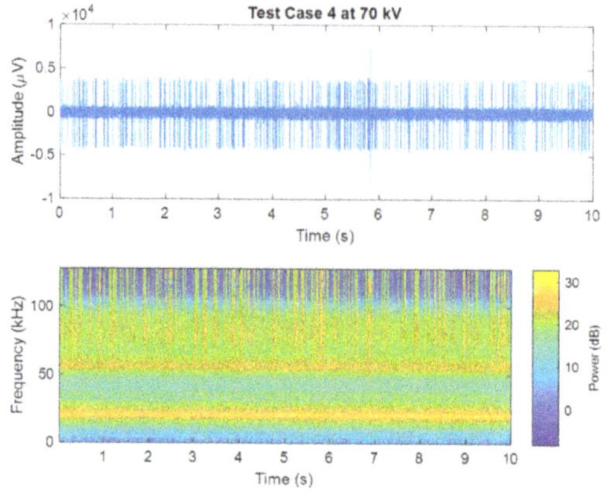

Figure 8. Amplitude and frequency spectrogram of the ultrasound signal for Test Case 4 at 70 kV.

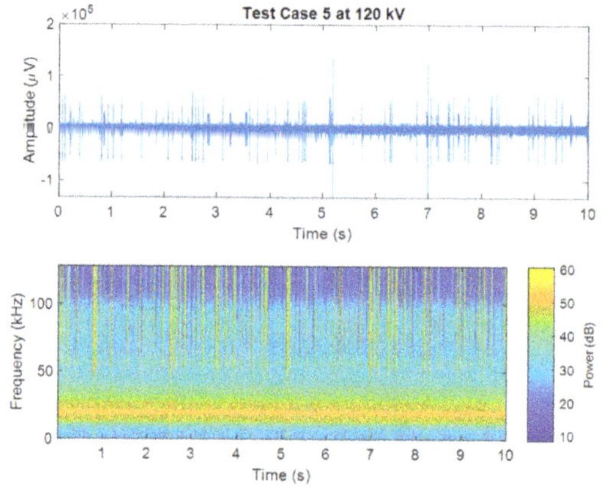

Figure 9. Amplitude and frequency spectrogram of the ultrasound signal for Test Case 5 at 120 kV.

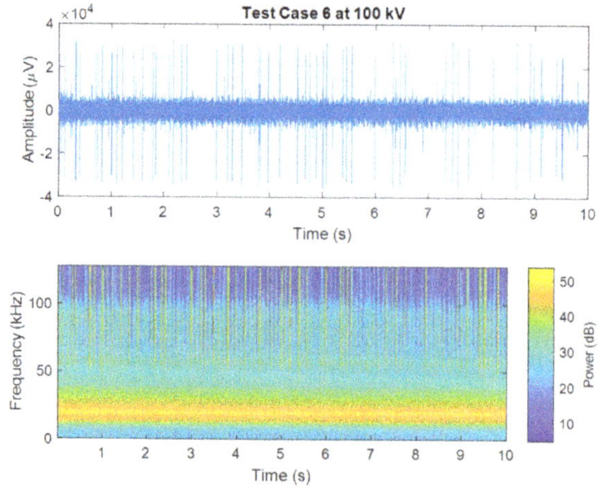

Figure 10. Amplitude and frequency spectrogram of the ultrasound signal for Test Case 6 at 100 kV.

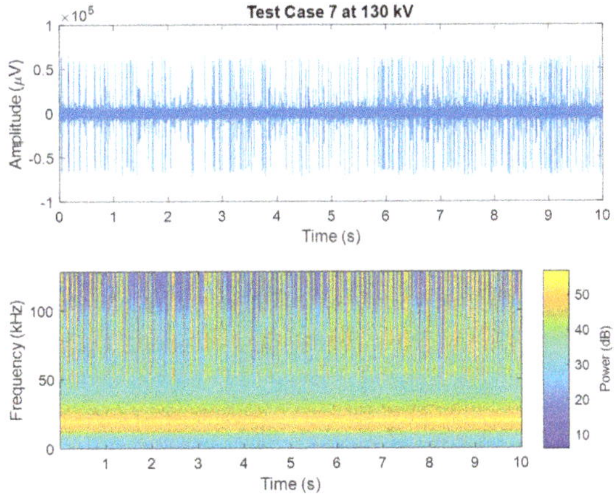

Figure 11. Amplitude and frequency spectrogram of the ultrasound signal for Test Case 7 at 130 kV.

The 10 s measurement time has been demonstrated to be adequate for capturing the occurrence of PDs. This duration also allows the operator to maneuver the digital ultrasound tester along the insulator chain without experiencing fatigue or shaking, thereby facilitating field measurements.

It can be seen that the development in the degradation of the insulation capacity of the HV glass insulators has a behavioral trend. This tendency is marked by a gradual increase in the amplitude of the ultrasound signal, ranging from values close to 500 μV at 10 kV (see Figure 5) and reaching values approximately equal to 50,000 μV for a voltage of 120 kV (see Figure 10). This represents an increase of 100 times in the amplitude of the ultrasound signal. The occurrence of high-frequency components can also be seen in the spectrogram, marked by the presence of frequencies in the order of 20 kHz for HV and relative insulation degradation.

4. Application of the Proposed Method

Based on the ultrasound measurements performed, as well as on the mathematical foundation provided in Section 2, a method is proposed for identifying and classifying PDs in HV glass insulators. This approach is based on the ZCR evaluation to identify anomalies in the ultrasound signal. If there is an anomaly, a fundamental frequency estimation technique is used to classify the type of the PD, as described in Section 2.

4.1. ZCR Evaluation for the Test Cases

Figure 12 shows the ZCR evaluation for the test cases.

The pattern depicted in Figure 12 clearly demonstrates the ability to detect PDs using the ZCR technique. It is essential to note that this pattern has undergone validation in a HV laboratory setting, and Section 5 provides further explanation for the reader's comprehension.

In Figure 12, a distinct behavior pattern emerges, wherein the ZCR remains relatively constant for low voltages. However, as the voltage increases within the range of 40 to 60 kV (depending on the specific case), the ZCR experiences a notable decrease. These decreases, with values below 0.5, are accompanied by discernible peaks in the acoustic raw signal, which are strongly correlated with PDs occurrences. This characteristic serves as a clear indication of anomalies present within the ultrasound signal, thus confirming the presence of PDs.

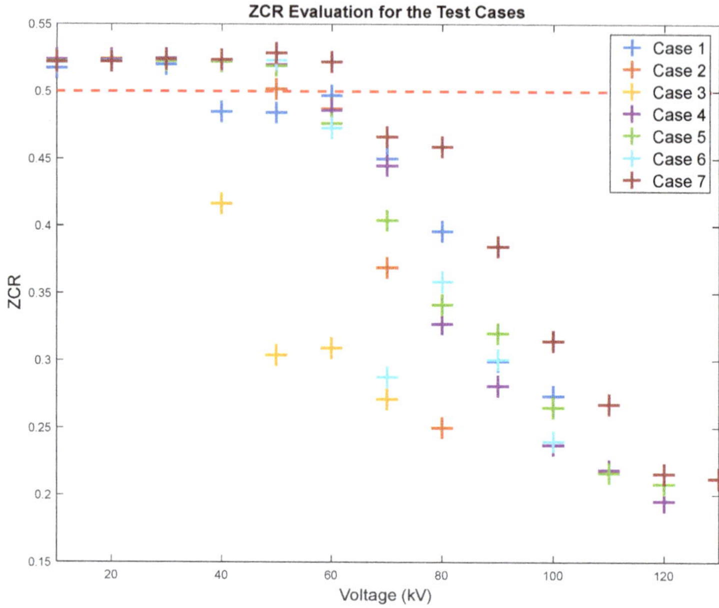

Figure 12. ZCR evaluation for the test cases.

This observation confirms the effectiveness and reliability of the proposed methodology in identifying and characterizing PDs. The ability to discern such abnormalities through the analysis of the ZCR and associated acoustic raw signals further reinforces the diagnostic capabilities of the technique.

4.2. Fundamental Frequency Estimation for the Test Cases

The fundamental frequency via the periodogram method for each test case is shown in Table 2. The "-" notation represents the lack of measurements, that is, the source overcurrent relay was triggered by a short-circuit.

Table 2. Fundamental frequency estimation.

Voltage (kV)	Fundamental Frequency Estimation (Hz)						
	Test Case 1	Test Case 2	Test Case 3	Test Case 4	Test Case 5	Test Case 6	Test Case 7
10	64.000 k	64.000 k	0.1667	23.362 k	23.359 k	56.933 k	64.000 k
20	11.938 k	0.1000	9.0617 k	23.363 k	23.361 k	23.361 k	64.000 k
30	59.9538	23.366 k	0.2666	23.360 k	23.359 k	23.358 k	64.000 k
40	10.429 k	11.957 k	23.856 k	23.361 k	60.0315	60.0003	64.000 k
50	60.0072	60.0068	16.846 k	23.359 k	23.355 k	59.9697	60.0334
60	60.0358	60.0033	20.218 k	60.0280	60.0092	60.0032	60.0321
70	60.0326	19.074 k	21.831 k	60.0220	60.0005	20.596 k	60.0225
80	59.9981	20.194 k	-	20.841 k	60.0037	59.9699	19.290 k
90	60.0329	-	-	60.0332	19.725 k	59.9692	9.9998
100	19.424 k	-	-	20.991 k	19.846 k	19.853 k	59.9972
110	-	-	-	19.317 k	19.901 k	-	59.9862
120	-	-	-	19.640 k	19.835 k	-	19.443 k
130					-	-	19.110 k

In yellow: frequencies in the range of 60 Hz (Corona PDs). In red: frequencies in range of 20 kHz (Superficial PDs).

The relationship between PD types and the estimated frequencies has been established through numerous HV laboratory tests, comparing the results of the proposed methodology with measurements of apparent electrical charges. This comparison has provided validation for the diagnostic approach, with further details discussed in Section 5.

As outlined in Table 2, the estimated frequencies around 60 Hz (indicated in yellow) signify the presence of corona discharges. Essentially, electric charges accumulate during the crest of the voltage waveform, leading to intermittent sparking that occurs at the same frequency as the voltage, i.e., 60 Hz. In this context, Figure 13 illustrates the comparison between the ultrasound signal from Test Case 1, where a voltage of 90 kV was applied, and a sine wave at 60 Hz.

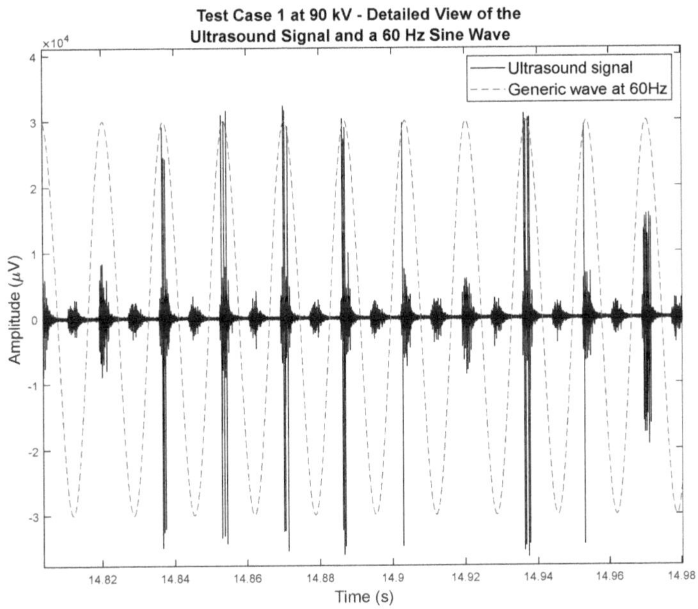

Figure 13. Ultrasound signal with the presence of corona discharges for Test Case 1 at 90 kV.

During the laboratory experiments, as the applied voltage increases, a fundamental frequency with values close to 20 kHz (indicated in red) emerges in the spectrogram. This frequency indicates the likelihood of superficial PDs occurrences, which are, usually, more damaging to equipment than corona discharges and have a frequency near the audible range (below 20 kHz). Superficial discharges are generally more detrimental to electrical equipment as they generate heat, emit ultraviolet radiation and form chemical compounds such as ozone and nitric acid. Additionally, surface-type PDs can lead to the phenomenon of "treeing channels" in more advanced stages. It is worth noting, however, that corona discharges also pose a significant threat to the integrity and performance of the insulation system as they indirectly contribute to the deterioration of dielectrics by generating ozone.

Finally, it is crucial to emphasize that, in this paper, the concept of superficial PDs incorporates both dry surface discharges, also known as dry band arcing (DBA), and wet surface discharges, when detecting faulty insulators.

4.3. *Algorithm for Classification of PDs*

The proposed approach calculates ZCR and, if it is greater than 0.5, the equipment is considered to be operating in normal conditions. However, an anomaly is detected for a ZCR lower than 0.5. If there is an anomaly, the fundamental frequency is estimated. If the frequency presents intensities approximately equal to 60 Hz—between 59 Hz and 61 Hz—the PD is characterized as a corona discharge. If the estimated frequency presents values close to 20 kHz—between 18 kHz and 22 kHz—the PD is identified as a superficial type. If none of these conditions are met, an inconclusive result is obtained and a new measure is required.

Figure 14 shows the method flowchart for identifying and classifying PDs based on ultrasound measurements.

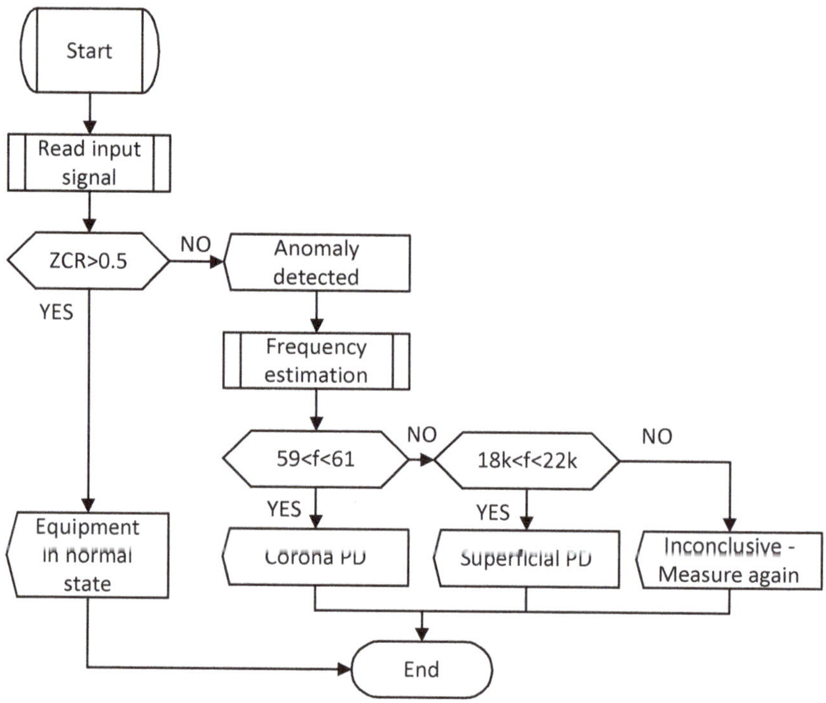

Figure 14. Methodology flowchart for identifying and classifying PDs based on ultrasound measurements.

The proposed method output for the cases is shown in Table 3.

Table 3. Results of the proposed method for classification of PDs.

Voltage (kV)	Results						
	Test Case 1	Test Case 2	Test Case 3	Test Case 4	Test Case 5	Test Case 6	Test Case 7
10	N	N	N	N	N	N	N
20	N	N	N	N	N	N	N
30	N	N	N	N	N	N	N
40	AI	N	AI	N	N	N	N
50	AC	N	AI	N	N	N	N
60	AC	AC	AS	AC	AC	AC	N
70	AC	AS	AS	AC	AC	AS	AC
80	AC	AS	-	AC	AC	AC	AS
90	AC	-	-	AS	AS	AC	AI
100	AS	-	-	AS	AS	AS	AC
110	-	-	-	AS	AS	-	AC
120	-	-	-	AS	AS	-	AS
130	-	-	-	-	-	-	AS

N: Normal (Green—No precautions); AI: anomaly inconclusive (Yellow—Attention); AC: anomaly corona (Orange—Attention); AS: anomaly superficial (Red—Caution).

In general, PDs undergo a transition from a normal behavior (N) to the presence of an anomaly (AI, AC, or AS) in the acquired ultrasonic signals as the voltage is increased. At lower voltage levels (between 60 and 90 kV), the anomaly is represented by corona discharges (AC), which then progress to surface discharges (AS) at higher voltages. The presence of surface discharges, characterized by the emission of acoustic signals with frequency components within the audible range, can indicate a stage of the phenomenon that is close to the formation of an electrical arc [20].

The identification of surface PDs for cases 3 and 6, subjected to voltages equal to 60 kV and 70 kV, respectively, can be attributed to the presence of pollution from saline solution or mud on all elements of the insulator chain, affecting the electrical conductivity of the medium. This may lead to DBA, which ceases if the generated heat completely dries out the pollution layer.

Identifying a surface PD for Case 7 at a voltage level of 80 kV may indicate the formation of a dry and localized flashover (a transient breakdown event where the insulation momentarily fails due to the combined effects of contamination and high electric field) that regressed to a corona discharge. It is important to note that the process of PD formation may exhibit instabilities, whereby it can abruptly cease, persist or progress in severity in an uncertain manner depending on changes in the electrical characteristics of the surrounding medium. An example of this is when the phenomenon progresses in severity when the wetting rate exceeds the drying rate of the pollution layers deposited on the dielectric surfaces.

Uncertainties and inconclusive cases (AI) arise from various factors that impact the occurrence and measurement of PDs. These factors encompass a range of influences, such as environmental conditions and external noise, which can affect the accuracy and reliability of PD assessments. Notably, environmental conditions play a significant role, with humidity emerging as a factor influencing the absorption of ultrasound waves by the air [21]. However, despite the inherent complexities and potential sources of uncertainty, the proposed method demonstrates the capability to identify abnormal conditions in electrical equipment, as evidenced by the results presented in Table 3. In such situations, it is highly recommended to establish a routine monitoring program for the equipment and conduct further measurement campaigns to gain a more accurate understanding of the underlying factors at play.

5. Validation of the Proposed Method

To validate the proposed methodology, a comparative analysis was conducted using apparent charge signals and phase-resolved partial discharge (PRPD) patterns. The comparison was made for a heavily contaminated insulator chain scenario where all three elements were affected, and the results are presented in Table 4. The attributes of interest in these kinds of measurements include the phase angle and polarity of PDs, the frequency and regularity of impulses, changes in magnitude relative to the test voltage and the ratio between the ignition and extinction voltages of the PDs, as stated in [2].

Figure 15 shows the spectrogram for a voltage of 60 kV applied in the insulator chain. The ZCR value is 0.5209, and the estimated fundamental frequency is 54.316 kHz, indicating a normal condition according to the proposed methodology. Similarly, the PRPD analysis and interpretation on apparent charge signals, as illustrated in Figure 16, reveal a situation of normality.

Conversely, an irregularity is observed, following the analysis of the acoustic signal depicted in Figure 17, with a ZCR and fundamental frequency amounting to 0.3856 and 59.9959, respectively. These attributes align with the corona discharge concept proposed in this study. Reinforcing this, Figure 18 displays apparent charge measurements for corona discharge [2]. Regular magnitude impulses can be seen, occurring near the voltage waveform peaks.

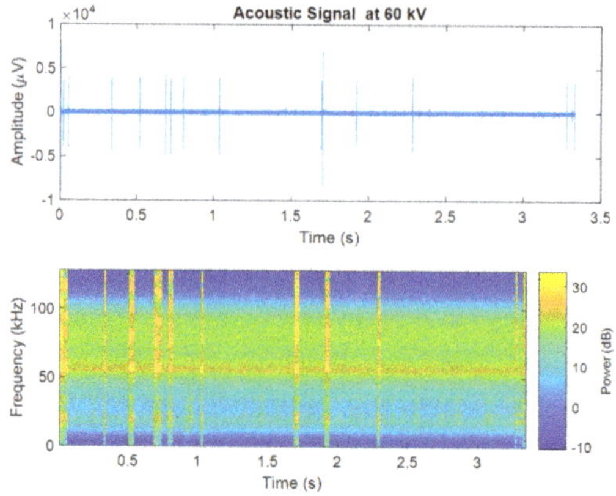

Figure 15. Amplitude and frequency spectrogram of the ultrasound signal for validation of the proposed method at 60 kV.

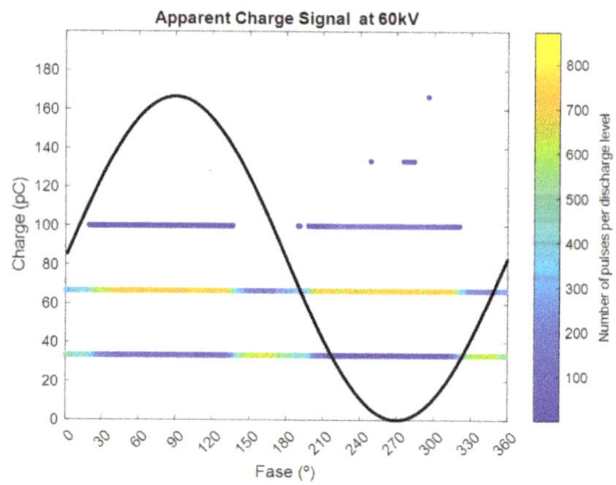

Figure 16. Apparent charge signal at 60 kV.

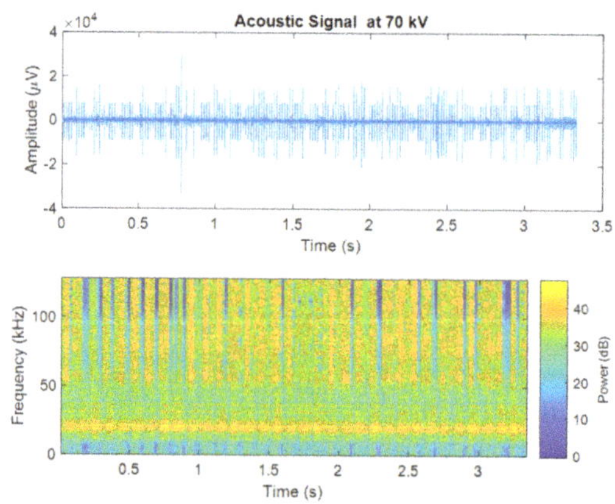

Figure 17. Amplitude and frequency spectrogram of the ultrasound signal for validation of the proposed method at 70 kV.

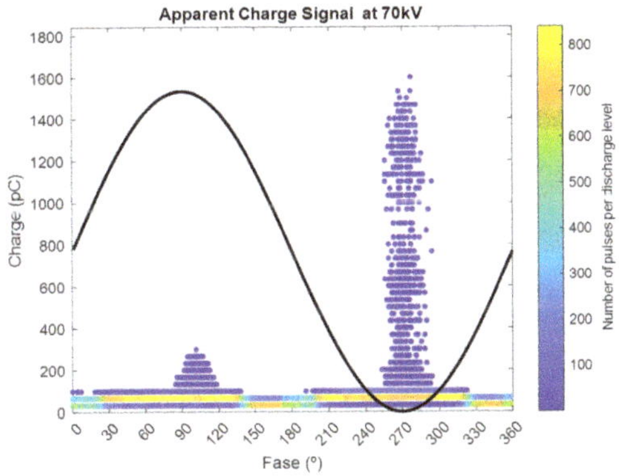

Figure 18. Apparent charge signal at 70 kV.

Another irregularity is discerned in the spectrogram presented in Figure 19 when the voltage level attains 80 kV. Under these circumstances, the ZCR and fundamental frequency yield values of 0.2713 and 18.316 kHz, respectively. Based on the methodology, these values imply a surface discharge. In this context, Figure 20 demonstrates impulses with increasing intensities from the zero-crossing point to the voltage signal peak (occasionally surpassing this value). This pattern is generally representative of superficial PDs [2].

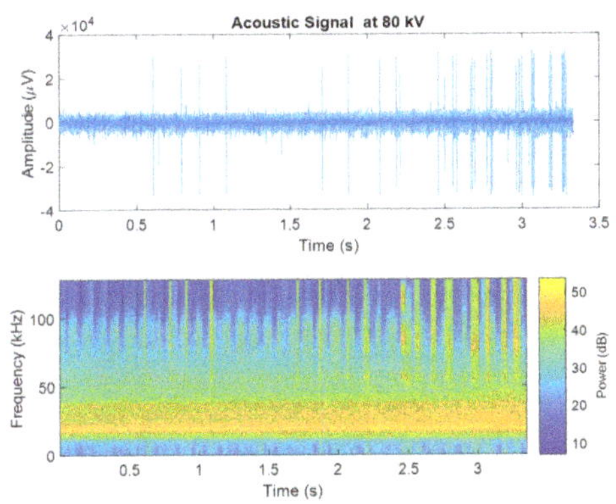

Figure 19. Amplitude and frequency spectrogram of the ultrasound signal for validation of the proposed method at 80 kV.

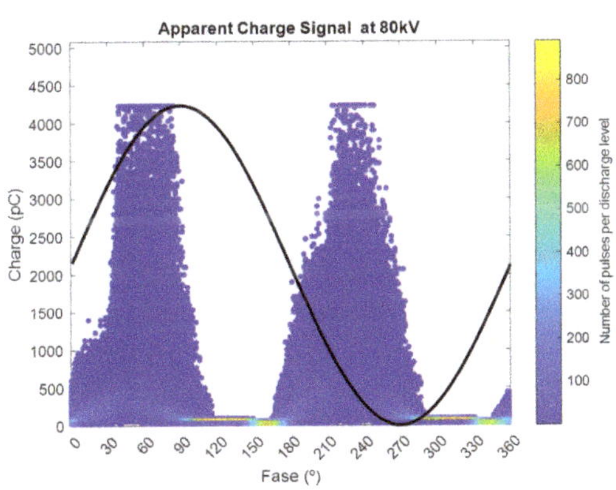

Figure 20. Apparent charge signal at 90 kV.

Table 4. Validation of the proposed method.

Voltage (kV)	60	70	80
ZCR	0.5209	0.3856	0.2713
Fundamental Frequency Estimation (Hz)	54.316 k	59.9959	18.316 k
Result of the Proposal	Normal	Anomaly Corona	Anomaly Superficial
Amplitude and Frequency Spectrogram	Figure 15	Figure 17	Figure 19
PRPD and interpretation	Normal	Corona	Superficial
PRPD Apparent Charge Signals	Figure 16	Figure 18	Figure 20

6. Measurements in Field

In order to assess the effectiveness of the technique in actual field conditions, ultrasound measurements were conducted at a 230 kV substation. It was discovered that there were two insulation chains with PDs, both located at the entrance of the substation. A visual inspection of Figures 21 and 22 reveals that the vertical insulation chains are noticeably dirtier compared to the horizontal ones. Therefore, it can be inferred that this condition is responsible for the occurrence of corona and superficial discharge phenomena, which were identified in the measurements and confirmed by the proposed technique. The results obtained from the method are presented in Table 5.

Figure 21. In this insulation chain of 230 kV, corona-type partial discharge was found to be present specifically in phase B.

Table 5. The outcomes achieved from the suggested approach in practical situations.

Measurement Number	1	2
Location	230 kV Sector, vertical insulation chain	230 kV Sector, vertical insulation chain
Phase	B	A
ZCR	0.3924	0.2831

Table 5. *Cont.*

Measurement Number	1	2
Fundamental Frequency Estimation (Hz)	60.01111	21.164 k
Results of the Proposal	Anomaly corona	Anomaly superficial
Figures	Figure 21	Figure 22
Amplitude and Frequency Spectrograms	Figure 23	Figure 24

Figure 22. In this insulation chain of 230 kV, superficial-type partial discharge was found to be present specifically in phase A.

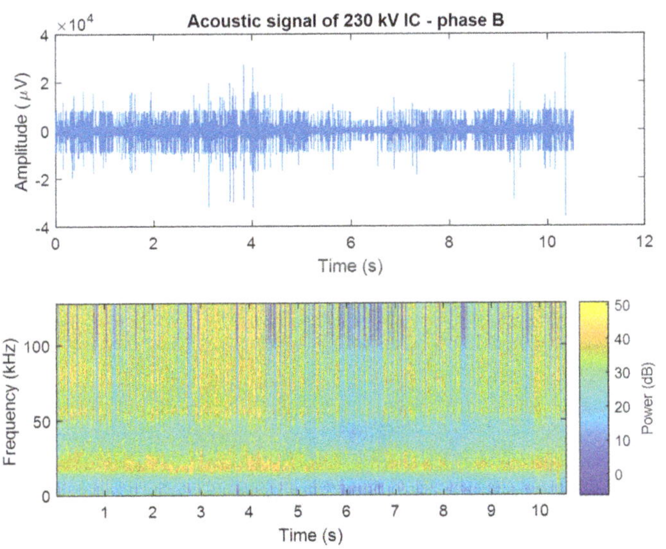

Figure 23. Amplitude and frequency spectrogram of the ultrasound signal captured from the insulation chain of Figure 21.

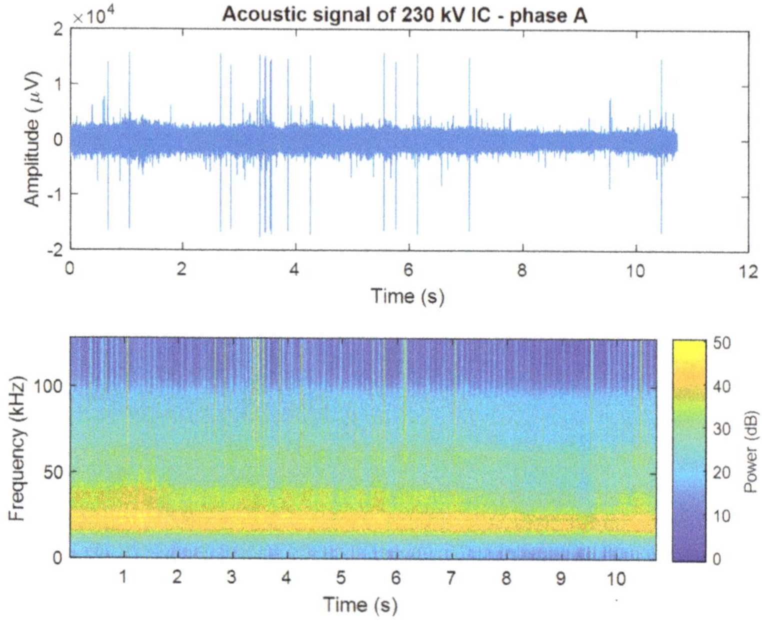

Figure 24. Amplitude and frequency spectrogram of the ultrasound signal captured from the insulation chain of Figure 22.

7. Conclusions

Identification and classification of PDs through acoustic inspection can be a powerful predictive maintenance tool. Raw acoustic signals, without any pre-processing and covering a wide frequency range (from 0 to 128 kHz), contain a large amount of information, both in the audible spectrum (<20 kHz) and in the ultrasound spectrum (>20 kHz). When appropriate techniques are applied, they can provide a more accurate diagnosis of PDs.

Seven test cases of artificial pollution on HV glass insulators were performed, producing a total of seven cases with different physical conditions and different voltage levels, such as mud build-up, conductive microfilm formation and damage. Thus, some conditions may be more favorable to the formation of PDs than others, as can be seen considering the different voltages achieved.

By using the zero-crossing rate, a trigger is created for detecting abnormal operating conditions, and the fundamental frequency estimate provides a diagnosis of the type of PD; thus, PDs can be classified in terms of severity and maintenance can be planned in advance, contributing to a higher level of power system safety and reliability at lower costs.

Although this study is based on experimental data from laboratory-tested insulator chains, the proposed method can be extended to other scenarios and equipment that require the diagnosis of partial discharges (provided they are external phenomena) through acoustic inspection.

It is noteworthy that environmental conditions influence the appearance of partial discharges, that is, measurements on different days can lead to different diagnoses. In this way, monitoring evolution regarding the type of PD can be a viable alternative to avoid making wrong decisions. By executing regular monitoring procedures and follow-up measurement campaigns, a more comprehensive comprehension of the underlying conditions can be attained, thereby enabling informed decision-making and proactive maintenance strategies.

Author Contributions: Conceptualization, K.M. and G.C.J.; methodology, K.M.; validation, K.M. and G.C.J.; formal analysis, K.M.; investigation, K.M.; resources, A.L.O.; data curation, A.M.M. and A.S.B.; writing—original draft preparation, K.M. and L.F.F.-G.; writing—review and editing, L.F.F.-G. and J.S.D.; supervision, A.L.O.; project administration, E.F.M.; funding acquisition, C.H.C. All authors have read and agreed to the published version of the manuscript.

Funding: This research was funded by CPFL Transmission through the project PD-05785-2107/2021—DE2107: "Methodology for aging assessment of insulation in instrument transformers and surge arresters through ultrasonic measurements of partial discharges," conducted under the Electric Sector Research and Development (R&D) Program, regulated by the Agência Nacional de Energia Elétrica (ANEEL) of Brazil. Partial support was also provided by the Coordenação de Aperfeiçoamento de Pessoal de Nível Superior—Brazil (CAPES/PROEX)—Finance Code 001.

Data Availability Statement: Not applicable.

Acknowledgments: The authors would like to acknowledge the R&D Program of ANEEL, for enabling the execution of this scientific research. The investigations and experiments described in this paper are associated with the project PD-05785-2107/2021—DE2107: "Methodology for aging assessment of insulation in instrument transformers and surge arresters through ultrasonic measurements of partial discharges", carried out by CPFL Transmission and UFSM, the Federal University of Santa Maria.

Conflicts of Interest: The authors declare no conflict of interest.

References

1. *IEC Standard 60060-1:2010*; High Voltage Test Techniques, Part 1: General Definitions and Test Requirements. International Electrotechnical Commission: Geneva, Switzerland, 2010.
2. Küchler, K. *High Voltage Engineering: Fundamentals, Technology, Applications*; Springer: Berlin/Heidelberg, Germany, 2018.
3. *IEC Standard 60060-2:2010*; High Voltage Test Techniques, Part 2: Measuring Systems. International Electrotechnical Commission: Geneva, Switzerland, 2010.
4. Lundgaard, L.; Runde, M.; Skyberg, B. Acoustic diagnosis of gas insulated substations: A theoretical and experimental basis. *IEEE Trans. Power Deliv.* **1990**, *5*, 1751–1760. [CrossRef]
5. Howells, E.; Norton, E.T. Detection of Partial Discharges in Transformers Using Acoustic Emission Techniques. *IEEE Trans. Power Appar. Syst.* **1978**, *PAS-97*, 1538–1549. [CrossRef]
6. He, Y. An overview of acoustic emission inspection and monitoring technology in the key components of renewable energy systems. *Mech. Syst. Signal Process.* **2021**, *148*, 107146. [CrossRef]
7. Maraaba, L.; Al-Soufi, K.; Ssennoga, T.; Memon, A.M.; Worku, M.Y.; Alhems, L.M. Contamination Level Monitoring Techniques for High-Voltage Insulators: A Review. *Energies* **2022**, *15*, 7656. [CrossRef]
8. Corso, M.P.; Perez, F.L.; Stefenon, S.F.; Yow, K.-C.; Ovejero, R.G.; Leithardt, V.R.Q. Classification of Contaminated Insulators Using k-Nearest Neighbors Based on Computer Vision. *Computers* **2021**, *10*, 112. [CrossRef]
9. Stefenon, S.F.; Zanetti Freire, R.; dos Santos Coelho, L.; Meyer, L.H.; Bartnik Grebogi, R.; Gouvêa Buratto, W.; Nied, A. Electrical Insulator Fault Forecasting Based on a Wavelet Neuro-Fuzzy System. *Energies* **2020**, *13*, 484. [CrossRef]
10. Stefenon, S.F.; Ribeiro, M.H.; Nied, A.; Mariani, V.C.; Coelho, L.D.; Leithardt, V.R.; Silva, L.A.; Seman, L.O. Hybrid Wavelet Stacking Ensemble Model for Insulators Contamination Forecasting. *IEEE Access* **2021**, *9*, 66387–66397. [CrossRef]
11. Nasir, A.; Al-geelani, M.; Piah, A.M.; Shaddad, R.Q. Characterization of acoustic signals due to surface discharges on H.V. Glass insulators using wavelet radial basis function neural networks. *Appl. Soft Comput.* **2012**, *12*, 1239–1246. [CrossRef]
12. Nasir, A.; Al-geelani, M.; Piah, A.M.; Bashir, N. A review on hybrid wavelet regrouping particle swarm optimization neural networks for characterization of partial discharge acoustic signals. *Renew. Sustain. Energy Rev.* **2015**, *45*, 20–35. [CrossRef]
13. Mahmoudi, J.; Moussavi, S.Z.; Naderi, P. Partial discharge diagnosis of ceramic pin insulators considering cost-worth analysis: Case study in a medium voltage feeder. *IEEE Trans. Dielectr. Electr. Insul.* **2017**, *24*, 2493–2502. [CrossRef]
14. Stefenon, S.F.; Seman, L.O.; Neto, N.F.; Meyer, L.H.; Nied, A.; Yow, K.C. Echo State network applied for classification of medium voltage insulators. *Int. J. Electr. Power Energy Syst.* **2022**, *134*, 107336. [CrossRef]
15. Giannakopoulos, T.; Pikrakis, A. *Introduction to Audio Analysis: A MATLAB Approach*; Elsevier: Amsterdam, The Netherlands; Academic Press: Cambridge, MA, USA, 2014.
16. Erman, L.D. An Environment and System for Machine Understanding of Connected Speech. Ph.D. Thesis, Stanford University, Stanford, CA, USA, 1974.
17. Guimarães, D.A. *Digital Transmission: A Simulation-Aided Introduction with VisSim/Comm*; Springer Science & Business Media: Berlin/Heidelberg, Germany, 2010.
18. Poularikas, A.D. *Understanding Digital Signal Processing with MATLAB and Solutions*; CRC Press: Boca Raton, FL, USA, 2017.
19. Electrovidro/Seves. *Isoladores de Suspensão de Vidro Temperado SEDIVER*; Electrovidro: Rio de Janeiro, Brasil, 2019; Available online: https://bit.ly/3vJSQWs (accessed on 1 May 2023).

20. Dehlinger, N.; Stone, G. Surface partial discharge in hydrogenerator stator windings: Causes, symptoms, and remedies. *IEEE Electr. Insul. Mag.* **2020**, *36*, 7–18. [CrossRef]
21. Lundgaard, L.E. Partial discharge. XIII. Acoustic partial discharge detection-fundamental considerations. *IEEE Electr. Insul. Mag.* **1992**, *8*, 25–31. [CrossRef]

energies

MDPI

Article

Numerical Modeling of PD Pulses Formation in a Gaseous Void Located in XLPE Insulation of a Loaded HVDC Cable

Paweł Mikrut * and Paweł Zydroń *

Department of Electrical and Power Engineering, AGH University of Krakow, al. Mickiewicza 30, 30-059 Krakow, Poland
* Correspondence: mikrut@agh.edu.pl (P.M.); pzydron@agh.edu.pl (P.Z.)

Abstract: Power cables are one of the key components of fast-growing HVDC transmission systems. The long-term reliability of HVDC cables is closely related to the occurrence of partial discharges (PDs) in their insulation systems. The article analyzes the conditions for the formation of PD pulses in gaseous voids located in the XLPE insulation of an HVDC cable. For this purpose, the MATLAB® procedure and the coupled electro-thermal simulation model implemented in COMSOL Multiphysics® software were used. The FEM model was used to study the effect of the applied voltage, the temperature field (created in the insulation of the loaded cable) and the location of the gaseous void (on cable radius) in the distribution and values of the electric field in the cable insulation. The model takes into account the influence of temperature and the electric field on the conductivity of the insulating material and relates the value of the PD inception field to the temperature/pressure of the gas inside the void. In the numerical simulation procedure, the time sequences of PDs arising in the gaseous defects of the HVDC cable insulation were analyzed, by observing changes caused by the increase in the temperature of the cable core. The model was used for a study of conditions for PD formation in models of three HVDC cables, for DC voltages from 150 kV to 500 kV. The critical dimensions of gaseous voids were also estimated for each of the analyzed cables, i.e., the dimension which, if exceeded, makes a void a source of PD.

Keywords: HVDC cables; partial discharges (PDs); XLPE; FEM simulation; coupled electro-thermal field

Citation: Mikrut, P.; Zydroń, P. Numerical Modeling of PD Pulses Formation in a Gaseous Void Located in XLPE Insulation of a Loaded HVDC Cable. *Energies* **2023**, *16*, 6374. https://doi.org/10.3390/en16176374

Academic Editor: Mario Marchesoni

Received: 21 June 2023
Revised: 30 August 2023
Accepted: 31 August 2023
Published: 2 September 2023

1. Introduction

Alternating current (AC) and direct current (DC) transmission systems are designed and implemented using overhead transmission lines and power cable lines [1–7]. Since the beginning of the power industry, the constantly growing demand of world societies and economies for electricity implies the need to increase the rated voltage of these systems. The choice of the type of system voltage (AC or DC) depends on the transmission distance and/or the location of generation sources and power consumption areas [6–8]. Economic considerations, taking into account investment and operating costs estimated over a long time horizon, play an important role in the decision on the selection of a transmission system [9,10]. In many cases, HVDC (and UHVDC) systems have better technical and economic performance than HVAC (UHVAC) systems. Currently, these systems are used for:

- Long-distance, high-capacity transmission of electric energy [1–3,7];
- The implementation of inter-system connections (usually for non-synchronized systems) [11,12];
- submarine electrical energy transmission [4].

The latter group includes the transmission of electricity from offshore wind farms, especially those located far from the shoreline, in deep water areas. In recent years, the number of such HVDC system implementations has been growing significantly in different regions of the world [10,13–15].

In each of the above-mentioned applications, the very high reliability of the electrical insulation systems is of key importance for the failure-free operation of the entire system. To achieve this goal, HVDC cable production technologies are being improved, based on analyses and model research for estimating the product lifetime, taking into account the factors influencing the rate of insulation aging. The basic stresses causing structural changes in the insulation, the formation and development of defects and the shortening of the effective lifetime of the insulation system include: thermal (T), electrical (E), ambient (A) and mechanical (M) stresses, collectively referred to as TEAM [16,17].

The effective lifetime of the cable insulation is estimated based on aging models that take into account both intrinsic and extrinsic aging processes [4,5,18–24]. In the case of HVDC cable lines, one of the potential causes of reduced reliability is the occurrence of various types of defects in the insulation of cables or cable joints. Some of them are the effects of imperfect technological production processes, whereas others arise as a result of errors made during the laying of the cable line or are the effect of various types of stresses during its operation. The required high quality and reliability of cable insulation can be significantly reduced due to extrinsic aging processes and the related occurrence of various types of defects in the insulation [25,26]:

- Gaseous voids;
- Delaminations;
- Protrusions;
- Contaminations;
- Electrical trees;
- Water trees.

The problems of the occurrences of these types of defects in insulation have been intensively studied, both during non-standard experimental tests and type tests, including long-term tests on model samples, prototypes and complete devices, as well as using various types of numerical simulation tools. Due to the specific nature of the problem resulting from the use of high DC voltage, the methods used for testing AC insulation are not always adequate for use in the analysis and testing of HVDC insulation.

A recognized but still current research problem is the development of electrical and water treeing processes in the polymer insulation of HV cables [27–30]. For example, the problems of the development of electrical trees in HVDC insulation in various operating conditions were analyzed in the research presented in article [31]. The influence of HVDC voltage polarity on the development of electrical trees and partial discharge parameters was also studied [30–33]. Simulation tools using the finite element method implemented in COMSOL software were used to study the conditions and processes of the development of water trees in the XLPE insulation of submarine cables [34,35].

The research described in this paper concerns the determination of the conditions for the formation of partial discharges in the gaseous voids of the XLPE insulation of HVDC cables and the analysis of the time sequences of PD pulses generated as a result of discharge inception. The problem of voids in the insulation of modern XPLE cables concerns the presence of defects in the polymer insulation structure of small size (~60 μm) or even smaller, where partial discharges (PDs) may occur [36–38]. The inception of PDs in internal voids in solid insulation is possible when the local electric field inside the void exceeds the value of the PD inception field [39,40]. For AC 50/60 Hz systems, the distribution of the electric field in the cable insulation depends on the permittivity (dielectric constant) of the insulation material [41,42]. In contrast, for DC systems, the distribution of the electric field in the cable insulation is determined by the volume conductivity/resistivity of the insulating material. Due to the dependence of this material parameter on temperature and local electric field values, the E field strength analysis in a DC cable must take into account the influence of these factors on the field distribution in cable insulation for steady and transient states [41–46].

The distribution of the E field strength in the HVDC cable is affected by the space charge accumulated in the insulation (both homocharges and heterocharges trapped in the bulk of a dielectric), the presence of which may increase or decrease the local field strength [47,48]. For this reason, it is required that new materials intended for extruded cable insulation have very low or even no ability to accumulate space charge [48–51]. For the developed HVDC XLPE 500 kV systems, a certain solution to the problem of space charge accumulation was reported, and confirmed in qualification tests up to 90 °C [52–54]. The test results presented in [54] show that the *field enhancement factor* (FEF), defined by the ratio of the E field strength in the DC-XLPE insulation, taking into account the effect of space charge, to the E field strength without this charge, was always less than 1.10 for 20 kV/mm and 50 kV/mm at 30 °C. In a longer period of time after the application of the DC voltage, it decreased to a value of about 1.05.

For design and operational purposes, extended analyses of temperature and electric field strength distributions in the insulation of HVDC power cables and their accessories are performed. The static and dynamic analyses take into account the geometry of the cable structure, the dielectric material parameters, the processes of space charge accumulation and decay, the effects of reversing the polarity of the DC voltage (in the case of using LCC technology), the presence of PD sources and the influence of DC voltage ripple [6–8,55–58]. One of the basic reasons for the analyses is the estimation of the permissible current-carrying capacity of DC cable lines due to the limit parameters of the insulating material [55,59]. Such analyses are also performed to determine the expected lifetime of the HVDC insulation of cable designed to operate in the planned electro-thermal conditions [60,61].

The article presents an analysis of the influence of the position of the gaseous void in the XLPE insulation of the HVDC cable on the conditions of PD formation and the generation of the PD pulse sequences. In the numerical simulations, made using the finite element method (FEM) in COMSOL Multiphysics® 6.0 software, the presence of an electric field, modified by the influence of the temperature field, in the XLPE insulation of the loaded HVDC cable was taken into account. On this basis, using a dedicated program implemented in MATLAB® R2021b software, conditions for PD formation and time sequences of PD pulses generated at DC voltage were analyzed. Based on the results of numerical simulations, the critical dimension of the gaseous voids was also estimated, depending on its location on the radius of the cable.

2. The Problem of PD Formation and the Numerical Model of the DC Cable

2.1. PD Formation in Gaseous Void—Overview

The formation of streamer PD pulses in a gaseous cavity located in a solid dielectric occurs when two required conditions are met simultaneously [39,40]:

(1) The electric field strength in the void must exceed the value of the discharge inception field strength;
(2) As a result of volume or surface processes, an initial electron must be present to trigger the development of the discharge process.

The PD inception field strength E_{inc} inside the gaseous void can be estimated using the following equation [39,40]:

$$E_{\text{inc}} = (E/p)_{\text{cr}} \cdot p\left[1 + B \cdot (p \cdot d)^{-n}\right] \tag{1}$$

where:

E_{inc}—PD inception field strength, V·m^{-1};
$(E/p)_{\text{cr}}$—critical electric field to pressure ratio, for air 25.2 [V·m^{-1}·Pa^{-1}];
p—gas pressure in void [Pa];
d—gaseous void dimension (the void diameter or height parallel to the applied E field) [m];
B, n—ionization process coefficients (for air B = 8.6 [m$^{1/2}$·Pa$^{1/2}$], n = 0.5).

Figure 1 shows the curves of the PD inception field strength E_{inc} as a function of the dimension of the gaseous void d, estimated for air at different temperatures (from 4 °C to 90 °C). The reduction in the dimension of the gaseous void enclosed in the dielectric is accompanied by an increase in the value of the E field strength required for PD inception.

When the size of the void is reduced from 1.0 mm to 10 μm (i.e., 100 times), the PD inception field strength increases about 5 times. As can be seen in Figure 1, an increase in temperature also increases the value of the PD inception field strength. Due to the properties of gases, for an un-vented void of constant volume, p/T = const., an increase in temperature causes a proportional increase in pressure and, consequently, also E_{inc} value. Both of the phenomena mentioned above (the effect of gaseous void dimension and gas temperature on E_{inc}) are of key importance for the analysis of the conditions for the formation of PD pulses in insulation defects of a current-carrying HVDC cable.

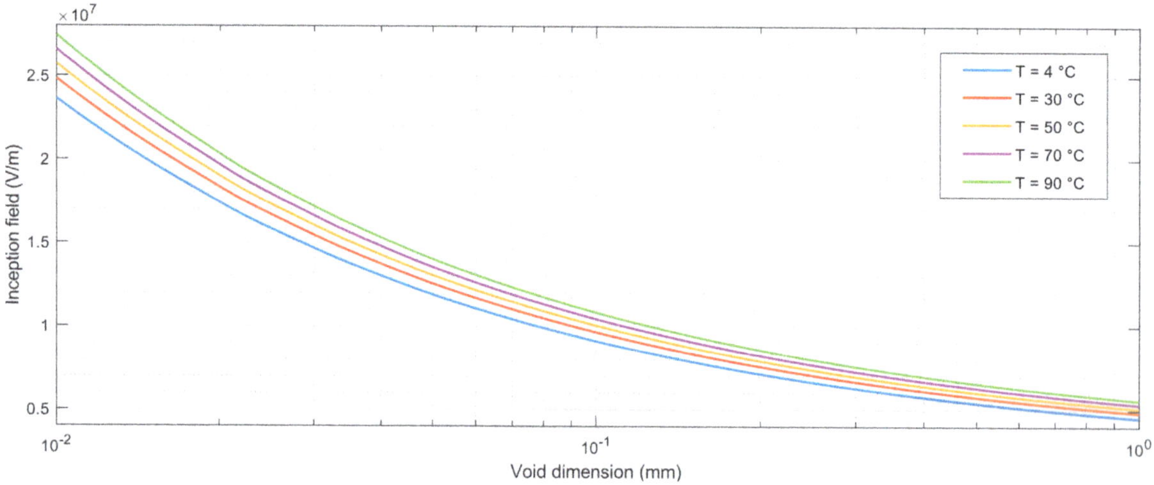

Figure 1. PD inception field vs. void dimension for different temperatures, from 4 °C up to 90 °C.

Fulfillment of the second condition necessary for PD inception requires the presence of a starting electron triggering the discharge development process. Due to the stochastic nature of the appearance of such an electron, a probability function (usually exponential) is defined for the lag time of the discharge occurrence relative to the moment when the field strength in the void reaches the E_{inc} value. An overview of the functions proposed by various authors for free electron supply and discharge probability can be found in [40]. In the PD source model implemented in the presented simulations, the discharge probability P_d was estimated based on the formula:

$$P_d = \begin{cases} 1 - \exp\left(\frac{\Delta t_{inc}}{\tau_{lag}}\right) & E_{void} \geq E_{inc} \\ 0 & E_{void} < E_{inc} \end{cases} \quad (2)$$

where:

E_{void}—electric field strength in a gaseous void, V·m^{-1};
τ_{lag}—PD inception lag time, s;
Δt_{inc}—the period of time counted from the moment when the field strength in the void E_{void} exceeded the E_{inc}, s.

The occurrence of PD inside a gas void does not result in a reduction of the field strength in the void to zero at the end of the discharge. Due to the not fully recognized conditions of PD extinction in a gaseous void enclosed in dielectric, various estimations of the PD extinction field strength E_{ext} were proposed in published works. One way is to define E_{ext} as a proportional part of the critical field strength [39,40,62–64]:

$$E_{ext} \approx \chi \cdot E_{cr} \quad (3)$$

where:

E_{ext}—the PD extinction field strength, V·m^{-1};
E_{cr}—the critical field strength, V·m^{-1};
χ—the PD extinction field coefficient, -.

The value of the χ coefficient may depend on the type of gas, the dielectric in which the void is enclosed and the processes on its surface, streamer polarization, temperature and field strength E inside the void, before the PD. Despite so many variable factors, approximate values of $\chi = 0.2$ for positive streamers and $\chi = 0.5$ for negative streamers [39,63] or their average value of $\chi = 0.35$ [64] are assumed for numerical PD simulations. An alternative method of estimating the value of E_{ext} is to assume a constant value of the residual field strength, which, based on various research studies, may be about 0.1 kV·mm^{-1} or about 1 kV·mm^{-1} [40].

2.2. Electric Field in Model HVDC Cable Insulation

HVDC cable usually has a complex, multi-layer coaxial construction. There is an electric field in the insulation of the energized cable, which is created between the semi-conductive screen on the cable core, with a high electric potential resulting from the applied voltage, and the semi-conductive screen on the insulation, which is at ground potential. The distribution of the DC electric field depends on the electrical conductivity σ of the dielectric. This material parameter of dielectrics is non-linearly dependent on temperature and electric field strength. To determine the distribution of the E field in the HVDC cable insulation, often an empirically determined function is used to estimate the value of electrical conductivity of a dielectric in specified conditions [42–45,48,51,55,57,59,60]:

$$\sigma(T,E) = \sigma_0 \cdot \exp(\alpha \cdot T) \cdot \exp(\beta \cdot E) \tag{4}$$

where:
σ_0—specific dielectric conductivity, S·m^{-1} (at E = 0.0 V·m^{-1} and T = 0 °C);
T—temperature, °C;
E—electric field strength, V·m^{-1};
α—temperature factor of conductivity, °C^{-1};
β—field factor of conductivity, V^{-1}·m.

Determining the appropriate values of coefficients α and β for numerical simulations may be a problem, because the published results of measurements performed for various XLPE samples show a large dispersion of their values. Reported values of the temperature factor of conductivity α range from 0.052 °C^{-1} to 0.180 °C^{-1} [45,65–67] and the variation of the field factor of conductivity β extends from 0.018×10^{-6} V^{-1}·m to 0.5×10^{-6} V^{-1}·m [65–67]. In addition, the measurements showed a dependence of β on temperature [66,67].

In order to take into account the influence of temperature and electric fields on conductivity in polymer insulation, a model function derived from the hopping theory of conduction in dielectrics is also often used [41,42,46,49,68–72]:

$$\sigma(T,E) = \mathrm{A} \cdot \exp\left(\frac{-\varphi \cdot q_e}{k_B \cdot T}\right) \cdot \frac{\sinh|\mathrm{B}|E||}{|E|} \tag{5}$$

where:
A, B— specific factors for the dielectric;
φ—thermal activation energy, eV;
q_e—elementary charge;
T—temperature, K;
E—electric field strength, V·m^{-1}.

Electrical conductivity is highly sensitive to temperature changes and less sensitive to electric field strength changes. For this reason, the temperature field existing in the insulating material has a dominant influence on the distribution of the electric field. Joule heat is produced in a cable conductor by the flow of operating, overload or fault current. This causes the temperature distribution in the insulation of the coaxial cable to be radial. For a loaded cable, the highest temperature is in the cable core and the lowest is at the outer shield. As a result, the inner layers of insulation, at the cable core, have a significantly higher conductivity than the outer layers. The increase in the conductivity of the inner layers of the dielectric material reduces the electric field stress in the vicinity of the cable core. For this reason, the high temperature of the core conductor can cause the inversion of the electric field strength distribution in the dielectric (in a radial direction). Then the electric field stress near the screen on the insulation may be higher than the stress near the cable core.

The second effect of the change in the radial temperature distribution in the insulation, caused by the heating of the cable core, is the modification of the PD inception field strength. An increase in gas temperature in a void of constant volume causes a proportional increase in pressure, which in turn leads to an increase in E_{inc} (see Section 2.1).

For numerical simulations of PD in HVDC cable insulation, a coupled electro-thermal 2D model was implemented in the COMSOL® program [73]. The model takes into account both mentioned effects of the temperature field influence on the conditions of PD formation in gaseous voids. The parameters of the model used for problem analysis take into account the individual properties of each material included in the layered structure of the HVDC cable with XLPE insulation.

In the numerical procedure of the electric field simulation used to determine the time sequence of the PD pulses, it was assumed that the test value of the DC voltage is not determined by a voltage step at the time $t = 0$ s, but is achieved by monotonically increasing the voltage at the cable core in accordance with the quarter-wave of the 50 Hz AC voltage from 0 V to peak value, i.e., during 5 ms. The initial distributions of the electric field and the space charge in the cable insulation are, therefore, determined mainly by the permittivity and the defined voltage transient at the cable core in relation to the outer shield on the insulation (i.e., between the inner and outer electrode of the concentric cable model). Space charges from other sources (e.g., trapped in the volume of insulating material) have not been modeled, so the transient processes of the E field and charge distributions occurring after the DC voltage is established are related to the influence of electrical conductivity, the value of which at a specific location on the cable radius depends on the temperature and electric field, according to Equation (4).

2.3. *Cable Model for E Field Analysis with Finite Element Method*

The objects of the numerical analysis were three models of the HVDC cables with XLPE insulation with rated voltages of 150 kV [65,73], 320 kV and 500 kV [53]. Due to the thickness of the XPLE insulation layer of each cable, the average electric field strength E_{avg} in the insulation is 18.75 kV·mm^{-1}, 16.00 kV·mm^{-1} and 21.74 kV·mm^{-1}, respectively.

A cross-section showing the multi-layer structure of the modeled cables is shown in Figure 2. Parameters of materials of subsequent layers, necessary for modeling thermal processes in the cable and the temperature field in XLPE insulation, are listed in Table 1. The geometric dimensions of the layers for each of the analyzed cables are defined in Table 2, where the radii of individual layers are listed. The inner radius of the insulation layer is x_{min} and the outer radius is x_{max} (Figure 2, Table 2). Table 3 summarizes the general parameters defining the properties of XLPE insulation and the air filling the void, as well as the thermal parameters of the cable environment (laying the cable in water with a temperature of 4 °C is assumed). In the simulations, time sequences of PD pulses in two spherical voids with a diameter of 0.5 mm each were determined. In all three analyzed cable cases, the first void is located near the cable core (center of the gaseous void on the inner insulation radius +1 mm), and the second one near the grounded screen on the insulation (center of the gaseous void on the outer insulation radius −1 mm), Table 4.

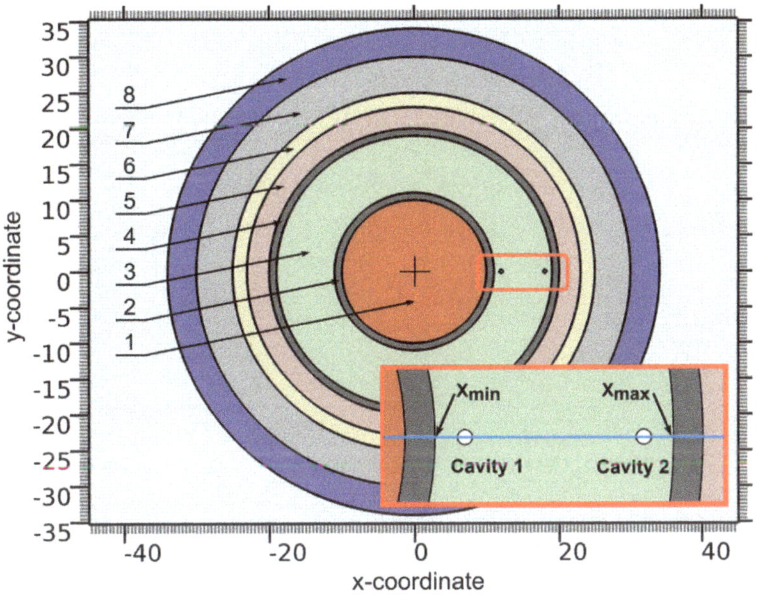

Figure 2. Geometry of HVDC cable model with two voids in XLPE insulation. Details of the insulation area with voids located on the cable radius (blue line) are shown in the red box. The cable layers are numbered according to their order in Tables 1 and 2.

Table 1. Parameters of materials for 150 kV, 320 kV and 500 kV cable models [65].

No	Cable Layer	k ($W \cdot m^{-1} \cdot K^{-1}$)	Cp ($J \cdot kg^{-1} \cdot K^{-1}$)	ρ ($kg \cdot m^{-3}$)
1	Core conductor	385	384	8900
2	Semicon layer	0.23	2050	1100
3	Insulation (XLPE)	0.32	2250	920
4	Semicon layer	0.23	2050	1100
5	Lead sheath	0.21	125	11,340
6	PE sheath	0.40	2300	950
7	Armoring	260	2300	2700
8	Outer serving	0.30	2350	950

Table 2. Dimensions of layers for 150 kV, 320 kV and 500 kV cable models.

No	Cable Layer	150 kV Cable r (mm)	320 kV Cable r (mm)	500 kV Cable r (mm)
1	Core conductor	10	28.8	33.5
2	Semicon layer (x_{min})	11	30.9	36
3	Insulation (x_{max})	19	50.9	59
4	Semicon layer	20	52.9	60
5	Lead sheath	23	55.8	64.5
6	PE sheath	25	58.3	70.5
7	Armoring	30	63.3	75.5
8	Outer serving	34	67.3	79.5

Table 3. Basic simulation parameters for HVDC cable models.

Parameter	Value	Unit
Initial conductivity of XLPE, σ_0 [65]	5.4×10^{-16}	$S{\cdot}m^{-1}$
Temperature factor of XLPE conductivity, α [65]	0.064	$°C^{-1}$
Field factor of XLPE conductivity, β [65]	6.7×10^{-8}	$V^{-1}{\cdot}m$
Conductivity of air in void [70,71]	10^{-16}	$S{\cdot}m^{-1}$
Conductivity of air in void during PD	10^{-3}	$S{\cdot}m^{-1}$
Dielectric constant of gas	1.0	-
Dielectric constant of XLPE	2.3	-
External heat transfer coefficient	30	$W{\cdot}m^{-2}{\cdot}K^{-1}$
External temperature	4	°C

Table 4. Simulation parameters for 150 kV, 320 kV and 500 kV cable models.

Parameter	150 kV Cable	320 kV Cable	500 kV Cable
Applied voltage, kV	150	320	500
1st void diameter, mm	0.5	0.5	0.5
1st void center *x*-coordinate, mm	12	31.9	37
2nd void diameter, mm	0.5	0.5	0.5
2nd void center *x*-coordinate, mm	18	49.9	58

3. Simulation Results—Case Studies for Three Model Cables

Figures 3–8 present, in the form of graphs, the results of the subsequent stages of the analysis of the conditions of the formation of PD pulses in the defined gaseous voids of three model HVDC cables. Distribution analyses of individual physical quantities were performed for thermal steady states at five different temperatures of the cable core: 4 °C, 30 °C, 50 °C, 70 °C and 90 °C. Figure 3 summarizes the temperature distributions in the insulation of individual cable models, which are the basis for the subsequent analyses.

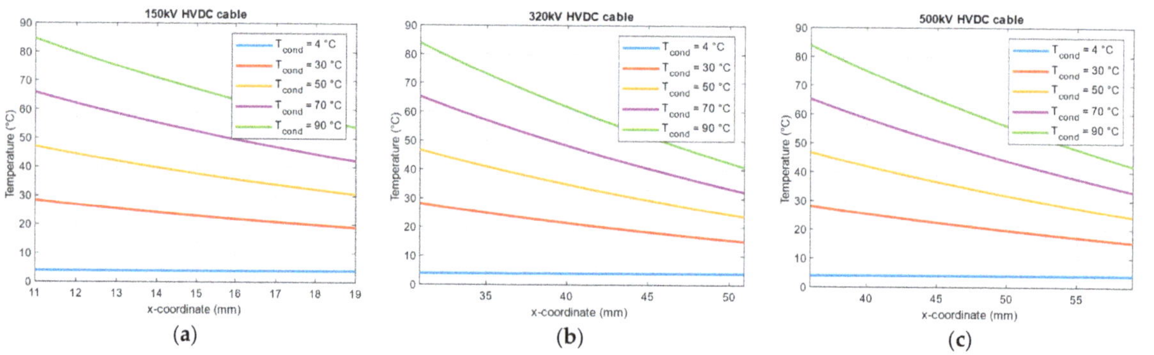

Figure 3. Distribution of temperature $T(x)$ across the cable insulation at 4 °C, 30 °C, 50 °C, 70 °C and 90 °C cable core temperatures for: (**a**) 150 kV HVDC cable; (**b**) 320 kV HVDC cable; (**c**) 500 kV HVDC cable.

The data visualized in Figure 4 shows the distributions of electrical conductivity in XLPE insulation on the radius of each of the cable models. The influence of temperature T and E field on conductivity σ was taken into account according to Equation (4), using the parameter values specified in Table 3.

Taking into account temperature distributions in the insulation of each cable model, simulations of E field strength distributions were carried out for a wide range of cable core temperatures (from 4 °C to 90 °C). As a result, distributions of the E field in cable insulation without voids were determined as a function of the x-coordinate on the radius of the cable and the temperature of the cable core (Figure 5). For all three 2D distributions, cross-sections for five selected temperatures were separated, showing the corresponding reference distributions of the E field as a function of the x-coordinate (Figure 6).

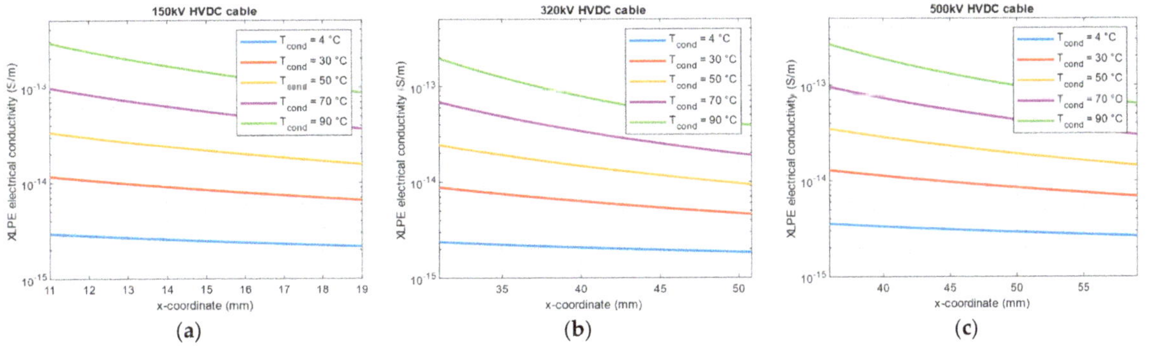

Figure 4. Distribution of electrical conductivity $\sigma(x)$ across the cable insulation at 4 °C, 30 °C, 50 °C, 70 °C, 90 °C cable core temperatures for: (**a**) 150 kV HVDC cable; (**b**) 320 kV HVDC cable; (**c**) 500 kV HVDC cable.

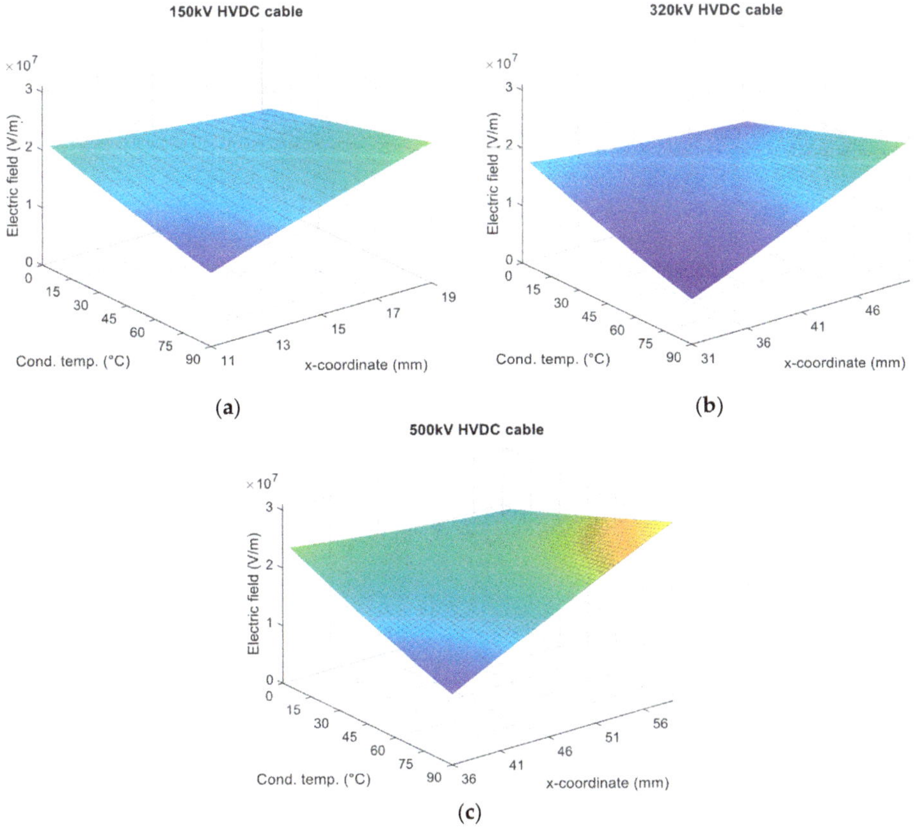

Figure 5. A 2D electric field distribution across the cable insulation at 4 °C to 90 °C cable conductor temperatures: (**a**) 150 kV HVDC cable; (**b**) 320 kV HVDC cable; (**c**) 500 kV HVDC cable.

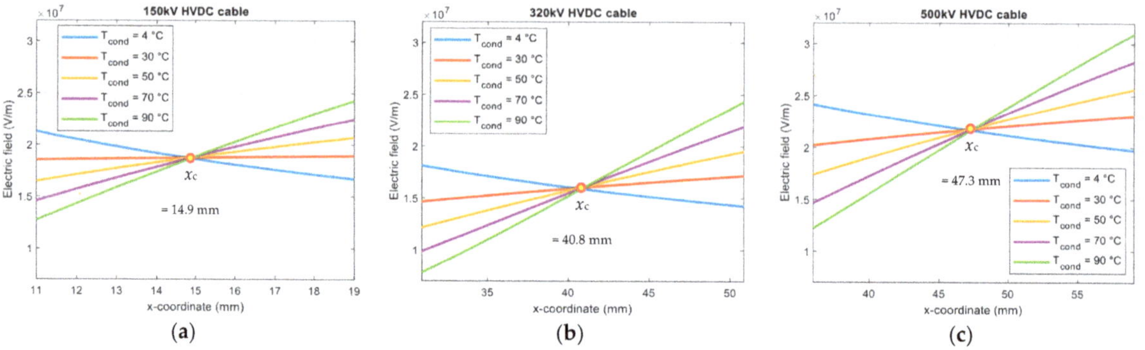

Figure 6. Electric field strength distribution $E(x)$ across the insulation at cable core temperatures of 4 °C to 90 °C: (**a**) 150 kV HVDC cable; (**b**) 320 kV HVDC cable; (**c**) 500 kV HVDC cable. At the x_c coordinate, the electric field strength E is almost constant.

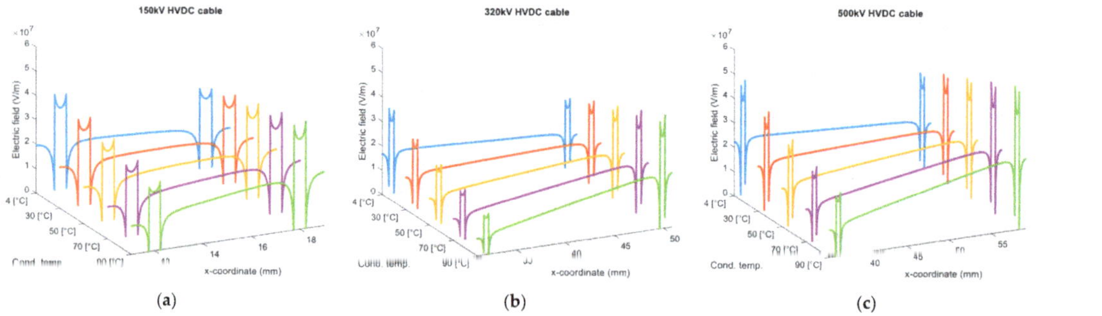

Figure 7. E field strength distribution in the cable insulation with two cavities at 4 °C, 30 °C, 50 °C, 70 °C and 90 °C conductor temperatures: (**a**) 150 kV HVDC cable; (**b**) 320 kV HVDC cable; (**c**) 500 kV HVDC cable. (Effects related to the formation of PD in the void have not been taken into account).

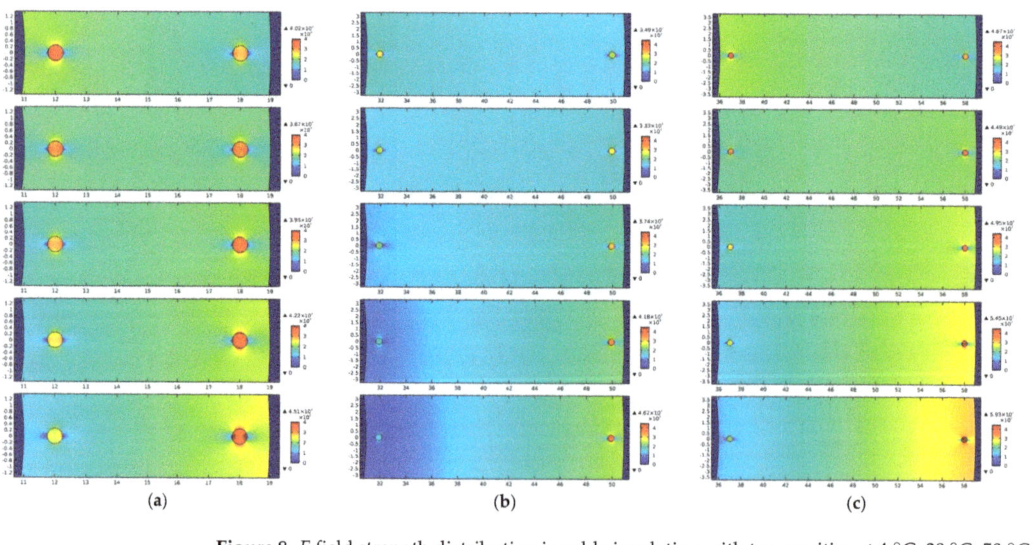

Figure 8. E field strength distribution in cable insulation with two cavities at 4 °C, 30 °C, 50 °C, 70 °C, 90 °C conductor temperature: (**a**) 150 kV HVDC cable; (**b**) 320 kV HVDC cable; (**c**) 500 kV HVDC cable. (Effects related to the formation of PD in the void have not been taken into account).

On the basis of numerically determined distributions of $T(x)$, $\sigma(x)$ and $E(x)$ for each cable, the values of these variables were determined for the boundary values of x coordinate, i.e., boundary temperatures (Table 5), boundary electrical conductivity (Table 6) and boundary E field strength in insulation without voids (Table 7). In addition, the conductance increase factor k_σ was defined as a ratio of the σ_{xmin} to σ_{xmax} (Table 6), and the differences ΔE between the E field strength at x_{min} and x_{max} coordinates (Table 7) were calculated. A negative value of ΔE means that the electric field inversion effect has occurred in insulation.

Table 5. Boundary temperatures in insulation of 150 kV, 320 kV and 500 kV cable models.

Cable Core Temperature	150 kV Cable			320 kV Cable			500 kV Cable		
	T_{xmin}, °C	T_{xmax}, °C	ΔT, °C	T_{xmin}, °C	T_{xmax}, °C	ΔT, °C	T_{xmin}, °C	T_{xmax}, °C	ΔT, °C
4 °C	4.0	4.0	0.0	4.0	4.0	0.0	4.0	4.0	0.0
30 °C	28.4	19.1	9.3	28.2	15.2	13.0	28.2	15.5	12.7
50 °C	47.1	30.7	16.4	46.7	23.8	22.9	46.7	24.3	22.4
70 °C	65.9	42.3	23.6	65.3	32.3	33.0	65.3	33.1	32.2
90 °C	84.6	53.9	30.7	83.9	40.9	43.0	83.9	42.0	41.9

Table 6. Boundary electrical conductivity of insulation for 150 kV, 320 kV and 500 kV cable models.

Cable Core Temperature	150 kV Cable			320 kV Cable			500 kV Cable		
	σ_{xmin}, S·m^{-1}	σ_{xmax}, S·m^{-1}	k_σ, -	σ_{xmin}, S·m^{-1}	σ_{xmax}, S·m^{-1}	k_σ, -	σ_{xmin}, S·m^{-1}	σ_{xmax}, S·m^{-1}	k_σ, -
4 °C	2.9×10^{-15}	2.2×10^{-15}	1.32	2.4×10^{-15}	1.8×10^{-15}	1.33	3.6×10^{-15}	2.7×10^{-15}	1.33
30 °C	1.2×10^{-14}	6.6×10^{-15}	1.82	8.8×10^{-15}	4.6×10^{-15}	1.91	1.3×10^{-14}	6.9×10^{-15}	1.88
50 °C	3.4×10^{-14}	1.6×10^{-14}	2.13	2.5×10^{-14}	9.3×10^{-15}	2.69	3.5×10^{-14}	1.4×10^{-14}	2.50
70 °C	9.9×10^{-14}	3.7×10^{-14}	2.68	6.9×10^{-14}	1.9×10^{-14}	3.63	9.6×10^{-14}	3.0×10^{-14}	3.20
90 °C	2.9×10^{-13}	8.7×10^{-14}	3.33	2.0×10^{-13}	3.9×10^{-14}	5.13	2.7×10^{-13}	6.4×10^{-14}	4.22

Table 7. Boundary E field strength in insulation of 150 kV, 320 kV and 500 kV cable models.

Cable Core Temperature	150 kV Cable			320 kV Cable			500 kV Cable		
	E_{xmin} kV·mm^{-1}	E_{xmax} kV·mm^{-1}	ΔE kV·mm^{-1}	E_{xmin} kV·mm^{-1}	E_{xmax} kV·mm^{-1}	ΔE kV·mm^{-1}	E_{xmin} kV·mm^{-1}	E_{xmax} kV·mm^{-1}	ΔE kV·mm^{-1}
4 °C	21.3	16.7	4.6	18.1	14.3	3.8	24.2	19.7	4.5
30 °C	18.5	18.9	−0.4	14.6	17.2	−2.6	20.2	23.0	−2.8
50 °C	16.5	20.7	−4.2	12.1	19.5	−7.4	17.4	25.6	−8.2
70 °C	14.6	22.5	−7.9	9.9	21.9	−12.0	14.7	28.3	−13.6
90 °C	12.7	24.3	−11.6	7.8	24.4	−16.6	12.2	30.9	−18.7

In the next step of the analysis, the E field strength distributions were determined on the radius of each of the model cables with the presence of spherical gaseous voids, with the dimensions and locations described in Table 4. The indicated results of the simulation of E field strength distributions vs. the x-coordinate are presented in Figure 7. The scale of the E field axis is identical in all three sub-figures, while the scale of the x-coordinate axis is adapted to the geometry of each analyzed cable. It should be noted that these E_{void} simulations do not take into account that the highest field strength in the void is limited by the PD generation process.

The last stage of the static analyses was to determine the 2D distributions of the E field strength in the insulation of each of the model cables, in the areas where spherical gaseous voids were located (Figure 8). The E field strength (expressed in V/m) was color-coded with a scale from 'cold' (lowest E field value) to 'hot' (highest E field value). The E_{void} values for both voids and the E_{cable} in XLPE for the same two radii were tabulated and field enhancement factor ($FEF = E_{void}/E_{cable}$) values were calculated (Table 8).

Table 8. *E* field in void and in XLPE simulation results for 150 kV, 320 kV and 500 kV cable models.

Cable Core Temperature	150 kV Cable						320 kV Cable						500 kV Cable					
	1st Void			2nd Void			1st Void			2nd Void			1st Void			2nd Void		
	E_{void} kV·mm^{-1}	E_{cable} kV·mm^{-1}	*FEF* -	E_{void} kV·mm^{-1}	E_{cable} kV·mm^{-1}	*FEF* -	E_{void} kV·mm^{-1}	E_{cable} kV·mm^{-1}	*FEF* -	E_{void} kV·mm^{-1}	E_{cable} kV·mm^{-1}	*FEF* -	E_{void} kV·mm^{-1}	E_{cable} kV·mm^{-1}	*FEF* -	E_{void} kV·mm^{-1}	E_{cable} kV·mm^{-1}	*FEF* -
4 °C	33.6	20.5	1.64	28.5	17.2	1.66	29.7	17.8	1.67	24.4	14.4	1.69	38.6	23.9	1.62	32.7	19.4	1.69
30 °C	30.7	18.6	1.65	31.2	18.9	1.65	25.0	14.8	1.69	28.5	17.1	1.67	33.5	20.4	1.64	37.3	22.9	1.63
50 °C	28.5	17.1	1.67	33.0	20.2	1.63	21.6	12.6	1.71	31.6	19.3	1.64	29.7	17.8	1.67	40.7	25.3	1.61
70 °C	26.3	15.8	1.66	35.0	21.6	1.62	18.2	10.5	1.73	34.7	21.4	1.62	25.9	15.3	1.69	44.1	27.8	1.59
90 °C	24.4	14.4	1.69	37.1	23.0	1.61	15.1	8.6	1.76	37.9	23.6	1.61	22.4	13.1	1.71	47.4	30.2	1.57

The red color indicates the highest and blue the lowest values of the *E* field strength in each pair of gaseous voids (1st or 2nd), or in the dielectric at void *x*-coordinates, or value of the *FEF*. The green color indicates nearly equal *E* field values in pairs of voids.

The radial temperature distribution existing in the insulation of the HVDC cable causes the intensity of the PD inception field in the gaseous void, depending on the location of the void on the radius of the cable. Assuming that the gas temperature in the void is equal to the temperature of the surrounding dielectric, the values of E_{inc} vs. *x*-coordinate were determined for the selected cable core temperatures of each cable (Figure 9). Using a dedicated numerical procedure implemented in the MATLAB® program, sequences of successive distributions of the *E* field strength were generated for the electro-thermal model of the cable simulated in the COMSOL® program [73]. The general procedure of PD simulation is presented in Figure 10. The first step is creating a geometric model, setting the materials and physical parameters, establishing boundary settings and meshing. Next, electric and thermal computation vs. time is performed. During the simulation, the PD initiation condition is checked. Discharge can occur when the electric field strength in the void exceeds the value of the PD inception field, Equation (1). Moreover, the stochastic nature of PD generation is modeled using Equation (2). Discharge probability P_d is calculated and compared with random number *R* between 0 and 1. If $P_d > R$, the PD initiation conditions are met. Partial discharge development is simulated by increasing the conductivity of the gas in the void from 10^{-16} S·m^{-1} to 10^{-3} S·m^{-1}. The flowing charge causes a decrease in the electric field inside the void. When the electric field in the gaseous void is reduced to the PD extinction field E_{ext}, discharge ends and the gas conductivity is reset to 10^{-16} S·m^{-1}. Individual PD charge is determined by integrating the current flowing through the void during discharge process. The entire procedure is repeated until the assumed simulation time is reached. In each simulation, the procedure was started 2000 s after applying the DC voltage.

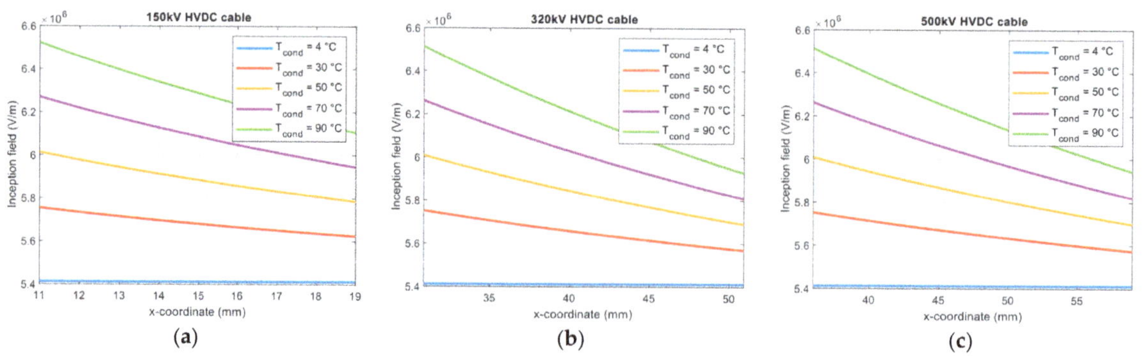

Figure 9. PD inception field in void of 0.5 mm diameter located in XLPE insulation of HVDC cable at 4 °C, 30 °C, 50 °C, 70 °C and 90 °C cable core temperatures for: (**a**) 150 kV HVDC cable; (**b**) 320 kV HVDC cable; (**c**) 500 kV HVDC cable.

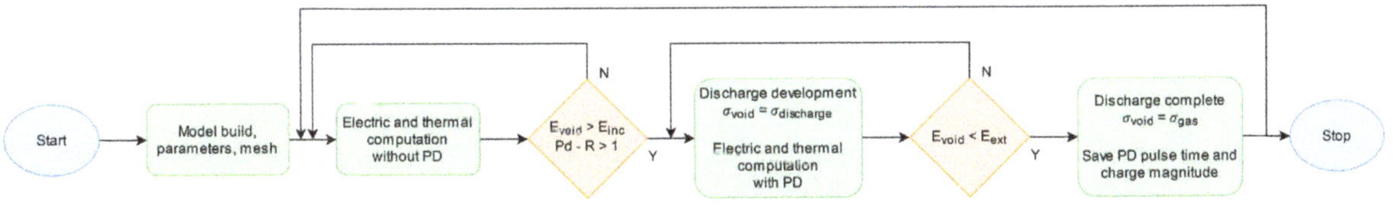

Figure 10. Block diagram of numerical procedure for determination of PD pulses generated in gaseous void in modeled HVDC cable insulation.

Effects related to the formation of PD pulses in both gaseous voids at field strengths reaching their individual E_{inc} values were taken into account. The time sequences of the PD pulses simulated in this way are shown in Figure 11.

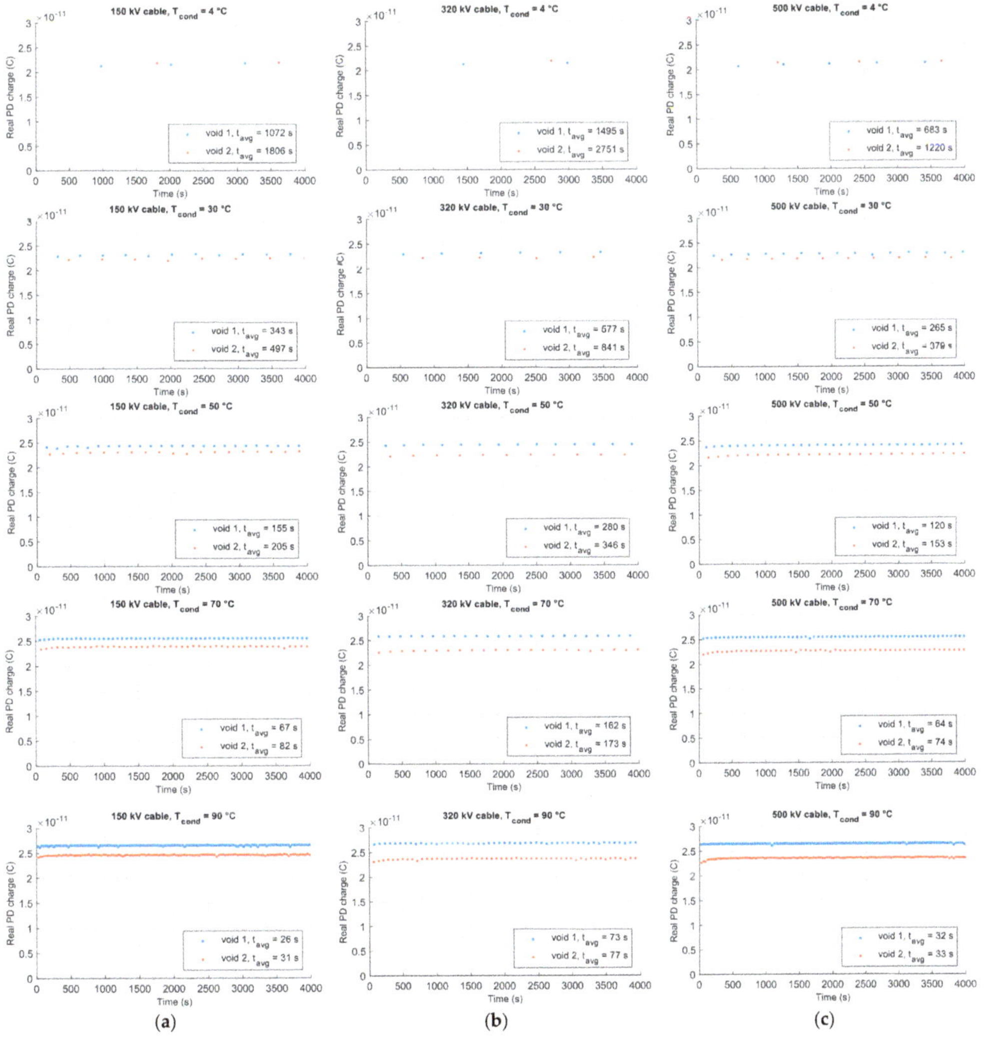

Figure 11. Simulated PD pulse sequences vs. time at 4 °C, 30 °C, 50 °C, 70 °C and 90 °C cable core temperatures for: (**a**) 150 kV HVDC cable; (**b**) 320 kV HVDC cable; (**c**) 500 kV HVDC cable.

The observed sequence of PD pulses in the cable insulation is the result of two independent discharge generation processes arising in the first and second gaseous voids (Figure 11). For each modeled cable (150 kV, 320 kV and 500 kV) and for selected cable core temperatures (4 °C, 30 °C, 50 °C, 70 °C and 90 °C) based on the simulated PD sequences, two basic statistical parameters were determined, characterizing the PD pulses at DC voltage (Table 9):

- Average PD magnitude q_{av}, pC;
- Average time interval between successive PD pulses Δt_{av}, s.

Table 9. Average PD charge q_{av} and average PD pulse time interval Δt_{av} for 150 kV, 320 kV and 500 kV cable models at selected cable core temperatures.

Cable Core Temperature	150 kV Cable				320 kV Cable				500 kV Cable			
	1st Void		2nd Void		1st Void		2nd Void		1st Void		2nd Void	
	q_{av} pC	Δt_{av} s	q_{av} pC	Δt_{av} s	q_{av} pC	Δt_{av} s	q_{av} pC	Δt_{av} s	q_{av} pC	Δt_{av} s	q_{av} pC	Δt_{av} s
4 °C	21.6	1072	21.8	1806	21.4	1495	21.9	2751	21.1	683	21.5	1220
30 °C	23.1	343	22.3	497	23.1	577	22.1	841	22.8	265	21.8	379
50 °C	24.3	155	23.1	205	24.4	280	22.3	346	24.1	120	22.2	153
70 °C	25.5	67	23.9	82	25.9	162	22.9	173	25.4	64	22.7	74
90 °C	26.5	26	24.6	31	26.8	73	23.6	77	26.5	32	23.6	33

4. Discussion

The presented results of thermal and electric field analyses for models of three HVDC cables with different rated voltages (150 kV, 320 kV and 500 kV), different surface areas of the cable core (300 mm^2, 2500 mm^2 and 3000 mm^2) and different structures of the outer layers covering the XLPE insulation are the basis for determining the conditions for the formation of PD pulses in internal gaseous voids.

The electric field in the insulation of the HVDC cable, due to the direct dependence on the electrical conductivity of the insulating material, is a field coupled (indirectly) with the thermal field present in the volume of the cable insulation. For current-carrying cables with a concentric design, the resulting radial temperature distribution in the cable also modifies radially the conductivity of the insulation material. This leads to a significant increase in the electrical conductivity of the dielectric along the radius of the cable, from the lowest conductivity at the outer radius of the insulation to the highest at the inner radius of the insulation (i.e., near the cable core with the highest temperature). Moreover, according to Equation (4) or (5), the conductivity of the dielectric is modified, also radially, by the presence of a DC electric field, which is influenced by the space charge present in the insulation volume. If the cable insulation is a modern XLPE polymer dedicated to HVDC applications (DC-XLPE), then the additional component of the E field produced by the space charge in steady state, a long time after applying a DC voltage, has a small value, even below 5% [54]. It follows that in such a state, the presence of space charge will only slightly affect the repetition rate of the PD pulses. However, the nature and trends of changes in PD parameters under the influence of the radial thermal field will be the same.

When estimating the conditions for the generation of the PD pulse sequence, the presence of a space charge in the insulation from sources other than the charge resulting from the application of the test voltage was not considered. More complex simulations may also take into account other charges present in the polymer structure, e.g., related to the presence of polar additives or impurities [74].

For the numerical simulations of steady states of the coupled thermo-electric field, performed in the COMSOL Multiphysics® 6.0 program, a constant value of the temperature of the external environment of the cable (4 °C) was assumed, while the temperature of the cable core was the modified parameter (from 4 °C to 90 °C). Obviously, for all three cables, similar insulation temperatures were obtained at the inner radius of the insulation $x_{\min}$ for each core temperature and the calculated differences are not greater than 0.7 °C for a conductor temperature of 90 °C (Figure 3 and Table 5). Due to the thickness of the XLPE insulation and the structure of the outer layers of the cables, the smallest decrease in temperature occurs in the insulation of the 150 kV cable, and the largest in the 320 kV cable. For example, when the temperature of the conductor is 70 °C, then the numerically estimated temperature differences for the inner and outer radius of insulation for the mentioned cables are 23.6 °C and 33 °C, respectively (Table 5). This fact has consequences for the conductivity distribution (Figure 4) and electric field strength (Figures 5 and 6) observed along the x-coordinate in the cable insulation. Already, at a conductor temperature of 30 °C, the electric field inversion effect occurs. The full ranges of variation of electric field strength E, caused by a change in conductor temperature, are:

(a) On the inner radius of the insulation $x_{\min}$

- For 150 kV cable, from 12.7 kV·mm^{-1} to 21.3 kV·mm^{-1};
- For 320 kV cable, from 7.8 kV·mm^{-1} to 18.1 kV·mm^{-1};
- For 500 kV cable, from 12.2 kV·mm^{-1} to 24.2 kV·mm^{-1}.

(b) On the outer radius of the insulation

- For 150 kV cable, from 16.7 kV·mm^{-1} to 24.3 kV·mm^{-1};
- For 320 kV cable, from 14.3 kV·mm^{-1} to 24.4 kV·mm^{-1};
- For 500 kV cable, from 19.7 kV·mm^{-1} to 30.9 kV·mm^{-1}.

Taking into account the above data, and Equation (1) and the data contained in Figure 1, it should be noted that the distributions of the field strength E for the cable model with two gaseous voids, presented in Figures 7 and 8, are not adequate to the real situation in the modeled cables. These distributions were only used to estimate the FEF values for the analyzed gaseous voids. The *FEF* values, obtained as a result of numerical simulations, ranging from 1.62 to 1.76 for all considered cases, are in good agreement with the results obtained in the 2D simulations presented in [71]. According to this paper, the more accurate results obtained from the 3D simulation are about 16% lower.

Due to the very high values of the DC field strength in the insulation of all three analyzed model cables (Figure 6), a 0.5 mm spherical gas void will be the source of PD in each of the considered cases (Figure 9). Of course, any gaseous void large enough to have a PD inception field strength (Figure 1) lower than the estimated electric field strength inside the void will be a potential PD source. This is a problem that is related to the influence of micro-defects of the XLPE polymer structure (with dimensions below 30 μm) on the rate of damage growth and the estimated lifetime of the insulation [37,38]. If the radial distributions of the electric field in HVDC cables (Figure 6) are related to the estimated values of the PD inception field E_{inc} as a function of the dimension d of the gaseous void (Figure 1; curves $E_{\mathrm{inc}} = f(d)$, parameterized by the gas temperature), then it is possible to estimate the critical dimension d_{cr} of the gaseous void (i.e., the minimum void size for which PD can arise for local electro-thermal field conditions) vs. the location of the gaseous void on the radius of the cable. Figure 12 compares the results of the d_{cr} estimation for the three analyzed HVDC cables. For these calculations, a simplifying assumption was made that for each gaseous void, the field enhancement factor *FEF* is 1.65 (see Table 8).

The results of the estimation of the critical dimension d_{cr} (Figure 12) indicate that for 150 kV and 500 kV cables in the full range of the analyzed temperatures, they are lower than 20 μm, and for the 320 kV cable only for the highest temperatures and x-coordinates close to the cable core are they are greater than this value. Three decades ago, the size of a gaseous void of 20 μm was the quality control level for defects of this type in the XLPE insulation of 500 kV cables [75]. Analyses of XLPE insulation lifetime models have shown, however, that the estimated critical sizes of microstructural defects are one order of magnitude smaller [37,38,51]. Damage to the cross-linked structure of XLPE at the microvoid–polymer interface caused by effective hot electrons may, under conditions of strong field stress, lead to the initiation of the electric treeing process and, as a result, shorten the life of the insulation [38,51]. The performed numerical simulations show that for 150 kV and 500 kV cables in a wide range of x-coordinates, and for a 320 kV cable in a narrower range, PD inception can occur in voids with dimensions of several micrometers, below 10 µm and down to 2 µm.

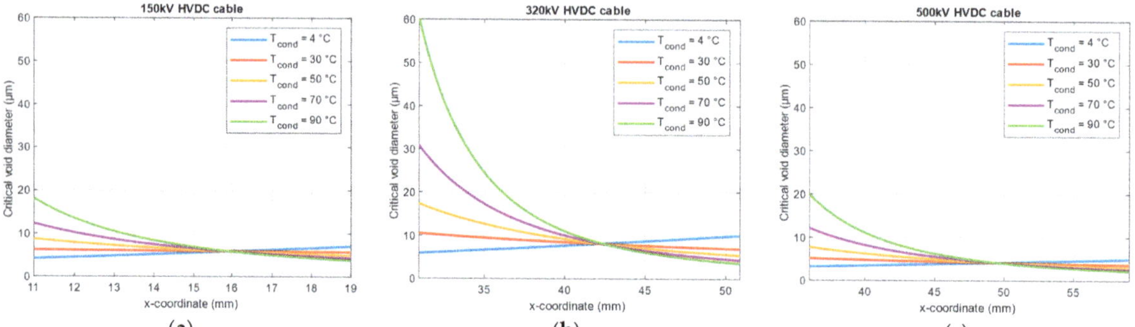

Figure 12. Critical diameter of gaseous void d_{cr} vs. x-coordinate in HVDC cable at 4 °C, 30 °C, 50 °C, 70 °C and 90 °C cable core temperatures for: (**a**) 150 kV HVDC cable; (**b**) 320 kV HVDC cable; (**c**) 500 kV HVDC cable.

The comparison of the results for all three cables shows that the design parameters, electrical and thermal, as well as the geometry of the cable have a significant impact on the critical dimension of the gaseous void. The 320 kV cable, with a more conservative design, has slightly lower electric field stress than the other two cables (Figures 5 and 6). For this reason, and due to the temperature field generated in the insulation of the loaded cable (Figure 3), this results in higher values of the critical dimension d_{cr} for this cable.

It should be noted that from the point of view of assessing the correctness and accuracy of simulation results carried out in the COMSOL® program, reducing the size of gaseous voids leads to a significant increase in the numerical complexity of the entire model and time-consuming calculations. High-resolution analysis of time sequences of PD pulses, generated in voids with sizes of the order of single micrometers, while taking into account the full geometry of the HVDC cable, requires the use of a computer hardware platform with very high computing performance and memory resources sufficient to allocate the extensive mesh model of the analyzed cable.

If the conditions necessary for PD inception are not met in a thermally stable model of cable insulation with a gaseous inclusion, the considered field problem is static. The presence of an active void in the insulation (i.e., one in which the conditions for the formation of PD are met) makes modeling a dynamic problem. The mechanism of discharge generation in the void is conditioned by the threshold value E_{inc}, exceeding which causes the creation of the PD pulse. Thus, a time-varying disturbance of the local electric field in the void and the dielectric around it is initiated. In circuit models, this corresponds to a short-term discharge of the void's electric capacitance and a slow process of its charging in a circuit with a long time constant. The parameters of the equivalent scheme elements of an insulating system with a void depend on the values of material constants (dielectric constant and resistivity/conductivity), which in turn are conditioned by temperature and electric field [39,76–78]. The magnitude of the PD from the void at a smooth DC voltage has a small dispersion compared to the PD at an AC voltage. When DC voltage ripple is present, then the PD charge dispersion is greater due to the influence of the statistically variable lag time.

In the HVDC cable model implemented by combining the functionality of MATLAB® and COMSOL®, it was possible to determine the sequence of PD pulses generated in the analyzed gaseous voids. Considering the variable nature of the radial distribution of the E field in the insulation of the HVDC cable and the presence of a point with the coordinate

x_c, for which the E field is always almost constant (Figure 6), the insulation volume can be divided into two zones:

- Zone I, closer to the cable core, limited by x_{min} and x_c radii;
- Zone II, closer to the outer screen on the insulation, limited by x_c and x_{max} radii.

The relationship between the volume of cable insulation in Zone II and the insulation in Zone I can be defined by the formula:

$$k_V = 100\frac{Volume\ of\ Zone\ II}{Volume\ of\ Zone\ I} = 100\frac{x_{max}^2 - x_c^2}{x_c^2 - x_{min}^2}[\%] \quad (6)$$

where:

k_V—zone volume factor, %;
x_{min}—inner radius of insulation, mm;
x_{max}—outer radius of insulation, mm;
x_c—radius of the insulation layer with (almost) constant E field, mm.

For the three analyzed cable models (150 kV, 320 kV and 500 kV) the k_V values are 138%, 130% and 132%, respectively. Assuming that the distribution of defects in the insulation is uniform, more than 40% of the gaseous voids are exposed to the conditions characteristic of Zone I, and almost 60% to the conditions of Zone II. In an unloaded cable, the E field distribution is 'normal, i.e., it is similar to the distribution in an AC cable and the highest field stress occurs at the inner radius of the insulation. This is the reason for the shorter average time interval between successive PD pulses Δt_{av} in voids located near the cable core (Figure 10 and Table 9, conductor temperature 4 °C). The heating of the conductor and the creation of a radial thermal field causes an inversion of the E field distribution (Figures 5 and 6). Then, the complex electro-thermal stresses in both zones change, affecting the conditions of PD formation in gaseous voids. An increase in the gas temperature also causes an increase in the PD inception field strength, and thus an increase in the PD charge. This effect is more pronounced in the area near the cable core (Figure 10 and Table 9). At the same time, due to a significant increase in conductivity, the PD repetition rate increases, in particular in sources located in areas with the highest temperature, i.e., in Zone I, close to the cable core. A similar effect of changing the parameters of the PD pulse sequence was also obtained in other studies based on field modeling of the gaseous void in XLPE insulation of HVDC cables [70,72]; however, the influence of the location of the defect on the PD time sequences was not analyzed. It is seen that the factor increasing the PD repetition rate is also the E field strength, reaching for loaded cable the highest values in Zone II, especially in the area close to the outer radius of the insulation. The factors causing the increase of the PD repetition rate, i.e., increase of conductivity and increase of the E field strength, occur simultaneously, but in the heated cable their radial distributions are mutually inverse. For this reason, the observed effect of increasing the PD repetition rate in HVDC cable is the result of the synergistic effect of both of these factors.

5. Conclusions

The problem presented in the article concerns the conditions for the generation of PD pulses and their time sequences in the XLPE insulation of HVDC cable in steady state. Elements of research originality, including numerical simulations and analyses, cover several different aspects of this topic:

1. Reference of the conditions of PD formation in XLPE insulation of 'cold' and 'hot' HVDC cables to the location of the gaseous void on the cable radius; these analyses concerned three different designs of HVDC cables, but their results can be generalized;
2. Simulating and analyzing the parameters of the sequence of PD pulses generated independently in two PD sources with different locations on the cable radius;

3. Distinguishing two zones in the HVDC cable where the conditions of PD formation in the 'cold' and 'hot' cable insulation are different due to the radial temperature distribution and 'normal' or 'inverted' electric field distribution;
4. Estimation of the critical dimension of the gaseous void depending on the position and thermal condition of the cable insulation, for each of the analyzed projects.

The planned further extension of the analyses will take into account the effects related to the accumulation of space charge in the insulation bulk and at interfaces, as well as the increase of *E* field stress in transient states, including those resulting from voltage polarity reversal.

Table 10 summarizes the conditions for the formation of PD in the gaseous void and the basic characteristics of the PD pulses and their time sequences in both distinguished zones of the XLPE insulation of the HVDC cable (Zone I and Zone II).

Table 10. Characteristics of PD in two zones of HVDC cable insulation.

Insulation Zone	Unloaded Cable ('Cold')	Loaded Cable ('Hot')
Zone I	Insulation temperature equal to environment temperature.	High insulation temperature, the highest at the cable core.
	Very low electrical conductivity of insulation material.	Higher electrical conductivity of the insulation material, highest near inner insulation radius.
	Very high electric field stress, highest near inner insulation radius (as in an AC cable).	Significant reduction of electric field stress, lowest stress near inner insulation radius.
	Very low repetition rate of PD pulses, but for voids with identical parameters higher than in Zone II (influence of higher *E* field stress).	Significantly higher repetition rate of PD pulses (several dozen times compared to a 'cold' cable), as a result of the combined action of higher temperature and increased *E* field stress.
	PD charges as in Zone II, with a small dispersion of values. In the case of DC voltage ripple, there is a greater variability of the PD magnitude due to the influence of lag time.	An increase in the PD magnitude (over a dozen percent), caused by an increase in E_{inc}, due to the higher gas pressure in the closed void.
	Due to the higher electric field strength, PD sources of smaller critical dimension than in Zone II may be also active.	An increase in the temperature of the cable core increases the critical dimension of the void near the core.
Zone II	Insulation temperature equal to environment temperature.	Increased insulation temperature, lowest on the outer radius of the insulation.
	Very low electrical conductivity of insulation material.	Higher electrical conductivity of the insulation material, but lower than in Zone I; lowest on the outer radius of the insulation.
	Reduced electric field stress, lower than in Zone I, lowest near the outer radius of the insulation (as in an AC cable).	Very high electric field stress, highest near outer insulation radius.
	Very low repetition rate of PD pulses.	Significantly higher repetition rate of PD pulses (several dozen times compared to a 'cold' cable), as a result of the combined action of higher temperature and increased *E* field stress.
	PD charges as in Zone I, with a small dispersion of values. In the case of DC voltage ripple, there is a greater variability of the PD magnitude due to the influence of lag time.	A slight increase in PD magnitude (several percent), caused by an increase in E_{inc}, due to the higher gas pressure in the closed void.
	Critical dimension of the gaseous void slightly larger than in Zone I.	The critical dimension of the gaseous void can be significantly smaller than in Zone I (depending on the temperature of the cable core).

Author Contributions: Conceptualization, P.Z.; methodology, P.Z. and P.M.; software, P.M.; validation, P.Z. and P.M.; resources, P.Z. and P.M.; data curation, P.M. and P.Z.; writing—original draft preparation, P.Z. and P.M.; writing—review and editing, P.Z. and P.M.; visualization, P.M.; supervision, P.Z. All authors have read and agreed to the published version of the manuscript.

Funding: This research received no external funding.

Data Availability Statement: Not applicable.

Conflicts of Interest: The authors declare no conflict of interest.

References

1. Humpert, C. Long distance transmission systems for the future electricity supply—Analysis of possibilities and restrictions. *Energy* **2012**, *48*, 278–283. [CrossRef]
2. Hammons, T.J.; Lescale, V.F.; Uecker, K.; Haeusler, M.; Retzmann, D.; Staschus, K.; Lepy, S. State of the art in ultrahigh-voltage transmission. *Proc. IEEE* **2012**, *100*, 360–390. [CrossRef]
3. Zhou, H.; Qiu, W.; Sun, K.; Chen, J.; Deng, X.; Qian, F.; Wang, D.; Zhao, B.; Li, J.; Li, S.; et al. (Eds.) *Ultra-High Voltage AC/DC Power Transmission*; Springer: Berlin/Heidelberg, Germany, 2019.
4. Worzyk, T. *Submarine Power Cables Design, Installation, Repair, Environmental Aspects*; Springer: Berlin/Heidelberg, Germany, 2009.
5. Mazzanti, G.; Marzinotto, M. *Extruded Cables for High-Voltage Direct-Current Transmission: Advances in Research and Development*; Wiley-IEEE Press: Hoboken, NJ, USA, 2013.
6. Jovcic, D.; Ahmed, K. *High-Voltage Direct-Current Transmission. Converters, Systems and DC Grids*; John Wiley & Sons: Hoboken, NJ, USA, 2015.
7. Liu, R. Long-distance DC electrical power transmission. *IEEE Electr. Insul. Mag.* **2013**, *29*, 37–46. [CrossRef]
8. Alassi, A.; Bañales, S.; Ellabban, O.; Adam, G.; MacIver, C. HVDC transmission: Technology review, market trends and future outlook. *Renew. Sustain. Energy Rev.* **2019**, *112*, 530–554. [CrossRef]
9. Li, Z.; Song, Q.; An, F.; Zhao, B.; Yu, Z.; Zeng, R. Review on DC transmission systems for integrating large-scale offshore wind farms. *Energy Convers. Econ.* **2021**, 2, 1–14. [CrossRef]
10. Acaroğlu, H.; Márquez, F.P.G. High voltage direct current systems through submarine cables for offshore wind farms: A life-cycle cost analysis with voltage source converters for bulk power transmission. *Energy* **2022**, *249*, 123713. [CrossRef]
11. ENTSO-E. *HVDC Links in System Operations, European Network of Transmission System Operators for Electricity*; Technical Paper; ENTSO-E: Brussels, Belgium, 2019.
12. Stan, A.; Costinaș, S.; Ion, G. Overview and assessment of HVDC current applications and future trends. *Energies* **2022**, *15*, 1193. [CrossRef]
13. Parol, M.; Robak, S.; Rokicki, Ł.; Wasilewski, J. Cable links designing in HVAC and HVDC submarine power grids—Selected issues. *Prz. Elektrotech.* **2019**, *95*, 7–13. [CrossRef]
14. Li, Y.; Liu, H.; Fan, X.; Tian, X. Engineering practices for the integration of large-scale renewable energy VSC-HVDC systems. *Glob. Energy Interconnect.* **2020**, *3*, 149–157. [CrossRef]
15. *Offshore Transmission Technology*; Report of ENTSO-E's Regional Group North Sea; European Network of Transmission System Operators for Electricity: Brussels, Belgium, 2011.
16. Stone, G.C.; Boulter, E.A.; Culbert, I.; Dhirani, H. *Electrical Insulation for Rotating Machines: Design, Evaluation, Aging, Testing, and Repair*, 1st ed.; Wiley-IEEE Press: Milwaukee, WI, USA, 2004.
17. *IEC 60505:2011*; Evaluation and Qualification of Electrical Insulation Systems. IEC: Genève, Switzerland, 2011.
18. Dissado, L.A.; Fothergill, J.C. *Electrical Degradation and Breakdown in Polymers*; Peregrinus Press: Chicago, IL, USA, 1992.
19. Jones, J.P.; Llewellyn, J.P.; Lewis, T.J. The contribution of field-induced morphological change to the electrical aging and breakdown of polyethylene. *IEEE Trans. Dielectr. Electr. Insul.* **2005**, *12*, 951–966. [CrossRef]
20. Marzinotto, M.; Mazzanti, G. The statistical enlargement law for HVDC cable lines part 1: Theory and application to the enlargement in length. *IEEE Trans. Dielectr. Electr. Insul.* **2015**, *22*, 192–201. [CrossRef]
21. Montanari, G.C.; Seri, P.; Dissado, L.A. Aging mechanisms of polymeric materials under DC electrical stress: A new approach and similarities to mechanical aging. *IEEE Trans. Dielectr. Electr. Insul.* **2019**, *26*, 634–641. [CrossRef]
22. Montanari, G.C.; Morshuis, P.; Seri, P.; Ghosh, R. Ageing and reliability of electrical insulation: The risk of hybrid AC/DC grids. *High Volt.* **2020**, *5*, 620–627. [CrossRef]
23. Naderiallaf, H.; Seri, P.; Montanari, G.C. Designing a HVDC insulation system to endure electrical and thermal stresses under operation. Part I: Partial discharge magnitude and repetition rate during transients and in DC steady state. *IEEE Access* **2021**, *9*, 35730–35739. [CrossRef]
24. Cambareri, P.; de Falco, C.; Rienzo, L.D.; Seri, P.; Montanari, G.C. Electric field calculation during voltage transients in HVDC cables: Contribution of polarization processes. *IEEE Trans. Power Deliv.* **2022**, *37*, 5425–5432. [CrossRef]
25. Lei, Z.; Song, J.; Tian, M.; Cui, X.; Li, C.; Wen, M. Partial discharges of cavities in ethylene propylene rubber insulation. *IEEE Trans. Dielectr. Electr. Insul.* **2014**, *21*, 1647–1659. [CrossRef]
26. Alshaikh Saleh, M.; Refaat, S.S.; Olesz, M.; Abu-Rub, H.; Guziński, J. The effect of protrusions on the initiation of partial discharges in XLPE high voltage cables. *Bull. Pol. Acad. Sci. Tech. Sci.* **2021**, *69*, e136037. [CrossRef]
27. Dissado, L.A.; Dodd, S.J.; Champion, J.V.; Williams, P.I.; Alison, J.M. Propagation of electrical tree structures in solid polymeric insulation. *IEEE Trans. Dielectr. Electr. Insul.* **1997**, *4*, 259–279. [CrossRef]
28. Chen, X.; Mantsch, A.R.; Hu, L.; Gubanski, S.M.; Blennow, J.; Olsson, C.O. Electrical treeing behavior of DC and thermally aged polyethylenes utilizing wire-plane electrode geometries. *IEEE Trans. Dielectr. Electr. Insul.* **2014**, *21*, 45–52. [CrossRef]
29. Ross, R. Inception and propagation mechanisms of water treeing. *IEEE Trans. Dielectr. Electr. Insul.* **1998**, *5*, 660–680. [CrossRef]

30. Li, J.; Zhao, X.; Yin, G.; Li, S.; Zhao, J.; Ouyang, B. The effect of accelerated water tree ageing on the properties of XLPE cable insulation. *IEEE Trans. Dielectr. Electr. Insul.* **2011**, *8*, 1562–1569. [CrossRef]
31. Fard, M.A.; Farrag, M.E.; McMeekin, S.; Reid, A. Electrical treeing in cable insulation under different HVDC operational conditions. *Energies* **2018**, *11*, 2406. [CrossRef]
32. Liu, H.; Zhang, M.; Liu, Y.; Xu, X.; Liu, A. Growth and partial discharge characteristics of DC electrical trees in cross-linked polyethylene. *IEEE Trans. Dielectr. Electr. Insul.* **2019**, *26*, 1965–1972. [CrossRef]
33. Liu, F.; Rowland, S.M.; Zheng, H.; Peesapati, V. Electrical tree growth in LDPE: Fine channel development during negative DC ramp down. *IEEE Trans. Dielectr. Electr. Insul.* **2022**, *29*, 1218–1220. [CrossRef]
34. Drissi-Habti, M.; Raj-Jiyoti, D.; Vijayaraghavan, S.; Fouad, E.-C. Numerical Simulation for Void Coalescence (Water Treeing) in XLPE Insulation of Submarine Composite Power Cables. *Energies* **2020**, *13*, 5472. [CrossRef]
35. Miceli, M.; Carvelli, V.; Drissi-Habti, M. Modelling electro-mechanical behaviour of an XLPE insulation layer for Hi-Voltage composite power cables: Effect of voids on onset of coalescence. *Energies* **2023**, *16*, 4620. [CrossRef]
36. Chen, G.; Davies, A.E. The influence of defects on the short-term breakdown characteristics and long-term dc performance of LDPE insulation. *IEEE Trans. Dielectr. Electr. Insul.* **2000**, *7*, 401–407. [CrossRef]
37. Mazzanti, G.; Montanari, G.C.; Civenni, F. Model of inception and growth of damage from microvoids in polyethylene-based materials for HVDC cables—Part 1: Theoretical approach. *IEEE Trans. Dielectr. Electr. Insul.* **2007**, *14*, 1242–1254. [CrossRef]
38. Mazzanti, G.; Montanari, G.C.; Civenni, F. Model of inception and growth of damage from microvoids in polyethylene-based materials for HVDC cables. 2. Parametric Investigation and Data Fitting. *IEEE Trans. Dielectr. Electr. Insul.* **2007**, *14*, 1255–1263. [CrossRef]
39. Niemeyer, L. A generalized approach to partial discharge modeling. *IEEE Trans. Dielectr. Electr. Insul.* **1995**, *2*, 510–528. [CrossRef]
40. Pan, C.; Chen, G.; Tang, J.; Wu, K. Numerical modeling of partial discharges in a solid dielectric-bounded cavity: A review. *IEEE Trans. Dielectr. Electr. Insul.* **2019**, *26*, 981–1000. [CrossRef]
41. Zhan, Y.; Chen, G.; Hao, M.; Pu, L.; Zhao, X.; Sun, H.; Wang, S.; Guo, A.; Liu, J. Comparison of two models on simulating electric field in HVDC cable insulation. *IEEE Trans. Dielectr. Electr. Insul.* **2019**, *26*, 1107–1115. [CrossRef]
42. Kumara, S.; Serdyuk, Y.V.; Jeroense, M. Calculation of electric fields in HVDC cables: Comparison of different models. *IEEE Trans. Dielectr. Electr. Insul.* **2012**, *28*, 1070–1078. [CrossRef]
43. Eoll, C.K. Theory of stress distribution in insulation of high-voltage DC cables: Part I. *IEEE Trans. Electr. Insul.* **1975**, *EI-10*, 27–35. [CrossRef]
44. Buller, F.H. Calculation of electrical stresses in DC cable insulation. *IEEE Trans. Power Appar. Syst.* **1967**, *PAS-86*, 1169–1178. [CrossRef]
45. Baferani, M.A.; Shahsavarian, T.; Li, C.; Tefferi, M.; Jovanovic, I.; Cao, Y. Electric field tailoring in HVDC cable joints utilizing electro-thermal simulation: Effect of field grading materials. In Proceedings of the 2020 IEEE Electrical Insulation Conference (EIC), Knoxville, TN, USA, 22 June–3 July 2020; pp. 400–404.
46. Qin, S.; Boggs, S. Design considerations for high voltage DC components. *IEEE Electr. Insul. Mag.* **2012**, *28*, 36–44. [CrossRef]
47. CIGRE Working Group D1.23. *Diagnostics and Accelerated Life Endurance Testing of Polymeric Materials for HVDC Application*; CIGRE Technical Brochure 636; CIGRE: Paris, France, 2015.
48. Hampton, R.N. Some of the considerations for materials operating under high-voltage, direct-current stresses. *IEEE Electr. Insul. Mag.* **2008**, *24*, 5–13. [CrossRef]
49. Li, Z.; Du, B. Polymeric insulation for high-voltage dc extruded cables: Challenges and development directions. *IEEE Electr. Insul. Mag.* **2018**, *34*, 30–43. [CrossRef]
50. Fu, M.; Chen, G.; Dissado, L.A.; Fothergill, J.C. Influence of thermal treatment and residues on space charge accumulation in XLPE for DC power cable application. *IEEE Trans. Dielectr. Electr. Insul.* **2007**, *14*, 53–64. [CrossRef]
51. Mazzanti, G. Issues and challenges for HVDC extruded cable systems. *Energies* **2021**, *14*, 4504. [CrossRef]
52. Reed, C.W. An assessment of material selection for high voltage DC extruded polymer cables. *IEEE Electr. Insul. Mag.* **2017**, *33*, 22–26. [CrossRef]
53. Maruyama, S.; Ishii, N.; Shimada, M.; Kojima, S.; Tanaka, H.; Asano, M.; Yamanaka, T.; Kawakami, S. Development of a 500 kV DC XLPE cable system. *Furukawa Rev.* **2004**, *25*, 47–52.
54. Murata, Y.; Sakamaki, M.; Abe, K.; Inoue, Y.; Mashio, S.; Kashiyama, S.; Matsunaga, O.; Igi, T.; Watanabe, M.; Asai, S.; et al. Development of high voltage DC XLPE cable system. *SEI Tech. Rev.* **2013**, *76*, 55–62.
55. Gu, X.; Zhu, T. Simulation of electrical field distribution of HVDC and EHVDC cable during load cycles in type test. In Proceedings of the 2022 IEEE Conference on Electrical Insulation and Dielectric Phenomena (CEIDP), Denver, CO, USA, 30 October–2 November 2022; pp. 83–86. [CrossRef]
56. Zhu, Y.; Yang, F.; Xie, X.; Cao, W.; Sheng, G.; Jiang, X. Studies on electric field distribution and partial discharges of XLPE cable at DC voltage. In Proceedings of the 2018 12th International Conference on the Properties and Applications of Dielectric Materials (ICPADM), Xi'an, China, 20–24 May 2018; pp. 562–565. [CrossRef]
57. Naderiallaf, H.; Seri, P.; Montanari, G.C. Investigating conditions for an unexpected additional source of partial discharges in DC cables: Load power variations. *IEEE Trans. Power Deliv.* **2021**, *36*, 3082–3090. [CrossRef]
58. Rizzo, G.; Romano, P.; Imburgia, A.; Ala, G. Partial discharges in HVDC cables—The effect of the temperature gradient during load transients. *IEEE Trans. Dielectr. Electr. Insul.* **2021**, *28*, 1767–1774. [CrossRef]

59. Huang, Z.Y.; Pilgrim, J.A.; Lewin, P.L.; Swingler, S.G.; Tzemis, G. Thermal-electric rating method for mass-impregnated paper-insulated HVDC cable circuits. *IEEE Trans. Power Deliv.* **2015**, *30*, 437–444. [CrossRef]
60. Diban, B.; Mazzanti, G.; Seri, P. Life-based geometric design of HVDC cables—Part I: Parametric analysis. *IEEE Trans. Dielectr. Electr. Insul.* **2022**, *29*, 973–980. [CrossRef]
61. Diban, B.; Mazzanti, G.; Seri, P. Life-based geometric design of HVDC cables—Part 2: Effect of electrical and thermal transients. *IEEE Trans. Dielectr. Electr. Insul.* **2023**, *30*, 97–105. [CrossRef]
62. Fruth, B.; Niemeyer, L. The importance of statistical characteristics of partial discharge data. *IEEE Trans. Electr. Insul.* **1992**, *27*, 60–69. [CrossRef]
63. Gutfleisch, F.; Niemeyer, L. Measurement and simulation of PD in epoxy voids. *IEEE Trans. Dielectr. Electr. Insul.* **1995**, *2*, 729–743. [CrossRef]
64. Callender, G.; Lewin, P.L. Modeling partial discharge phenomena. *IEEE Electr. Insul. Mag.* **2020**, *36*, 29–36. [CrossRef]
65. Tefferi, M. Characterization of Conduction Properties of DC Cable Dielectric Materials. Ph.D. Thesis, University of Connecticut, Storrs, CT, USA, 2019.
66. Khalil, M.S.; Gastli, A. Investigation of the dependence of DC insulation resistivity of ultra-clean polyethylene on temperature and electric field. *IEEE Trans. Power Deliv.* **1999**, *14*, 699–704. [CrossRef]
67. Weedy, B.M.; Chu, D. HVDC extruded cables—Parameters for determination of stress. *IEEE Trans. Power Appar. Syst.* **1984**, *PAS-103*, 662–667. [CrossRef]
68. Boggs, S.; Damon, D.H.; Hjerrild, J.; Holboll, J.T.; Henriksen, M. Effect of insulation properties on the field grading of solid dielectric DC cable. *IEEE Trans. Power Deliv.* **2001**, *16*, 456–461. [CrossRef]
69. Vu, T.T.N.; Teyssedre, G.; Le Roy, S. Electric field distribution in HVDC cable joint in non-stationary conditions. *Energies* **2021**, *14*, 5401. [CrossRef]
70. He, M.; Hao, M.; Chen, G.; Chen, X.; Li, W.; Zhang, C.; Wang, H.; Zhou, M.; Lei, X. Numerical modelling on partial discharge in HVDC XLPE cable. *COMPEL—Int. J. Comput. Math. Electr. Electron. Eng.* **2018**, *37*, 986–999. [CrossRef]
71. He, M.; Chen, G.; Lewin, P.L. Field distortion by a single cavity in HVDC XLPE cable under steady state. *High Volt.* **2016**, *1*, 107–114. [CrossRef]
72. He, M.; Hao, M.; Chen, G.; Li, W.; Zhang, C.; Chen, X.; Wang, H.; Zhou, M.; Lei, X. Numerical study of influential factors on partial discharges in HVDC XLPE cables. *COMPEL—Int. J. Comput. Math. Electr. Electron. Eng.* **2018**, *37*, 1110–1117. [CrossRef]
73. Mikrut, P.; Zydron, P. Partial discharge modeling in a gaseous inclusion located on the radius of a loaded HVDC XLPE cable. In Proceedings of the XVII Conference Progress in Applied Electrical Engineering (PAEE 2023), Kościelisko, Poland, 26–30 June 2023.
74. Khalil, M.S. International research and development trends and problems of HVDC cables with polymeric insulation. *IEEE Electr. Insul. Mag.* **1997**, *13*, 35–47. [CrossRef]
75. Kaminaga, K.; Ichihara, M.; Jinno, M.; Fujii, O.; Fukunaga, S.; Kobayashi, O.; Watanabe, K. Development of 500-kV XLPE cables and accessories for long-distance underground transmission lines. V. Long-term performance for 500-kV XLPE cables and joints. *IEEE Trans. Power Deliv.* **1996**, *11*, 1185–1194. [CrossRef]
76. Fromm, U.; Kreuger, F.H. Statistical behaviour of internal partial discharges at DC voltage. *Jpn. J. Appl. Phys.* **1994**, *33*, 6708–6715. [CrossRef]
77. Morshuis, P.H.F.; Smit, J.J. Partial discharges at DC voltage: Their mechanism, detection and analysis. *IEEE Trans. Dielectr. Electr. Insul.* **2005**, *12*, 328–340. [CrossRef]
78. Florkowski, M. Influence of insulating material properties on partial discharges at DC voltage. *Energies* **2020**, *13*, 4305. [CrossRef]

energies

MDPI

Article

Investigations into the Influence of Matrix Dimensions and Number of Iterations on the Percolation Phenomenon for Direct Current

Pawel Zukowski [1], Paweł Okal [2], Konrad Kierczynski [2], Przemyslaw Rogalski [2], Sebastian Borucki [3], Michał Kunicki [3] and Tomasz N. Koltunowicz [2,*]

1 Department of Economy, Vincent Pol University in Lublin, 2, Choiny Str., 20-816 Lublin, Poland; zukowski50pawel@gmail.com

2 Department of Electrical Devices and High Voltage Technology, Lublin University of Technology, 38D, Nadbystrzycka Str., 20-618 Lublin, Poland; p.okal@pollub.pl (P.O.); k.kierczynski@pollub.pl (K.K.); p.rogalski@pollub.pl (P.R.)

3 Department of Electric Power Engineering and Renewable Energy, Faculty of Electrical Engineering, Automatic Control and Informatics, Opole University of Technology, 76, Proszkowska Str., 45-758 Opole, Poland; s.borucki@po.edu.pl (S.B.); m.kunicki@po.edu.pl (M.K.)

* Correspondence: t.koltunowicz@pollub.pl; Tel.: +48-81-538-47-13

Abstract: The paper presents studies of the site percolation phenomenon for square matrixes with dimensions $L = 55$, 101 and 151 using the Monte Carlo computer simulation method. The number of iterations for each matrix was 5×10^6. An in-depth analysis of the test results using the metrological approach consisting of determining the uncertainty of estimating the results of iterations with statistical methods was performed. It was established that the statistical distribution of the percolation threshold value is a normal distribution. The coefficients of determination for the simulation results in approximations of the percolation threshold using the normal distribution for the number of iterations 5×10^6 are 0.9984, 0.9990 and 0.9993 for matrixes with dimensions 55, 101 and 151, respectively. The average value of the percolation threshold for relatively small numbers of iterations varies in a small range. For large numbers of iterations, this value stabilises and practically does not depend on the dimensions of the matrix. The value of the standard deviation of the percolation threshold for small numbers of iterations also fluctuates to a small extent. For a large number of iterations, the standard deviation values reach a steady state. Along with the increase in the dimensions of the matrix, there is a clear decrease in the value of the standard deviation. Its value is about 0.0243, about 0.01 and about 0.012 for matrixes with dimensions 55, 101 and 151 for the number of iterations 5×10^6. The mean values of the percolation threshold and the uncertainty of its determination are $(0.5927046 \pm 1.1 \times 10^{-5})$, $(0.5927072 \pm 7.13 \times 10^{-6})$ and $(0.5927135 \pm 5.33 \times 10^{-6})$, respectively. It was found that the application of the metrological approach to the analysis of the percolation phenomenon simulation results allowed for the development of a new method of optimizing the determination and reducing the uncertainty of the percolation threshold estimation. It consists in selecting the dimensions of the matrix and the number of iterations in order to obtain the assumed uncertainty in determining the percolation threshold. Such a procedure can be used to simulate the percolation phenomenon and to estimate the value of the percolation threshold and its uncertainty in matrices with other matrix shapes than square ones.

Keywords: metrological approach; uncertainty of measurement; percolation phenomenon; percolation threshold; Monte Carlo method; computer simulation

Citation: Zukowski, P.; Okal, P.; Kierczynski, K.; Rogalski, P.; Borucki, S.; Kunicki, M.; Koltunowicz, T.N. Investigations into the Influence of Matrix Dimensions and Number of Iterations on the Percolation Phenomenon for Direct Current. *Energies* **2023**, *16*, 7128. https://doi.org/10.3390/en16207128

Academic Editor: Mario Marchesoni

Received: 5 September 2023
Revised: 10 October 2023
Accepted: 16 October 2023
Published: 18 October 2023

1. Introduction

The phenomenon of percolation comes from chemistry in connection with the gelation of polymers during World War II [1]. Percolation theory is used in the description of various systems and phenomena, including the spread of epidemics [2], the reliability of computer networks [3], the spread of fires [4], molecular biology [5,6], materials science [7] and the flow of electricity by conductive and non-conductive mixtures [8]. Currently, the percolation phenomenon is widely used to describe DC and AC conductivity in metal–dielectric nanocomposites. This is the so-called hopping conductivity based on the quantum mechanical phenomenon of electron tunnelling [9–12]. Research on percolation has been growing rapidly since the 1950s. Currently, over 10,000 articles with the term "percolation" have been registered in the Web of Science Core Collection database. Moreover, there are many articles that have reported percolation studies without using the term in their titles. The phenomenon of percolation (Latin percolatio—a straining or filtering through) is a critical phenomenon. It describes the type 2 phase transitions due to the continuous change in parameters. It is an idea that is difficult to define precisely, although it is quite easy to describe qualitatively. The first article by Flory-Stockmayer [13] on the percolation theory was published in 1952 [14]. However, S. R. Broadbent and J. M. Hammersley [15] are the pioneers of the theory of percolation phenomena. In physical terms, the definition of percolation is quite general. In a medium made up of at least two phases, the elements of one phase, called dispersed, which are independent, under certain conditions form a more complex structure that changes the macroscopic properties of the entire medium. This is a consequence of a change in the concentration of the dispersed phase. The percolation theory includes models of flat and spatial random processes and the effects of the variable range of their mutual interactions and is used to describe systems with stochastic geometry and topologically disordered bodies [16]. This idea is illustrated by the network that was created by Broadbent and Hammersley as the first, simple, stochastic model [15]. Such a network consists of n-dimensions and can be both regular and irregular. It has a finite number of nodes, the distribution of which is determined by the type of the considered network and may be random or determined by the type of network. In such a model, nodes are assigned to two states, open or closed. The volume, number or concentration of open nodes is determined by x. They are distributed randomly with the probability p. For a small number of open nodes in relation to closed nodes, it is impossible to create the so-called percolation channel between two opposite planes of the network, i.e., a continuous connection of open nodes. This is due to the fact that the number of nodes making up the channel, even for the shortest straight line, is insufficient to connect the network planes. Both the nodes and the clusters they create as a result of aggregation are isolated. Then the percolation probability denoted by $P(p)$ is 0. With an increase in the number of connections, density, filling or concentration in disordered systems, the percolation threshold suddenly occurs. There is a value of x for which the probability of percolation is non-zero since the number of nodes making up the percolation channel is sufficient to create a continuous connection. This value is denoted x_c and is called the percolation threshold. Further increasing the value of x causes the formation of larger clusters, expansion of the resulting channel or the creation of new ones. The appearance of the percolation threshold determines the creation of a so-called infinite percolation cluster. Such a cluster takes an irregular geometric shape. Percolation phenomena can be divided into two main groups based on the types of connections, i.e., bond percolation and site percolation [17,18]. In bond percolation, open and closed states are assigned to connections between nodes, while in site percolation, these states are assigned to nodes. Both are models belonging to discrete percolation. This term is used to describe percolation theory models in which the media are discrete sets, such as sets of regular lattice points, or more generalised examples such as graphs. In addition to the site and bond percolation models, there are others that are a variation of these two types. An example is the percolation of two-phase continuous systems (Swiss cheese model) called continuous percolation. This model differs from site percolation in that the objects that make up the nodes are of any shape and size. In the case of three-dimensional networks, they can be cubes, canes or spheres, and in two-dimensional models, there are circles, ellipses, segments, etc. These objects are randomly placed in a non-conductive environment and are part of the system forming a percolation channel [19,20].

There is also a model of stirred percolation [21], polychromatic percolation [22] or bootstrap percolation [23]. The phenomenon of percolation is also analysed in terms of two main types of networks: random networks and those with translational symmetry. Over the past few decades, an enormous amount of work has been completed to find the exact and approximate percolation thresholds for different networks. The exact thresholds are known only for some two-dimensional networks. Examples are square networks (4^4), in which the threshold for bond percolation is exactly 0.5, or triangular networks (3^6), where the site percolation threshold is also 0.5 These values have been proven analytically [24]. It was possible because these networks can be divided into a self-similar pattern, i.e., after a triangle–triangle transformation, the system remains the same. Two-dimensional self-modality means that all fully triangulated networks (e.g., triangle, union jack, double cross, double martini and double asanoha, or double 3–12 and Delaunay triangulation) have a site percolation threshold of 0.5. In other cases, it is not possible, which resulted in the construction of models.

The model that describes the behaviour of fluids in porous media was the most primary and developed at the earliest [25]. It was presented as an analogy network of capillary tubes, which reflected the geometry of the rock sample, thanks to which parameters such as permeability, capillary pressure, capillary growth rate and saturation level were estimated. This model, even though it was based on the percolation theory, provided a lot of information about capillary flow without contributing much to the study of the percolation phenomenon. To model the percolation phenomenon in order to determine the percolation threshold, the most obvious one was the flow of electric current in a two-dimensional model of a square network composed of conductive and insulating elements [26]. Electric current will begin to flow between the electrodes at opposite edges of the matrix when the contacting conductive spheres create a conductive channel called a percolation channel, i.e., a continuous connection between the electrodes.

It is intuitively known that there is a critical value of the conductive sphere concentration for which such a channel will be created. For a small number of spheres $n < L$ it is impossible to create a percolation channel between two electrodes. This is because the number of spheres forming the channel is insufficient even for the shortest straight line connecting the opposite sides of the network. Both the nodes and the clusters they create, i.e., groups of adjacent conductive nodes formed as a result of aggregation, are isolated. When analysing a square matrix with dimensions $L \cdot L$, the minimum number of conducting spheres to reach the percolation threshold is L. However, there is a very small probability of a random distribution of spheres, which will result in a system that connects two opposite edges of the matrix in a straight line. There is also a number of conductive spheres for which the percolation probability $P(p) = 1$. For a given network example, this number is $(L^2 - L + 1)$. This means that the percolation threshold is in the sphere concentration range $L \leq x_c \leq (L^2 - L + 1)$. Thus, it cannot be determined based on a single iteration only. This problem needs to be identified and solved statistically. In order to achieve statistics with a high confidence coefficient, a given type of network should be repeatedly modelled. In the pre-computer era, this was technically very difficult to do. With the advent of the computer age, when the limitations of computational tools ceased to be a barrier, the phenomenon of percolation began to be analysed using the Monte Carlo method. The possibility of using the Monte Carlo method in the study of percolation resulted in the creation of numerous models. The most frequently studied models were two-dimensional networks with translational symmetry. Such a network was transformed into a random one by occupying points (vertices) or bonds (edges) with a statistically independent probability p. Depending on the method of the network obtaining, models of bond percolation or knot percolation, respectively, were created. These models assume that the occupied places or bonds are completely random (Bernoulli percolation). The percolation thresholds for many networks were estimated, incl. honeycombs (6^6) [27,28], Kagome (6, 3, 6, 3) [29,30], rhombitrihexagonal (3, 4, 6, 4) [31], maple leaf (3^4, 6) [31,32], non-planar chimera [33], subnetworks [34], many other three-dimensional [26] and even for dimensions larger than 3 [35]. One of the first statistical calculations of the node percolation threshold for a square matrix appeared in the 1960s with a result of 0.569(13)% [36]. The simulations were made for a square matrix with dimensions L = 78, and number of iterations n = 30. From the nineties of the last century until now, the intensification of research and increase in computational possibilities have provided more accurate results of the calculations. The publications give the values 0.59274605079210(2) [30], 0.59274601(2) [28], 0.59274621(13) [37], 0.59274621(33) [38]. The simulation results in the above works differ by no more than one ten-thousandth of a percent. In the publication [37] for the calculation of the site percolation threshold using the Monte Carlo method, a statistical calculation algorithm was assumed, which provides the result with a very large number of iterations and a much lower demand for computing power. The calculations were performed for the matrix with the size L = 256 for a very large number of iterations, amounting to 3×10^8. The estimation result was p_c = 0.59274621(13).

Unfortunately, most of the publications quoted above do not provide such model parameters as the dimensions of matrixes used in the simulations and the number of iterations. Moreover, the recommendations of the document [39], concerning the metrological analysis of the measurement results, which are undoubtedly the values of the percolation threshold and the uncertainty of their estimation, were not applied. According to this document and its earlier versions, to estimate the numerical values based on random variables, obtained with repeated repetition of measurements using statistical methods, the measurement uncertainty of type A is used [39]. This means that at first it should be specified if the obtained sequence of results corresponds to a normal distribution. According to this document, the full (correct) measurement in this case consists of the measured value and the measurement uncertainty. The measured value in the case of multiple repetitions of the measurement is equal to the arithmetic mean value of the subsequent measurement results. The measurement uncertainty is a function of the standard deviation and the number of iterations. In the work, the metrological approach for estimating the measurement uncertainty was used to analyse the results of the Monte Carlo simulation of the percolation phenomenon for many iterations. Such an analysis, together with other model parameters, such as the use of different dimension matrixes and the variation of the number of iterations, allows us to obtain new data on the percolation phenomenon in the so far analysed square matrixes.

The aim of the work is:

- Simulation of the site percolation phenomenon using the Monte Carlo method with the use of inverted simulation logic for square matrixes with dimensions L = 55, 101 and 151 for a large number of iterations 5×10^6;
- Determination of the node's initial coordinates and their distribution in space, generated in each iteration, and on their basis confirmation of the software generating non-conductive nodes in the matrixes correct operation;
- Determination of the percolation threshold value for each iteration;
- Determination of the probability statistical distribution of the percolation threshold value occurrence;
- Determination of the average value of the percolation threshold, standard deviation and the uncertainty of the percolation threshold estimation dependence on the number of iterations;
- The impact of the matrix dimensions on the values of the percolation threshold, standard deviation and the uncertainty of the percolation threshold estimation analysis.

2. Methods

Exact percolation thresholds, determined analytically, are known only for some two-dimensional networks. This is the square network, denoted as (44), for which the node percolation threshold is exactly 0.5, and the triangular network (36), where the node percolation threshold value is also 0.5 [24]. In other types of networks, for example, the square network for node percolation and the triangular network for bond percolation, the percolation threshold can only be determined by computer simulation. Therefore, the object of the Monte Carlo simulation was the square network for node percolation.

The most common percolation model is to define a regular network, such as a quadratic one, and transform it into a random network by randomly "occupying" sites (vertices) or bonds (edges) with a statistically independent probability p [24,40,41]. When a critical probability value p_c is reached, the so-called long-range connectivity appears for the first time and a percolation channel is formed, allowing current to flow. The value of p_c is called the percolation threshold. The percolation of nodes in a two-dimensional square network is the most studied percolation model by simulation methods. It is a special type of network because the square network has the property of self-duality [42].

The first step of the Monte Carlo computer simulation was to define the space in which the problem would be analysed. In this case, it is an L^2 square matrix whose nodes are described in the Cartesian plane. Despite the simplification of the model to two dimensions, the simulation calculations are performed sequentially, so the choice of the L parameter determines the time of the performed simulation.

Next, the logic of the model under study was determined. According to percolation theory, in the simulation, all K nodes of the quadratic network belong to the set E:

$$E = \left\{ K(x,y) : x,y \in L^2 \right\}, \tag{1}$$

$$G = \phi. \tag{2}$$

A sequential permutation without repetition of K nodes on the set E was then performed, which were assigned to subsets of the set G such that:

1. If the set $G = \emptyset$, the node K forms a subset of the set G.
2. If node K was the nearest node in the vertical, horizontal or diagonal direction of a node from the set G, it was added to the subset containing that node.
3. If the node K was the nearest node in the vertical, horizontal or diagonal direction of more than one point from the set G and these points belong to different subsets of the set G then the subsets are replaced by their conjunction with the added point K.

Regarding percolation, all subsets in the set G are clusters composed of neighbouring nodes of the network.

The above conditions define the reverse direction of percolation, i.e., the determination of the concentration N for which at least one subset will form a critical cluster [43]. This will be the concentration for which the last percolation channel of the network is broken. This means that the initial state of the matrix is the state in which all nodes of the matrix are occupied by conducting elements. In the case of a square matrix, the conductive nodes are in contact with their neighbouring nodes, arranged in vertical and horizontal directions. This state of the matrix results in current flow when electrodes are connected to a current source, contacting the conductive elements of the matrix at its top and bottom edges. The next simulation steps involve randomly eliminating individual conductive elements and replacing them with non-conductive nodes. As the number of non-conductive elements increases, clusters appear, until finally the removal of a single conductive element, which results in the complete disconnection of the top electrode from the bottom electrode and the interruption of current flow. This means that the percolation threshold x_c is reached. Previous studies [44] show that the expected percolation threshold for a two-dimensional square lattice is about 0.59. This means that the use of conduction element elimination is an optimisation action. In the case of adding conductive nodes, the number of nodes is about 0.59. In the case of elimination, about 0.41 nodes must be removed. This means that the algorithm to remove conductive elements requires approximately 1.44 times fewer calculations.

The algorithm and simulation software for this thesis was developed in the Python programming language in the Unix system environment. The calculations were based on multiplicity theory because the computer performs assignment processes much faster than computational processes. The only variable was the dimension of the square lattice L.

Figure 1 shows a schematic diagram of the computer software algorithm developed to simulate the percolation phenomenon using the Monte Carlo method. In the first step of the simulation, the computer created a set of square network nodes with coordinates of dimension $L{\cdot}L$. Then, using a pseudo-random number generator, the set was randomly mixed. The pseudo-random number generator implemented for the simulation is based on a basic function that generates a random regular floating-point number in the semi-open range [0,1). The main generator is the "Mersenne Twister" algorithm [45] which generates floating point numbers with a precision of 53 bits in the range 219937-1 and is a basic implementation in the C programming language. The MT19937 variant used has a high degree of uniform distribution which means that the periodic dependence between successive values of the output sequence is negligible. It also satisfies numerous statistical randomness tests, including diehard tests. It satisfies most, but not all, of the more rigorous randomness tests (e.g., TestU01 Crush) [45]. The "Mersenne Twister" algorithm is used in many Monte Carlo studies [46–49].

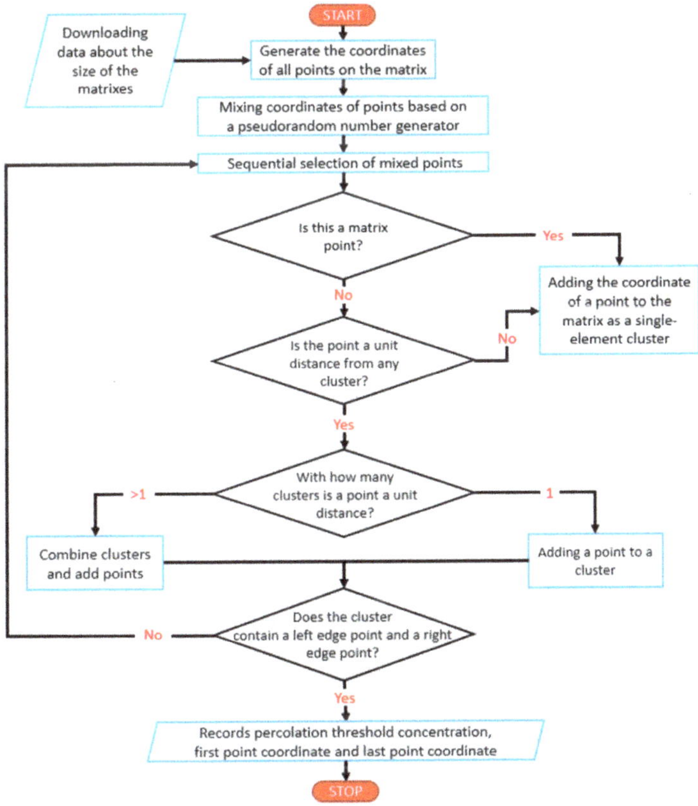

Figure 1. Schematic of the simulation algorithm.

The software then performed a sequential selection of a non-conductive nodes set. The first node drawn formed a one-element set. Each subsequent non-conductive node selected was subject to a membership check of the existing sets, i.e., if the selected node was within a unit distance in the vertical, horizontal or diagonal direction from any node in the existing set. It was then attached to this set to form a cluster. Otherwise, the node formed a one-element set. The exception was when the selected node belonged to more than one cluster. Then it was added to the combined clusters. Once a non-conducting node was added to an existing cluster, that cluster was tested to see if it contained points with coordinates $x = 0$ and $x = L$. This condition determined whether the cluster formed a continuous connection between the right and left edges of the matrix. Meeting these conditions means that a so-called "infinite cluster" is formed from non-conducting nodes and the percolation threshold is reached. The algorithm at this stage stored the coordinates of the first drawn node, the coordinates of the non-conducting node interrupting the last percolation channel and the calculated percolation threshold concentration and terminated its operation. The algorithm was run again until a preset number of iterations was obtained, which was 5×10^6. Simulations were performed for matrixes of size $L = 55$, 101 and 151.

The most relevant results in the first stage of the simulation are the coordinates of the first non-conductive nodes drawn in each iteration. They show the quality of the developed algorithm. In the case of a large number of iterations, the distribution of the first non-conductive node number over the area of the square matrix should be regular.

The initial analysis of the first drawn node distribution in each of the 5×10^6 iterations involved the creation of two-dimensional histograms or so-called heat maps [50]. These represent a graphical interpretation of multidimensional data, in which the individual values contained in the matrix are represented by a colour scale. Figures 2–4 show the corresponding colour for each node of the matrices for networks of dimensions $L = 55$, $L = 101$ and $L = 151$. The colours represent the number of the first non-conductive nodes in each node of the matrix.

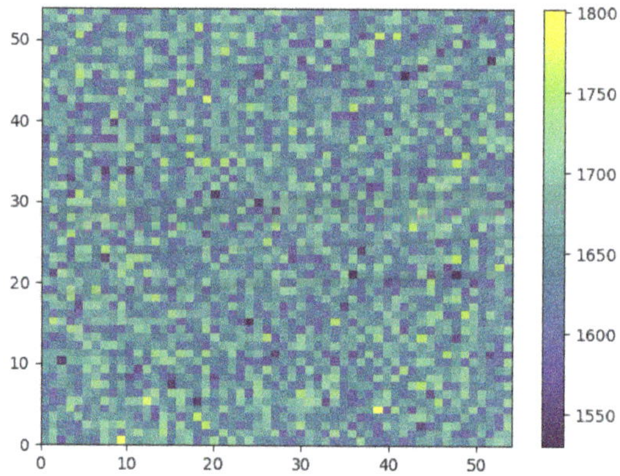

Figure 2. Two-dimensional histogram of first non-conductive nodes dependent on the coordinates of the nodes for matrix dimension $L = 55$ and 5×10^6 iterations.

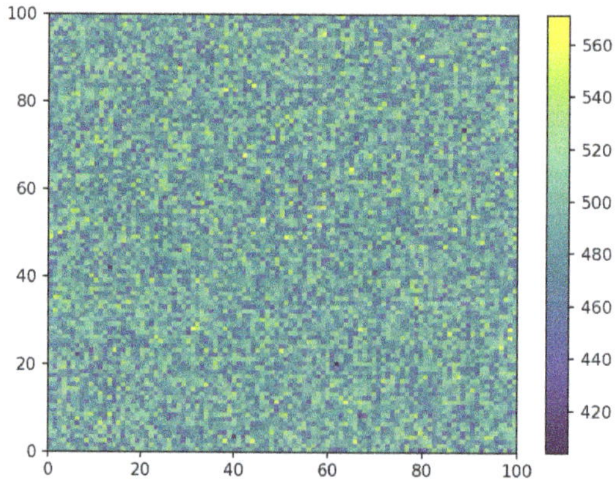

Figure 3. Two-dimensional histogram of first non-conductive nodes dependent on the coordinates of the nodes for matrix dimension $L = 101$ and 5×10^6 iterations.

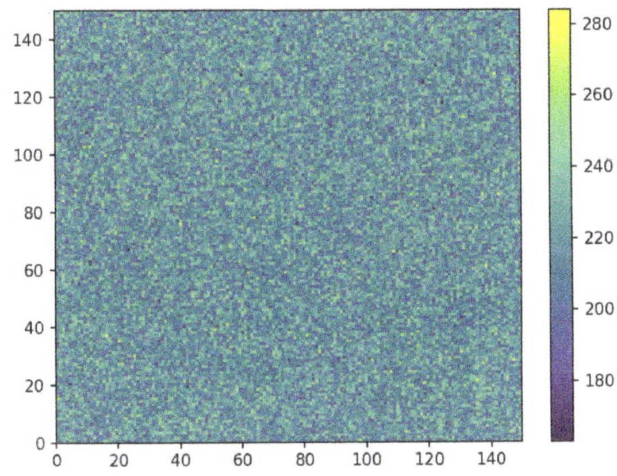

Figure 4. Two-dimensional histogram of first non-conductive nodes dependent on the coordinates of the nodes for matrix dimension $L = 151$ and 5×10^6 iterations.

Analysis of Figures 2–4 did not reveal any significant inhomogeneities in the distribution. To estimate the level of homogeneity, graphs were drawn for characteristic sections of the plane. As is known, a square matrix has a fourth-order symmetry axis passing perpendicular to the matrix through its centre. For the simulation, matrixes with odd L-dimensions of 55, 101 and 151 were used to obtain a node in the centre of the matrix through which the axis of symmetry passes. This allowed us to analyse the distribution of the first non-conductive node number along lines, perpendicular to their side edges, passing through the centres of the matrixes. Figure 5 shows a summary of the obtained results.

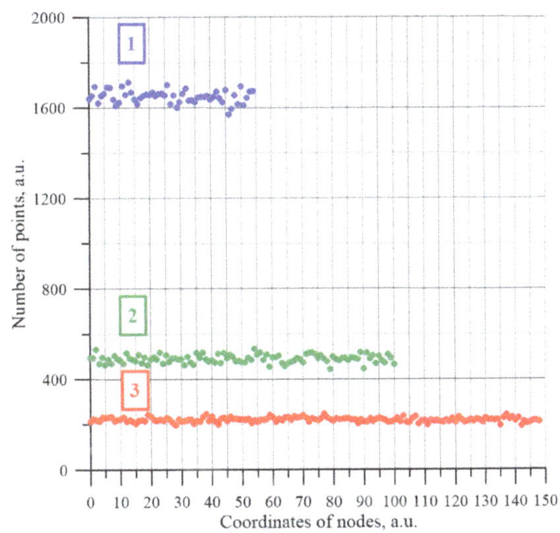

Figure 5. Number of first non-conductive nodes along the horizontal and vertical lines passing through the centre of the matrix for 5×10^6 iterations. Dimensions of matrixes L: 1—55, 2—101, 3—151.

It can be seen from Figure 5 that for all the matrixes analysed, the distributions are constant values. To estimate the homogeneity of the distributions shown in Figure 5, the standard deviation values, calculated from the equation, were used:

$$\sigma = \sqrt{\frac{\sum\limits_{k=1}^{n} (x_k - \overline{X})^2}{n-1}}, \tag{3}$$

where: σ—standard deviation, $\overline{X}$—mean value, n—number of iterations, x_k—results for iteration number k.

The standard deviation value is used in metrology to determine the type A uncertainty of measurement [39], given by the equation:

$$u = \frac{\sigma}{\sqrt{n}} = \sqrt{\frac{\sum\limits_{k=1}^{n} (x_k - \overline{X})^2}{n(n-1)}}. \tag{4}$$

Uncertainty of measurement is used to estimate numerical values, obtained by repeated measurements, by statistical methods. In this case, the measurement result consists of the measured value and measurement uncertainty. The measurand value in the case of repeated measurement is the arithmetic mean value of subsequent measurement results, while the uncertainty of measurement is given by Equation (4). The result of a measurement can be written in the form:

$$X \approx (\overline{X} \pm u). \tag{5}$$

Relative uncertainty, as defined [39], is used to compare the measurement uncertainty and the measured value:

$$\delta = \frac{u}{\overline{X}}. \tag{6}$$

The mean value of the first non-conductive nodes in each matrix node for matrixes with $L = 55$ is 1652.89, $L = 101$ is 490.15 and $L = 151$ is 219.29. Calculated from the simulation results, the type A uncertainty of measurement by Equation (4) and the relative uncertainties by Equation (6) for first nodes number are: for matrix with $L = 55$ $u = 8.28 \times 10^{-4}$ and $\delta = 5.01 \times 10^{-6}$, matrix $L = 101$ $u = 1.63 \times 10^{-3}$ and $\delta = 3.32 \times 10^{-6}$, for matrix $L = 151$ $u = 2.94 \times 10^{-3}$ and $\delta = 1.34 \times 10^{-5}$. Based on these results, it can be concluded that the distribution in space of the first randomly selected non-conductive nodes is regular. This indicates the correct operation of the software developed to simulate the percolation phenomenon.

3. Simulation of the Percolation Phenomenon for a Large Number of Iterations

The most important simulation result was the determination of the removed conductive node number, after which the percolation threshold occurs. In the algorithm used, the last percolation channel is interrupted when the randomly drawn node marked red in Figures 6–8 becomes non-conductive. This creates an "infinite" non-conductive cluster that connects the two side edges of the square matrix. This node thus cuts off the current flow in the vertical direction. Figures 6–8 below show exemplary visualisations of simulation results for matrixes with dimensions $L = 55$, 101 and 151. In the figures, black points are conducting nodes, which in the first phase of the analysis constitute 100% of the matrix nodes. Then the conducting nodes are randomly replaced by non-conducting nodes. The arrangement is defined such that the percolation edges constitute the top and bottom edges of the matrix so that the percolation threshold will be reached at the moment of interrupting the last percolation channel connecting these edges. Due to the inverted simulation logic used to optimise the analysis process, the percolation threshold is determined for the concentration in which the adjacent non-conductive nodes are connected in such a way that a non-conductive channel is created connecting the left and right edges of the tested system.

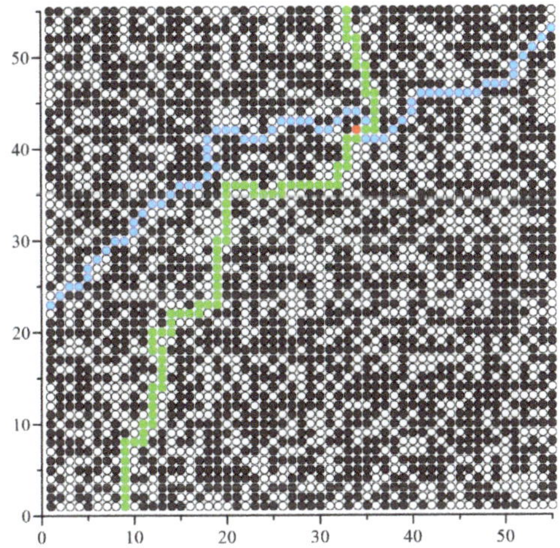

Figure 6. Example simulation result for an L = 55 square matrix. Black indicates conductive nodes, white indicates non-conductive nodes, red indicates the node interrupting the last percolation channel, blue indicates the "infinite" non-conductive cluster, and green indicates the last percolation channel.

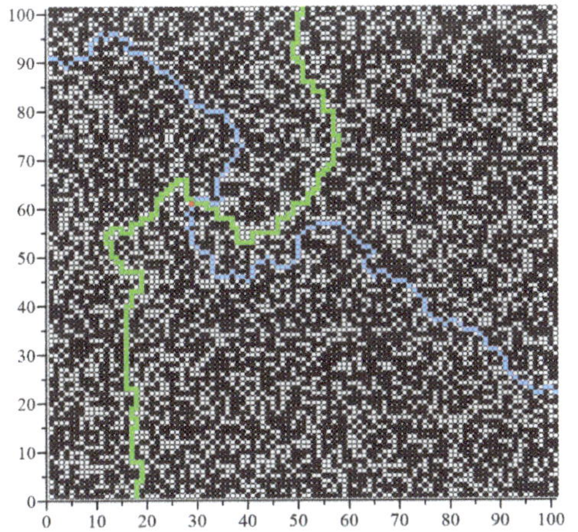

Figure 7. Example simulation result for an L = 101 square matrix. Black indicates conductive nodes, white indicates non-conductive nodes, red indicates the node interrupting the last percolation channel, blue indicates the "infinite" non-conductive cluster, and green indicates the last percolation channel.

During the last iteration, a non-conductive node was drawn, marked in red, whose change in property from conductive to non-conductive determines the interruption of the last percolation channel. After drawing this node, the percolation threshold is calculated from the ratio of conducting nodes remaining in the matrix to all network nodes.

When n independent observations have been made, for determining the percolation threshold value x_{ci} (i = 1, 2, . . . n), and when this value varies randomly, the best achievable estimate of the expected value $\overline{X}$ is the arithmetic mean. In this case, the type A method of determining standard uncertainty should be used to determine the uncertainty of the percolation threshold estimate [39]. In order to determine whether the percolation threshold values obtained as a result of the simulation are a random variable, histograms of the percolation threshold values distributions and their comparison with the normal distribution function for iterations number from 10^3 to 5×10^6 for matrixes with dimensions L = 55, 101 and 151 were performed, as recommended in [39]. In the simulation, the result of percolation threshold determination in each iteration is the independent variable, which is drawn from the same population. The expected value and standard deviation σ can be calculated for them. According to the central limit theorem [51], the sequence of random variables in the form of normalised U_n values converges to a standard normal distribution for the number of iterations going to infinity and is described as:

$$U_n = \frac{\frac{1}{n}\sum_{i=1}^{n}(x_i - \overline{X})}{\frac{\sigma}{\sqrt{n}}}, \tag{7}$$

where: σ—standard deviation, $\overline{X}$—expected value, n—number of iterations.

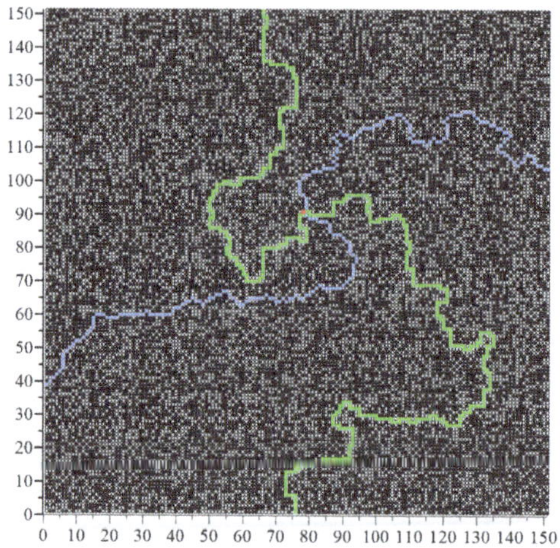

Figure 8. Example simulation result for an L = 151 square matrix. Black indicates conductive nodes, white indicates non-conductive nodes, red indicates the node interrupting the last percolation channel, blue indicates the "infinite" non-conductive cluster, and green indicates the last percolation channel.

Figures 9–11 show the histograms of the normalised U_n values dependence for the number of iterations $n = 5 \times 10^3$, 5×10^4 and 5×10^6 for a matrix of dimension L = 151. The normal distributions calculated for the mean values and standard deviations, which have been determined on the basis of the simulations, are also shown in these figures.

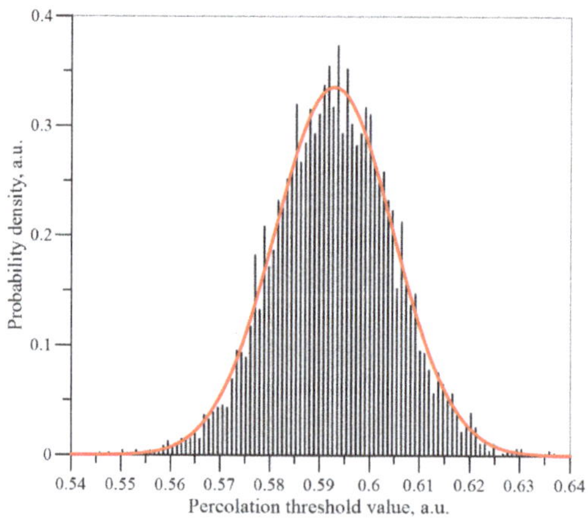

Figure 9. Histogram of the percolation threshold values distribution for an iteration number of 5×10^3 for a matrix with L = 151 and a plot of the normal distribution—solid line.

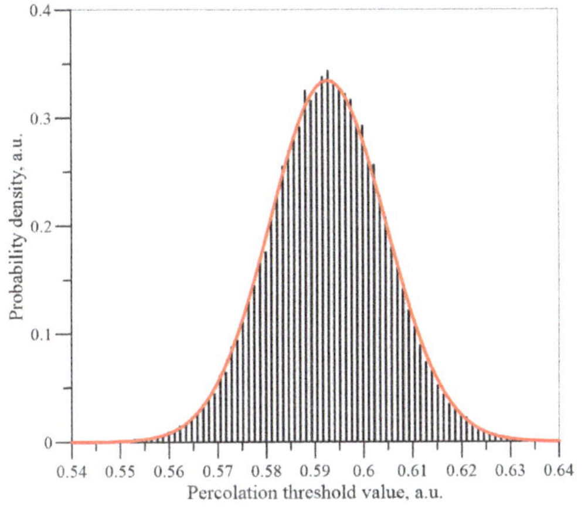

Figure 10. Histogram of the percolation threshold values distribution for an iteration number of 5×10^4 for a matrix with L = 151 and a plot of the normal distribution—solid line.

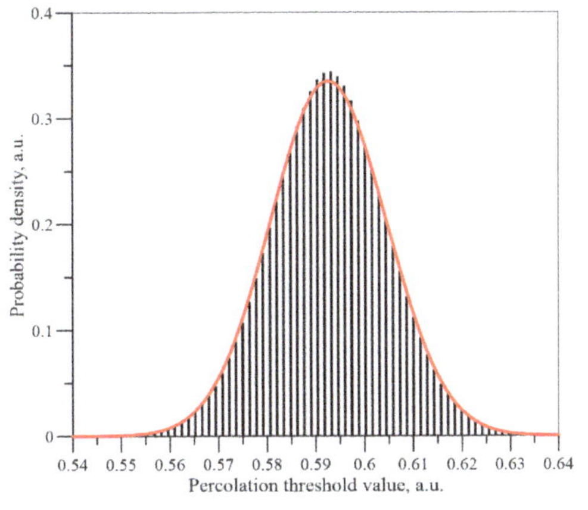

Figure 11. Histogram of the percolation threshold values distribution for an iteration number of 5×10^6 for a matrix with L = 151 and a plot of the normal distribution—solid line.

It can be seen from Figures 9–11 that, as the number of iterations increases, the results are closer and closer to a normal distribution. Similar simulations and calculations were performed for matrixes with a dimension of 55 and 101 for iteration numbers up to 5×10^6. The results for the three matrixes are shown in Figure 12 for iteration numbers 5×10^6. As can be seen from Figure 12, as the dimensions of the matrix increase, the expected value (position of the maximum) remains constant, while the standard deviation value (width of the distribution) decreases.

The percolation threshold values obtained were compared with the density value of the normal distribution for the given values of $\overline{X}$ and σ depending on the number of iterations and, as recommended in [39], R^2 coefficients of determination were calculated as functions of the regression results using least squares estimation. Table 1 summarises the coefficients of determination values for matrixes with L = 55, 101, 151, depending on the number of iterations.

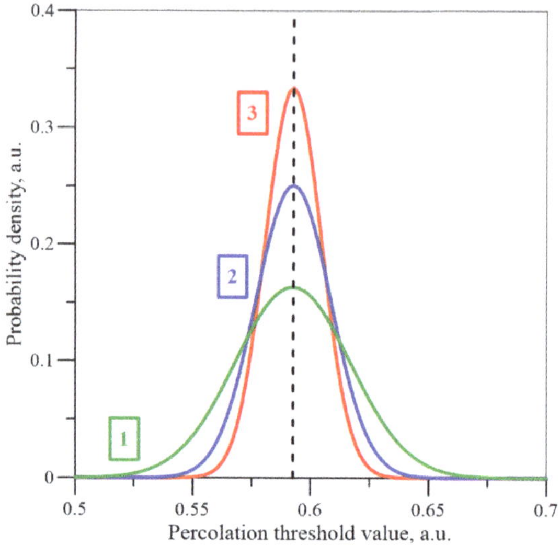

Figure 12. Normal distributions of percolation threshold values for matrixes with *L*: 1—55, 2—101, 3—151 for the number of iterations 5×10^6. Vertical dashed line—maximum value.

Table 1. Dependence of the coefficient of determination on the number of iterations for matrixes with dimensions L = 55, 101, 151.

Number of Iterations *n*, a.u.	R^2, a.u. L = 55	R^2, a.u. L = 101	R^2, a.u. L = 151
10^3	0.9285	0.8886	0.9067
2×10^3	0.9505	0.9699	0.9392
5×10^3	0.9763	0.9837	0.9830
10^4	0.9824	0.9921	0.9899
2×10^4	0.9853	0.9952	0.9974
5×10^4	0.9931	0.9963	0.9987
10^5	0.9985	0.9967	0.9986
2×10^5	0.9912	0.9971	0.9988
5×10^5	0.9954	0.9988	0.9989
10^6	0.9979	0.9984	0.9991
2×10^6	0.9981	0.9984	0.9990
5×10^6	0.9984	0.9990	0.9993

As can be seen from Table 1, the R^2 coefficients of determination already for an iteration number of 5×10^4 are greater than 0.99 and very close to unity. Their values for an iteration number of 5×10^6 are 0.9984, 0.9990 and 0.9993 for matrixes of 55, 101 and 151, respectively. The close to unity values of the coefficients of determination demonstrate the good quality of the simulation results approximation using normal distributions. This means that the probability distribution of nodes interrupting the last percolation channel is a normal distribution and the percolation threshold value is a random variable. This allows the type A method of determining the uncertainty given by Equation (4) [39] to be used to determine the uncertainty in the estimation of percolation thresholds later in the article.

The dependencies of the normal distribution basic parameters—the mean value of the percolation threshold and the standard deviation on the number of iterations *n*—were determined. Successive values of the iteration number for which calculations were made were determined using the equation:

$$n = \left\lceil 1000 \cdot 10^{0.02m} \right\rceil, \tag{8}$$

where: n—integer part of a number, enclosed in square brackets, $m = 0, 1, 2, 3$ ... This choice of iteration number allows them to be evenly distributed on a logarithmic scale.

Figure 13 shows the dependence of the average percolation threshold value on the number of iterations for matrixes of dimensions 55, 101 and 151. The figure shows every fifth number of iterations value. It can be seen from the figure that for a relatively small iteration number, the percolation threshold values change over a small range, not exceeding 0.0007. For iteration numbers of 5×10^4 and above, the percolation threshold values change in the range below 10^{-4} and gradually stabilise. This means that, for large numbers of iterations, the percolation threshold value hardly depends on the dimensions of the matrix.

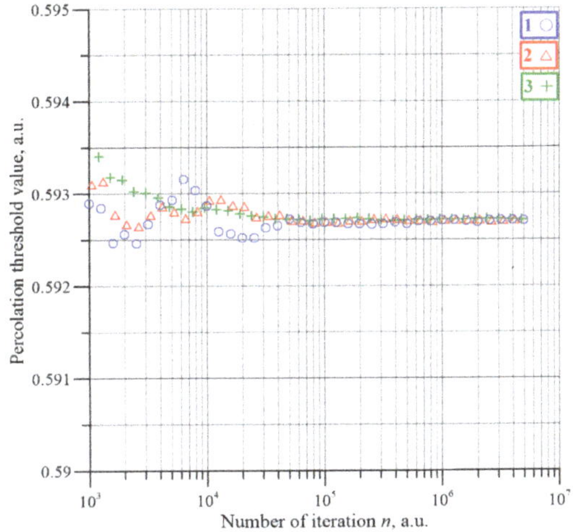

Figure 13. Dependence of the percolation threshold value on the number of iterations for matrixes with L dimensions: 1—55, 2—101, 3—151. The figure shows every fifth number of iterations.

Figure 14 shows the waveforms of standard deviation as a function of the iteration number. For a small number of iterations, some variation in the standard deviation values is apparent. Once the number of iterations exceeds approximately 5×10^4 and above, the standard deviation values practically stabilise.

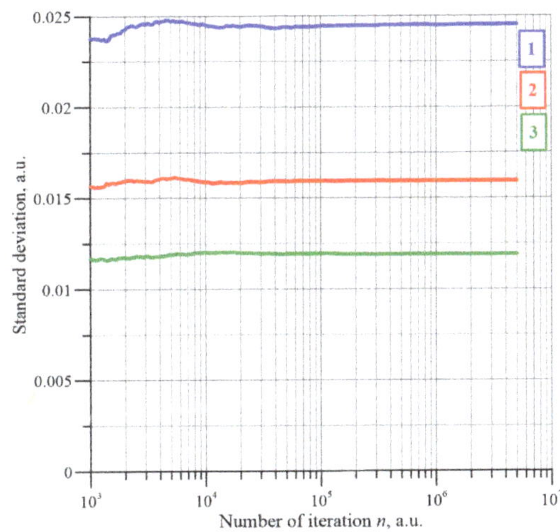

Figure 14. Dependence of the standard deviation on the number of iterations for matrixes with L dimensions: 1—55, 2—101 and 3—151.

It should be noted that Figure 14 shows that the value of the standard deviation for a large number of iterations decreases as the dimensions of the matrix increase. For iteration number values of 5×10^6, the following standard deviation values were obtained: 0.024526293 for a matrix with $L = 55$, 0.015941405—$L = 101$, 0.011915908—$L = 151$. Based on the standard deviation values obtained for the number of iterations 5×10^6, the percolation threshold estimation values and the relative uncertainty for matrixes with dimensions $L = 51, 101, 151$ were calculated using Equations (4) and (6). The calculation resulted in the following percolation thresholds: $x_c(L = 55) \approx (0.5927046 \pm 1.1 \times 10^{-5})$, relative uncertainty $\delta(L = 55) \approx 1.9 \times 10^{-5}$, $x_c(L = 101) \approx (0.5927072 \pm 7.13 \times 10^{-6})$, $\delta(L = 101) \approx 1.2 \times 10^{-5}$, $x_c(L = 151) \approx (0.5927135 \pm 5.33 \times 10^{-6})$, $\delta(L = 151) \approx 8.99 \times 10^{-6}$. From a comparison of the results, shown in Figures 13 and 14, for large iteration numbers the percolation threshold values are almost independent of the matrix dimensions. The standard deviations for large iteration numbers hardly depend on the number of iterations, while the values of standard deviation decrease as the dimensions of the matrix increase. Figures 9–14 show that increasing the number of iterations to very large values for a matrix of specific dimensions leaves the normal probability distribution of percolation threshold values virtually unchanged. This distribution is described by two parameters—the mean value of the percolation threshold and the standard deviation.

From Equation (4), it follows that the uncertainty in the percolation threshold estimation is proportional to the value of the standard deviation and inversely proportional to the square root of the iteration number. Figure 14 shows that the range of relatively low iteration number values should not be used to estimate the percolation threshold determination uncertainty. This is due to two factors. Firstly, the largest fluctuations in standard deviation values occur in the range of iteration numbers below 10^4. Secondly, in this area, the square root values are relatively small compared to the value for the maximum iteration number of 5×10^6. This means that the uncertainty values for an iteration number range below 10^4 are unsatisfactory and should be disregarded.

Figure 15 shows the dependence of the percolation threshold determination uncertainty on the number of iterations for matrixes of dimensions 55, 101 and 151, calculated according to Equation (4) for a range of iterations numbers from 10^4 to 5×10^6. This figure also shows the approximation waveforms in the form of power functions.

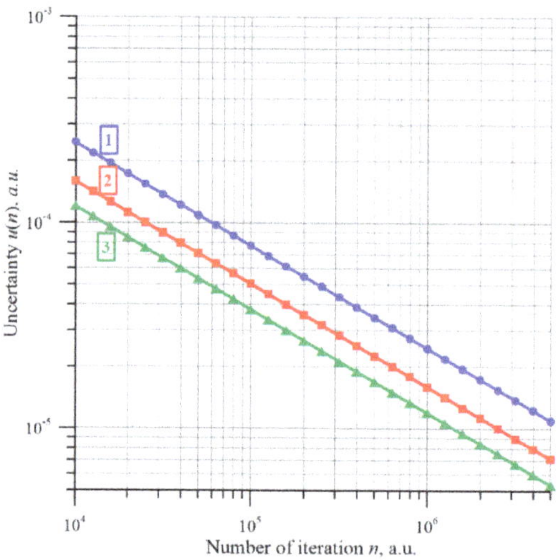

Figure 15. Dependence of the percolation threshold estimation uncertainty $u(n)$ on the number of iterations n for matrixes of dimension L: 1—55, 2—101 and 3—151. The figure shows every fifth number of iterations.

The following conclusions emerge from the analysis of Figure 15. The R^2 determination coefficients for the approximation waveforms are very close to unity and are 0.999981761, 0.99997486 and 0.99995301 for matrixes with L = 55, 101 and 151, respectively. The differences between them and unity are no more than 0.00007815. This indicates a very good approximation of the simulation results. As is well known, the dependence of uncertainty on the number of iterations is described, by definition, by the equation [39], which can be presented in the form:

$$u = \sigma n^{-0.5}, \tag{9}$$

where: σ—standard deviation, n—number of iterations.

From the approximations, shown in Figure 15, it follows that the experimental dependencies $u_D(n)$ are described by the equation:

$$u_D = \sigma_D n_D^{-K}, \tag{10}$$

where: u_D—uncertainty of the percolation threshold estimate, σ_D—experimental value of the standard deviation, n—number of iterations, k—power factor.

For matrixes with L = 55, 101 and 151, the following approximation equations were obtained:

$$u_D(L = 55, n) \approx 0.02497738 n^{-0.50173554}, \tag{11}$$

$$u_D((L = 101, n) \approx 0.0157898 n^{-0.49929933}, \tag{12}$$

$$u_D(L = 151, n) \approx 0.01153863 n^{-0.49735083} \tag{13}$$

Equations (11)–(13) show that the experimental power coefficients are very close to the value—0.5. The largest difference between the power value in Equation (9) and the coefficients from the approximating Equations (11)–(13) occurs for the matrix with L = 151 and is only about 0.00264917. For the other matrixes, the differences are even smaller.

In Equations (11)–(13), the numerical coefficients in front of the iterations number n are the standard deviation values, determined from the simulation results approximations of the type A uncertainty dependence on the number of iterations. These values were also calculated directly from 5×10^6 iterations (see Figures 13 and 15). The comparison of pairs of standard deviation values determined directly and from the approximation gives the following differences between them: $\Delta\sigma(L = 55) = 0.00045110$, $\Delta\sigma(L = 101) = 0.00015160$, $\Delta\sigma(L = 151) = 0.00037728$. The comparison of these differences shows that the values of standard deviations determined directly and from the approximation are very close to each other. This means that the dependencies of the percolation threshold value uncertainty estimation on the number of iterations, obtained by approximation of the simulation results acquired with the Monte Carlo method perfectly agree with the definition of type A uncertainty [39]. This is evidenced by both the power values, very close to −0.5, and the values of standard deviations. Comparing the results of the approximations with Equation (9), it should be noted that for a matrix of specific dimensions, the uncertainty in the estimation of the percolation threshold is a function of the number of iterations only, and the value of the standard deviation σ does not depend on the number of iterations for $n \geq 104$. This means that a reduction in the uncertainty of the percolation threshold determination for a matrix of certain dimensions is only possible by increasing the number of iterations.

It can be seen from Figure 15 that an increase in the matrix dimensions decreases the uncertainty, leaving the power relationship unchanged. This is associated, according to Equations (11)–(13), with a decrease in the standard deviation, as the matrix dimensions increase, which is also visible in Figure 14. The reduction in the standard deviation value established in the study with an increase in the matrix dimensions allowed us to propose a method to reduce the uncertainty in the percolation threshold estimation not only by increasing the number of iterations but also by increasing the dimensions of the matrix.

For example, to reduce the uncertainty in the estimation of the percolation threshold by an order of magnitude for a matrix with $L = 55$, that is, from 1.1×10^{-5} to 1.1×10^{-6}, requires, according to Equation (9), a hundredfold increase in the number of iterations from 5×10^6 to 5×10^8. Figure 15 shows that the reduction of the uncertainty value along with the increase in matrix dimensions makes it possible to apply the uncertainty reduction method, consisting of increasing the dimensions of the matrix. For example, the uncertainty in estimating the percolation threshold for a matrix with $L = 55$ and an iteration number of 5×10^6 is 1.1×10^{-5}. The same uncertainty for the matrix with $L = 151$ will be obtained after about 4.166 times the smaller number of iterations of about 1.2×10^6.

This means that based on the data obtained in the work, it is possible to optimise the determination of percolation threshold estimation uncertainty based on the use of the metrological approach. Optimisation consists of selecting the dimensions of the matrix and the number of iterations in order to obtain the assumed uncertainty of the percolation threshold determination. The proposed optimisation procedure can be used to simulate the percolation phenomena of both nodes and bonds in a matrix with shapes and symmetry types other than square matrixes.

4. Conclusions

In the paper, a study of the node percolation phenomenon for square matrixes with dimensions $L = 55$, 101 and 151 using Monte Carlo computer simulation was performed. The number of iterations for each matrix was 5×10^6. An in-depth analysis of the test results using the metrology approach of determining the uncertainty of estimating the results of repeated iterations using statistical methods was performed.

It was found that the mean value of the percolation threshold for relatively small iteration numbers varies over a small range. For large iteration numbers, the mean value stabilises and is practically independent of the matrix dimensions. The standard deviation values of the percolation threshold for small iteration numbers also fluctuate in a small range. For large iteration numbers, the standard deviation value reaches a steady state. Increasing the dimensions of the matrix for a large number of iterations results in a decrease in the standard deviation steady-state value, the percolation threshold estimation uncertainty and the relative uncertainty. The obtained percolation threshold values for matrixes with dimensions $L = 51$, 101, 151 and number of iterations 5×10^6 are ($0.5927046 \pm 1.1 \times 10^{-5}$), ($0.5927072 \pm 7.13 \times 10^{-6}$) and ($0.5927135 \pm 5.33 \times 10^{-6}$), respectively.

The reduction In the standard deviation value established in the paper by increasing the dimensions of the matrix allows us to develop a way to reduce the uncertainty of the percolation threshold estimation not only by increasing the iterations number but also by increasing the dimensions of the matrix. For example, for a matrix with $L = 55$ and an iterations number of 5×10^6, the uncertainty of the percolation threshold estimation estimate is about 1.0968×10^{-5}, while for a matrix with $L = 151$ a similar value can be obtained for an iterations number 4.166 times smaller, about 1.2×10^6.

The application of a metrology approach to the analysis of percolation simulation results allowed the development of a new optimisation method for determining the uncertainty of percolation threshold estimation. The optimisation consists of selecting the dimensions of the matrix and the number of iterations in order to obtain an assumed uncertainty in the determination of the percolation threshold. Such a procedure can be applied to simulate the percolation phenomenon and to estimate the value of the percolation threshold and its uncertainty in matrixes with other network shapes than squares.

Author Contributions: Conceptualisation, P.Z. and P.O.; methodology, P.Z., P.O., S.B. and T.N.K.; software, P.O. and K.K.; validation, P.Z. and P.O.; formal analysis, P.Z., P.O., K.K., P.R., S.B., M.K. and T.N.K.; investigation, P.O., K.K., P.R. and M.K.; resources, P.O. and T.N.K.; data curation, P.O., K.K., S.B. and T.N.K.; writing—original draft preparation, P.Z., P.O., K.K. and T.N.K.; writing—review and editing, T.N.K.; visualisation, P.O. and T.N.K.; supervision, P.Z.; project administration, P.O. and T.N.K.; funding acquisition, P.Z., P.O., K.K., P.R. and T.N.K. All authors have read and agreed to the published version of the manuscript.

Funding: This research was supported by a subsidy from the Ministry of Education and Science (Poland) for the Lublin University of Technology as funds allocated for activities in the scientific disciplines of Automation, Electronics, Electrical Engineering and Space Technologies (grants: FD-20/EE-2/702, FD-20/EE-2/703, FD-20/EE-2/705, and FD-20/EE-2/707).

Data Availability Statement: Not applicable.

Conflicts of Interest: The authors declare no conflict of interest.

References

1. Stauffer, D.; Aharony, A. Percolation. In *Encyclopedia of Physical Science and Technology*; Academic Press: New York, NY, USA, 2003; pp. 655–669. [CrossRef]
2. Browne, C.A.; Amchin, D.B.; Schneider, J.; Datta, S.S. Infection Percolation: A Dynamic Network Model of Disease Spreading. *Front. Phys.* **2021**, *9*, 171. [CrossRef]
3. Li, D.; Zhang, Q.; Zio, E.; Havlin, S.; Kang, R. Network Reliability Analysis Based on Percolation Theory. *Reliab. Eng. Syst. Saf.* **2015**, *142*, 556–562. [CrossRef]
4. Beer, T.; Enting, I.G. Fire Spread and Percolation Modelling. *Math. Comput. Model.* **1990**, *13*, 77–96. [CrossRef]
5. Brunk, N.E.; Twarock, R. Percolation Theory Reveals Biophysical Properties of Virus-like Particles. *ACS Nano* **2021**, *15*, 12988–12995. [CrossRef]
6. Brunk, N.E.; Lee, L.S.; Glazier, J.A.; Butske, W.; Zlotnick, A. Molecular Jenga: The Percolation Phase Transition (Collapse) in Virus Capsids. *Phys. Biol.* **2018**, *15*, 056005. [CrossRef]
7. Vionnet-Menot, S.; Grimaldi, C.; Ryser, P.; Maeder, T.; Strässler, S. Strain Modulation of Transport Criticality in RuO_2-Based Thick-Film Resistors. *Appl. Phys. Lett.* **2004**, *85*, 5619–5621. [CrossRef]
8. Ghanbarian, B.; Hunt, A.G. Modelling Flow and Transport in Variably Saturated Porous Media: Applications from Percolation Theory and Effective-Medium Approximation. In *Modelling of Flow and Transport in Fractal Porous Media*; Elsevier: Amsterdam, The Netherlands, 2021; pp. 79–117. [CrossRef]
9. Żukowski, P.; Kierczyński, K.; Kołtunowicz, T.N.; Rogalski, P.; Subocz, J. Application of elements of quantum mechanics in analysing AC conductivity and determining the dimensions of water nanodrops in the composite of cellulose and mineral oil. *Cellulose* **2019**, *26*, 2969–2985. [CrossRef]
10. Pogrebnjak, A.; Ivashchenko, V.; Maksakova, O.; Buranich, V.; Konarski, P.; Bondariev, V.; Zukowski, P.; Skrynskyy, P.; Sinelnichenko, A.; Shelest, I.; et al. Comparative measurements and analysis of the mechanical and electrical properties of Ti-Zr-C nanocomposite: Role of stoichiometry. *Measurement* **2021**, *176*, 109223. [CrossRef]
11. Komarov, F.F.; Zukowski, P.; Kryvasheyeu, R.M.; Munoz, E.; Koltunowicz, T.N.; Rodionova, V.N.; Togambaeva, A.K. Effects of surfactant and fabrication procedure on the electrical conductivity and electromagnetic shielding of single-walled carbon nanotube films. *Phys. Status Solidi Appl. Mater. Sci.* **2015**, *212*, 425–432. [CrossRef]
12. Koltunowicz, T.N. Dielectric properties of $(CoFeZr)_x(PZT)_{(100-x)}$ nanocomposites produced with a beam of argon and oxygen ions. *Acta Phys. Pol. A* **2014**, *125*, 1412–1414. [CrossRef]
13. Flory, P.J. Molecular Size Distribution in Three Dimensional Polymers. II. Trifunctional Branching Units. *J. Am. Chem. Soc.* **1941**, *63*, 3091–3096. [CrossRef]
14. Fujita, H. A Theory of Percolation for the Case of Two Solutes. *J. Phys. Chem.* **1952**, *56*, 949–953. [CrossRef]
15. Broadbent, S.R.; Hammersley, J.M. Percolation Processes: I. Crystals and Mazes. *Math. Proc. Camb. Philos. Soc.* **1957**, *53*, 629–641. [CrossRef]
16. Saberi, A.A. Recent Advances in Percolation Theory and Its Applications. *Phys. Rep.* **2015**, *578*, 1–32. [CrossRef]
17. Chayes, L.; Schonmann, R.H. Mixed Percolation as a Bridge between Site and Bond Percolation. *Ann. Appl. Probab.* **2000**, *10*, 1182–1196. [CrossRef]
18. Hammersley, J.M. A Generalization of McDiarmid's Theorem for Mixed Bernoulli Percolation. *Math. Proc. Camb. Philos. Soc.* **1980**, *88*, 167–170. [CrossRef]
19. Lorenz, C.D.; Ziff, R.M. Precise Determination of the Critical Percolation Threshold for the Three-Dimensional "Swiss Cheese" Model Using a Growth Algorithm. *J. Chem. Phys.* **2001**, *114*, 3659–3661. [CrossRef]
20. Youm, K.H.; Lee, S.I. AC Conductivity and Dielectric Constant in a Two-Dimensional Swiss Cheese Percolation System. *Solid. State Commun.* **1991**, *79*, 1069–1072. [CrossRef]
21. Bug, A.L.R.; Gefen, Y. Diffusion in a Stirred, Percolating System. *Phys. Rev. A* **1987**, *35*, 1301–1310. [CrossRef]
22. Wanzeller, W.G.; Cucchieri, A.; Mendes, T.; Krein, G. Percolation of Monte Carlo Clusters. *Braz. J. Phys.* **2004**, *34*, 247–250. [CrossRef]
23. Holroyd, A.E. Sharp Metastability Threshold for Two-Dimensional Bootstrap Percolation. *Probab. Theory Relat. Fields* **2003**, *125*, 195–224. [CrossRef]
24. Shante, V.K.S.; Kirkpatrick, S. An Introduction to Percolation Theory. *Adv. Phys.* **1971**, *20*, 325–357. [CrossRef]
25. Larson, R.G.; Scriven, L.E.; Davis, H.T. Percolation Theory of Two Phase Flow in Porous Media. *Chem. Eng. Sci.* **1981**, *36*, 57–73. [CrossRef]
26. Kirkpatrick, S. Percolation and Conduction. *Rev. Mod. Phys.* **1973**, *45*, 574–588. [CrossRef]

27. Djordjevic, Z.V.; Stanley, H.E.; Margolina, A. Site Percolation Threshold for Honeycomb and Square Lattices. *J. Phys. A Math. Gen.* **1982**, *15*, L405–L412. [CrossRef]
28. Jacobsen, J.L. High-Precision Percolation Thresholds and Potts-Model Critical Manifolds from Graph Polynomials. *J. Phys. A Math. Theor.* **2014**, *47*, 135001. [CrossRef]
29. Sykes, M.F.; Essam, J.W. Exact Critical Percolation Probabilities for Site and Bond Problems in Two Dimensions. *J. Math. Phys.* **1964**, *5*, 1117–1127. [CrossRef]
30. Jacobsen, J.L. Critical Points of Potts and O(N) Models from Eigenvalue Identities in Periodic Temperley-Lieb Algebras. *J. Phys. A Math. Theor.* **2015**, *48*, 454003. [CrossRef]
31. Scullard, C.R.; Jacobsen, J.L. Bond Percolation Thresholds on Archimedean Lattices from Critical Polynomial Roots. *Phys. Rev. Res.* **2019**, *2*, 012050. [CrossRef]
32. Suding, P.N.; Ziff, R.M. Site Percolation Thresholds for Archimedean Lattices. *Phys. Rev. E* **1999**, *60*, 275–283. [CrossRef]
33. Melchert, O.; Katzgraber, H.G.; Novotny, M.A. Site- and Bond-Percolation Thresholds in $K_{n,n}$-Based Lattices: Vulnerability of Quantum Annealers to Random Qubit and Coupler Failures on Chimera Topologies. *Phys. Rev. E* **2016**, *93*, 042128. [CrossRef] [PubMed]
34. Haji-Akbari, A.; Ziff, R.M. Percolation in Networks with Voids and Bottlenecks. *Phys. Rev. E* **2009**, *79*, 021118. [CrossRef] [PubMed]
35. Torquato, S.; Jiao, Y. Effect of Dimensionality on the Continuum Percolation of Overlapping Hyperspheres and Hypercubes. II. Simulation Results and Analyses. *J. Chem. Phys.* **2012**, *137*, 074106. [CrossRef] [PubMed]
36. Dean, P. A New Monte Carlo Method for Percolation Problems on a Lattice. *Math. Proc. Camb. Philos. Soc.* **1963**, *59*, 397–410. [CrossRef]
37. Newman, M.E.J.; Ziff, R.M. Efficient Monte Carlo Algorithm and High-Precision Results for Percolation. *Phys. Rev. Lett.* **2000**, *85*, 4104. [CrossRef]
38. De Oliveira, P.M.C.; Nóbrega, R.A.; Stauffer, D. Corrections to Finite Size Scaling in Percolation. *Braz. J. Phys.* **2003**, *33*, 616–618. [CrossRef]
39. JCGM. Evaluation of Measurement Data-Guide to the Expression of Uncertainty in Measurement Évaluation Des Données de Mesure-Guide Pour l'expression de l'incertitude de Mesure. *Int. Organ. Stand. Geneva* **2008**, *50*, 134.
40. Sahini, M.; Sahimi, M. *Applications Of Percolation Theory*; CRC Press: Boca Raton, FL, USA, 1994.
41. Bollobás, B.; Riordan, O. Sharp Thresholds and Percolation in the Plane. *Random Struct. Algorithms* **2006**, *29*, 524–548. [CrossRef]
42. Grimmett, G. Percolation. In *Grundlehren der Mathematischen Wissenschaften*; Springer: Berlin/Heidelberg, Germany, 1999; Volume 321, ISBN 978-3-642-08442-3.
43. Newman, C.M.; Schulman, L.S. Infinite Clusters in Percolation Models. *J. Stat. Phys.* **1981**, *26*, 613–628. [CrossRef]
44. Dean, P.; Bird, N.F. Monte Carlo Estimates of Critical Percolation Probabilities. *Math. Proc. Camb. Philos. Soc.* **1967**, *63*, 477–479. [CrossRef]
45. Matsumoto, M.; Nishimura, T. Mersenne Twister. *ACM Trans. Model. Comput. Simul.* **1998**, *8*, 3–30. [CrossRef]
46. Lee, M.J. Complementary Algorithms for Graphs and Percolation. *Phys. Rev. E* **2007**, *76*, 027702. [CrossRef]
47. Monteaguodo, J.E.P.; Lage, P.L.C. Cross-Properties Relations in 3D Percolation Networks: I. Network Characteristic Length Determination. *Transp. Porous Media* **2005**, *61*, 143–156. [CrossRef]
48. Oskouyi, A.B.; Mertiny, P. Monte Carlo Model for the Study of Percolation Thresholds in Composites Filled with Circular Conductive Nano-Disks. *Procedia Eng.* **2011**, *10*, 403–408. [CrossRef]
49. Hyytia, E.; Virtamo, J.; Lassila, P.; Ott, J. Continuum Percolation Threshold for Permeable Aligned Cylinders and Opportunistic Networking. *IEEE Commun. Lett.* **2012**, *16*, 1064–1067. [CrossRef]
50. Wilkinson, L.; Friendly, M. The History of the Cluster Heat Map. *Am. Stat.* **2009**, *63*, 179–184. [CrossRef]
51. Bauer, H. *Measure and Integration Theory*; Walter de Gruyter: Berlin, Germany, 2001; ISBN 978-3-11-016719-1.

energies

Article

Dielectric Breakdown Strength of PDMS Elastomers after Mechanical Cycling †

Emmanuel Taine [1,2], Thomas Andritsch [1,*], Istebreq A. Saeedi [1] and Peter H. F. Morshuis [3]

[1] The Tony Davies High Voltage Laboratory, University of Southampton, Southampton SO17 1BJ, UK
[2] R&D Laboratory, SBM Offshore, 06510 Le Broc, France
[3] Solid Dielectric Solutions, 2311 SG Leiden, The Netherlands
* Correspondence: t.andritsch@soton.ac.uk

† This paper is an extended version of our paper published in 2022 IEEE 4th International Conference on Dielectrics (ICD), Palermo, Italy, 3–7 July 2022; pp. 21–24, doi:10.1109/ICD53806.2022.9863500.

Abstract: PDMS-based composites such as silicone elastomers are commonly found in high-voltage engineering, especially in outdoor insulation as coatings or structural elements or at interfaces between network elements, such as cable sealing ends (CSE). They are also promising prospects for dielectric elastomer generators (DEG), which are retrieving electrostatic energy from large strain amplitudes. The upper limit of energy conversion from these transducers is determined by the dielectric breakdown strength (DBS). Therefore, developing reliable systems that operate under high electric fields and variable repeated strains requires a thorough understanding of the mechanisms behind electrical breakdown and its coupling to mechanical cycling. In this study, the effect of Mullins damage and mechanical fatigue on silicone elastomers has been investigated. An electro-mechanical instability model that considers cyclic softening allows for predicting the evolution of the breakdown strength depending on the loading history. The results highlight the importance of the "first cycle," where up to a 30% reduction in the mean DBS was measured. However, subsequent mechanical fatigue only marginally contributes to the degradation, which is a promising perspective for the long-term performance of any silicone elastomer as long as the precise impact of the first cycle is known.

Keywords: breakdown test; dielectric elastomer; electro-mechanical instability; low-cycle fatigue; Mullins effect; PDMS

Citation: Taine, E.; Andritsch, T.; Saeedi, I.A.; Morshuis, P.H.F. Dielectric Breakdown Strength of PDMS Elastomers after Mechanical Cycling. *Energies* **2023**, *16*, 7424. https://doi.org/10.3390/en16217424

Academic Editor: Tomasz Norbert Kołtunowicz

Received: 3 October 2023
Revised: 27 October 2023
Accepted: 31 October 2023
Published: 3 November 2023

1. Introduction

Polydimethylsiloxane (PDMS) based elastomers can be used in soft transducers, which are suitable for a wide range of potential applications such as sensors, soft robotics, energy harvesting, and biomedical devices. They consist of an elastomeric film, which is coated on both sides with compliant electrodes to form a stretchable capacitor. A number of DEG devices were proposed [1] with a particular interest in novel wave energy converters [2–4] due to the potentially high conversion efficiency at large strain amplitudes at low frequencies of ocean waves. The convertible energy of DEG scales with the strain amplitude, dielectric permittivity, and the applied electric field. Despite its low permittivity, PDMS is a popular choice for DEG due to its high stretchability, good aging resistance, and high electrical resistivity. Its versatility and ease of processing also allow the formulation of nanocomposites in order to tailor the mechanical and electrical properties to several electrical applications. For DEG, the dielectric breakdown strength bounds the ultimate convertible energy, and that physical limit is of primary importance because of its quadratic contribution to the energy output. The DBS of dielectric elastomers has been widely investigated, and a variety of parameters are found to influence the electrical limits. Among them, elastomer stiffness [5–7] and amount of pre-stretch [8–10] can determine the onset of electrical failure. The breakdown strength of PDMS is also reported to change with the thickness of the dielectric layer [11–13], the shape of the electrodes [12,13], and their size [10,14]. The choice of PDMS in high-voltage applications is often driven by a particular environment that requires stretchability. Although strain and electric field result from different mechanical and electrical sources, cross-interactions can build up, resulting in premature failures. These synergetic effects are scarcely explored in the literature, and this study aims to better understand the effect of mechanical loading on the DBS. Electro-mechanical instabilities are acknowledged as the dominant failure mechanisms in dielectric elastomers; therefore, the relation between mechanical properties and electrical limits is of particular importance in these soft materials subjected to extreme loadings.

Meanwhile, elastomers are highly nonlinear materials for which the mechanical response varies with time, temperature, strain rate, or load history. A particular feature of highly filled elastomers is known as the Mullins effect, which corresponds to a reduction in mechanical stress on the second and subsequent mechanical loadings. This softening has various interpretations, including damage in the elastomer matrix, filler network alteration, or rubber–filler interface changes [15]. For the specific case of silica-filled PDMS, this reduction is typically attributed to the disentanglement of adjacent chains [16] or detachment/slippage on the filler surface of chains having reached their limit of extensibility [17]. Additionally, under cyclic mechanical loadings, filled elastomers are prone to additional softening, which is often considered the result of stress relaxation [18]. Alternatively, the Mullins effect and cyclic softening could be related to one single process caused by sliding and friction between polymeric chains and fillers [19].

The relation between mechanical properties and electric breakdown has been widely investigated. When a dielectric elastomer film is subjected to an electric field, the electrostatic pressure is thinning the membrane and increasing the internal mechanical stress. When the voltage increases further and exceeds a critical threshold V_c, the equilibrium between electrostatic and mechanical stresses becomes unstable, causing huge, localized strains (Figure 1). This phenomenon is known as pull-in instability or electro-mechanical instability (EMI), which can cause an electrical breakdown if the resulting strain or electric field exceeds the material's intrinsic limits [20]. EMI is acknowledged as the main cause of dielectric breakdown in the absence of defects. An early description of EMI was proposed by Stark and Garton for stiff polymers with a model relating the breakdown strength to permittivity and Young's modulus [21]. Extensions of this model have been proposed to account for non-linear elasticity in polymers [22] and highly stretchable elastomers [6,11,23]. Therefore, the softening induced by the Mullins effect is expected to change the dielectric breakdown strength, and experimental results on silicone support this assumption [24].

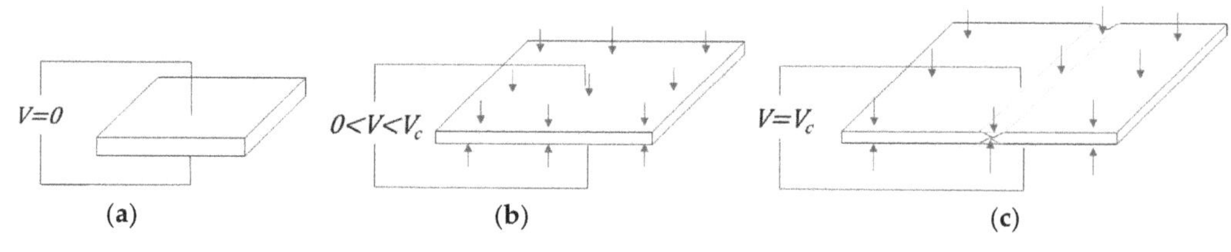

Figure 1. (**a**) Dielectric elastomer at rest, (**b**)thinned by an electric field, (**c**) and electro-mechanical instability (**c**). The electro-static pressure is schematically represented by red arrows.

The first part of this study further discusses the crucial role of the first cycle and extends the preliminary results presented in [7]. The long-term performance of dielectric elastomers remains an exploration field. The influence of millions of squared drive voltage applications on PDMS dielectric actuators has been investigated with regard to electrode performance [25] and dielectric breakdown strength [9]. However, degradation of the dielectric properties after large mechanical cyclic loadings remains scarcely investigated. In the second part of this study, a series of low-cycle fatigue tests were carried out at various stretch amplitudes, and then electrical breakdown tests were performed on the fatigued specimens. Based on the observed softening, an electro-mechanical instability model was proposed, which allows the prediction of the reduction in breakdown voltage from the mechanical damage accumulated.

2. Materials and Methods

The two dielectric elastomers have been prepared for analysis from two components, platinum catalyzed, liquid silicone rubbers (LSRs) with a shore hardness of 50 ShA and 70 ShA, respectively. Variation in the silica content between the two formulations gives different mechanical responses and Mullins damages under cyclic loadings. These LSRs were diluted in a volatile silicone fluid using a turbine mixer under a vacuum environment (~100 mbar). The resulting mixtures were coated on a polyester carrier film using a roll-to-roll coating process entirely enclosed in a clean room environment (ISO 8). A first crosslinking was performed at 110 °C for a duration of 10 min, then the dielectric elastomers were peeled from their polyester carriers, and a final post-curing was performed at 120 °C for a duration of 15 h to complete the crosslinking reaction and eliminate the volatile residuals.

Samples were cut from the PDMS thin films in a rectangular shape of 300 mm length and 200 mm width (Figure 2a). For the mechanical loading, the samples were placed in a tensile test machine (Zwick/Roell Z100, Ulm, Germany) controlled in displacement at a fixed strain rate of 3%·s^{-1}. Samples were held into pneumatic clamps to prevent slippage in the jaws under load. The tensile stretch was defined as $\lambda_1 = L/L_0$ with L, the distance between the clamps (Figure 2b) and L_0 to the initial length of the sample (L_0 = 220 mm). Incompressibility of the elastomers $\lambda_1\lambda_2\lambda_3 = 1$ yields an equal stretch in the direction of the width and thickness $\lambda_2 = \lambda_3 = \lambda_1^{-\frac{1}{2}}$.

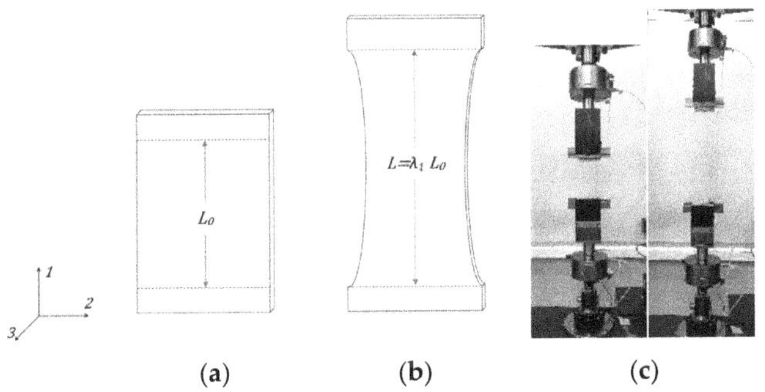

Figure 2. Tensile test sample: (**a**) unloaded; (**b**) stretched to λ_1; (**c**) corresponding experimental setup. Axes of the coordinate system are denoted as 1, 2, and 3.

The nominal tensile stress was defined as $T_1 = F/A_0$, with F the force measured by the load cell and A_0 the original cross-sectional area. The machine was loading the sample up to a pre-defined value of maximal stretch λ_{1max} before releasing completely the mechanical stress before the electrical test, individual samples were evaluated at different levels of maximal stretch for the two formulations.

Additionally, the effect of multiple cycles on the dielectric breakdown strength has been investigated for the 70 ShA elastomer. Individual samples were elongated to different pre-defined values of maximal stretch (respectively, $\lambda_{1max} = 1.3$, $\lambda_{1max} = 1.7$, and $\lambda_{1max} = 2.4$). Then, samples were fully released following a triangular displacement-controlled pattern at a crosshead speed of 6 mm/s. This sequence constitutes one mechanical cycle, which was repeated 1000 times before submitting the samples to the DBS evaluation. For the higher stretch amplitude evaluated ($\lambda_{1max} = 2.4$), intermediate breakdown strength measurements were performed after 10 and 100-cycle repetitions, respectively.

After mechanical cycling, the dielectric membranes are installed on an in-house developed breakdown tester, which allows measuring the DBS automatically in multiple locations on the tensile test sample. The PDMS dielectric elastomer film is deposited on a 304 stainless steel plate having a mirror finish polishing on which a DC power supply is connected (Heinzinger PNC 30 kV, Rosenheim, Germany). Ethanol is used to ease the sample installation and to prevent trapped air at the interface between the sample and the high-voltage electrode. Before starting any measurement, a rest time of 3 h is respected to ensure the desorption of solvent residuals. For each testing location, the film thickness is preliminary measured using an Eddy current displacement sensor (Micro-Epsilon EddyNCDT DT3100/EPS08, Ortenburg, Germany). The sensor is integrated into a cylindrical holder with a vertical offset between the sensor head and the lower surface of the cylinder, preventing contact with the measuring object (Figure 3a). The sensor holder is automatically positioned on the stainless-steel plate at the future locations of the breakdown measurements, and this calibration step gives the value of the offset δ_0. The same measurement is repeated at the surface of the film to obtain the local dielectric thickness $d_0 = \delta - \delta_0$. Repeatability of the thickness measurement was found to be 1 µm and mean membrane thicknesses were 138 ± 5 µm for the 70 ShA and 105 ± 3 µm for the 50 ShA. The corresponding thickness distributions (gathering the measurements of all the tested specimens) are represented in Figure 3c.

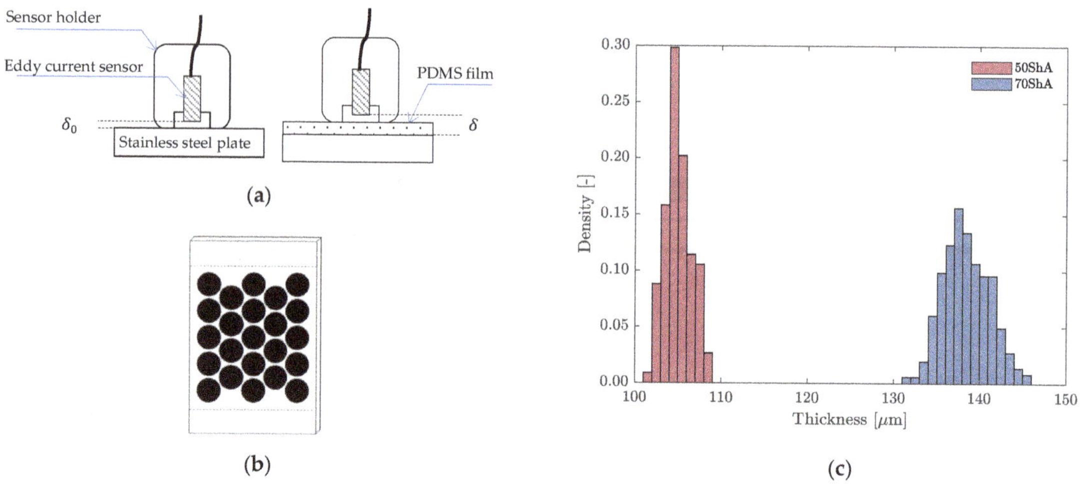

Figure 3. (**a**) Schematic representation of the thickness measurement using the Eddy current sensor, (**b**) location of measurement points on the tensile test sample, (**c**) distribution of thickness measurements for the two elastomers evaluated.

Then, the elastomer film is immersed in a silicone fluid to prevent flashover. For the ground electrode, a 40 mm diameter cylinder of stainless steel 304 is used (Figure 4a), for which all edges have been rounded to a 3.2 mm radius, such as limiting electric field enhancement at that location [13,26]. This electrode is actuated by a 3-axis motor to the first testing location, where the cylinder is deposited on the dielectric elastomer, and only the mass of the cylinder contributes to the contact pressure at the film interface. Subsequently, a positive DC voltage V is applied across the dielectric elastomer using a ramp of 500 V/s. Assuming that the thickness is uniform across the tested location, the electric field E across a linear membrane is obtained from the initial dielectric thickness as $E = V/d_0$. The breakdown voltage V_{BD} is detected from a sharp increase in the current measurement, and the corresponding DBS is calculated from (1) [27].

$$E_{BD} = V_{BD}/d_0 \tag{1}$$

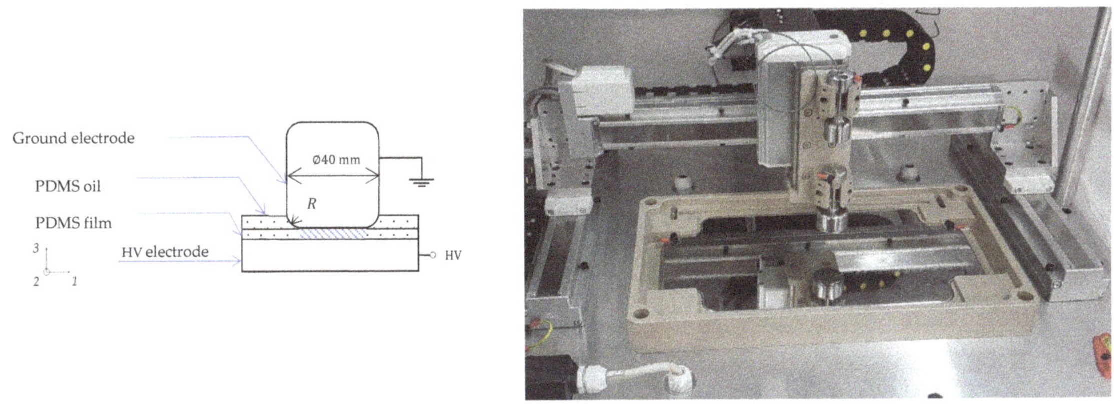

(a) (b)

Figure 4. (**a**) Schematic of the breakdown setup; (**b**) automated multipoint DBS tester.

After lifting the ground electrode, the motor automatically advances to the next testing location and repeats the breakdown process there. Using only the stretched surface and ignoring the part of the sample that was previously clamped in the tensile test machine, a maximum of 23 breakdown points can be measured (Figure 3b). Samples were stored and tested in a climate-controlled room, where the temperature and relative humidity were maintained at 23 °C and 50%, respectively.

Experimental breakdown results of every single mechanical testing scenario are fitted with a two-parameter Weibull distribution. The probability density function of a Weibull distribution is given by (2), where η is the scale parameter and β the shape parameter. The mean of the Weibull distribution is given by (3), with Γ being the gamma function [28].

$$f(E) = \frac{\beta E^{\beta-1}}{\eta^{\beta}} e^{-\left(\frac{E}{\eta}\right)^{\beta}} \tag{2}$$

$$\overline{E} = \eta \Gamma\left(\frac{1}{\beta} + 1\right) \tag{3}$$

For each material, a virgin unstretched sample was tested to attain a reference point and evaluate the effect of the subsequent mechanical loadings.

3. Model

3.1. Hyperelastic Model

The mechanical properties of the two elastomers have been modeled using an Ogden–Roxburgh hyperelastic model. The detailed methodology and results were introduced in a previous work [7]. In this approach, the Mullins effect is accounted for through a damage parameter D, which evolves with the maximal strain energy density W_{max} reached in the mechanical loading history of the elastomer [29]:

$$D = 1 - \frac{1}{r} erf\left[\frac{W_{max} - \widetilde{W}}{m + bW_{max}}\right] \tag{4}$$

where r, m, and b are material parameters, $erf(x)$ is the error function, W_{max} is the evolving maximum strain energy density reached throughout the deformation history and $\widetilde{W}$ the strain energy density function of a virgin material. Under a uniaxial test condition applied along direction i, the Cauchy stress (5) and the nominal stress (6) are represented as follows:

$$\sigma_i = D\mu\left(\lambda_i{}^{\alpha} - \lambda_i{}^{-\frac{\alpha}{2}}\right) \tag{5}$$

$$T_i = \frac{\sigma_i}{\lambda_i} \tag{6}$$

Figure 5 shows the experimental tensile test results of the two elastomers and the Ogden–Roxburgh model obtained from the material parameters of Table 1.

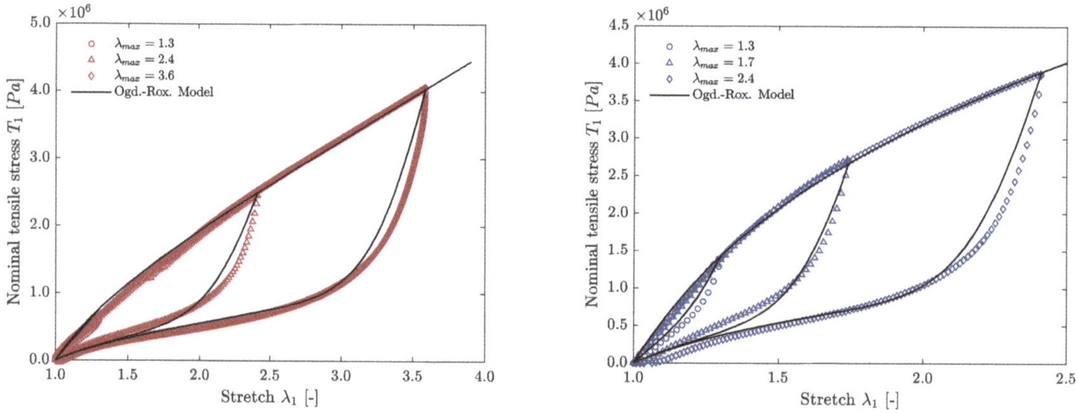

(a) (b)

Figure 5. Mullins effect and strain softening behavior modeled with an Ogden–Roxburgh model: (**a**) 50 ShA material; (**b**) 70 ShA material.

Table 1. Ogden and Ogden–Roxburgh parameters [7].

Parameter	Symbol	50 ShA	70 ShA	Unit
Ogden	μ	1.01×10^6	2.58×10^6	Pa
	α	2.1	1.61	–
Ogden–Roxburgh	r	1.4	1.35	–
	m	7.0×10^5	3.4×10^5	–
	b	0.2	0.26	–

3.2. Electro-Mechanical Model

When subjected to an external voltage V, a dielectric elastomer experiences a Maxwell compressive stress σ_E that reduces its thickness to $d = \lambda_3 d_0$. The equation of this electrostatic compressive stress is given by (7) with ε_0 is the vacuum permittivity and ε_r is the relative permittivity of the dielectric elastomer. This equation only holds if the permittivity is stretch-independent. In previous works [30,31], a variation of the permittivity with the applied stretch was reported for acrylic elastomers, whereas the permittivity of natural rubber was found insensitive to stretch. In our study, a stretch-independent relative permittivity of $\varepsilon_r = 2.7$ was measured, and that value is used in our model.

$$\sigma_E = \varepsilon_0 \varepsilon_r \left(\frac{V}{d}\right)^2 \tag{7}$$

At equilibrium, the mechanical stress (5) balances the electrostatic stress (6), and the relation between voltage and compressive stretch is given by (8). Voltage-induced actuation depends on the historical loading through the value of the damage parameter D. In this approach, the friction at the interface between the dielectric layer and electrodes is considered negligible, and the stiffness of the surrounding inactive material (not covered by electrodes) is assumed to have a marginal contribution on the actuation stretch of the active region.

$$V = \lambda_3 d_0 \sqrt{-\frac{D\mu}{\varepsilon_0 \varepsilon_r}\left(\lambda_3{}^{\alpha} - \lambda_3{}^{-\frac{\alpha}{2}}\right)} \tag{8}$$

For a virgin material, the damage parameter D is inactive (taking the value 1), and λ_3 decreases (corresponding to compressive stretch) when the voltage increases (going from right to left in Figure 6b,d). This trend is observed until a maximum critical point $\{\lambda_c; V_c\}$ at which no additional increase in the voltage is required to cause further deformation. This yields extreme deformations, causing an electrical breakdown because of the monotonic decrease in the voltage curve beyond that critical point. The model predicts a critical voltage V_c = 15 kV for the virgin unloaded 50 ShA and V_c =28 kV for 70 ShA used in this study.

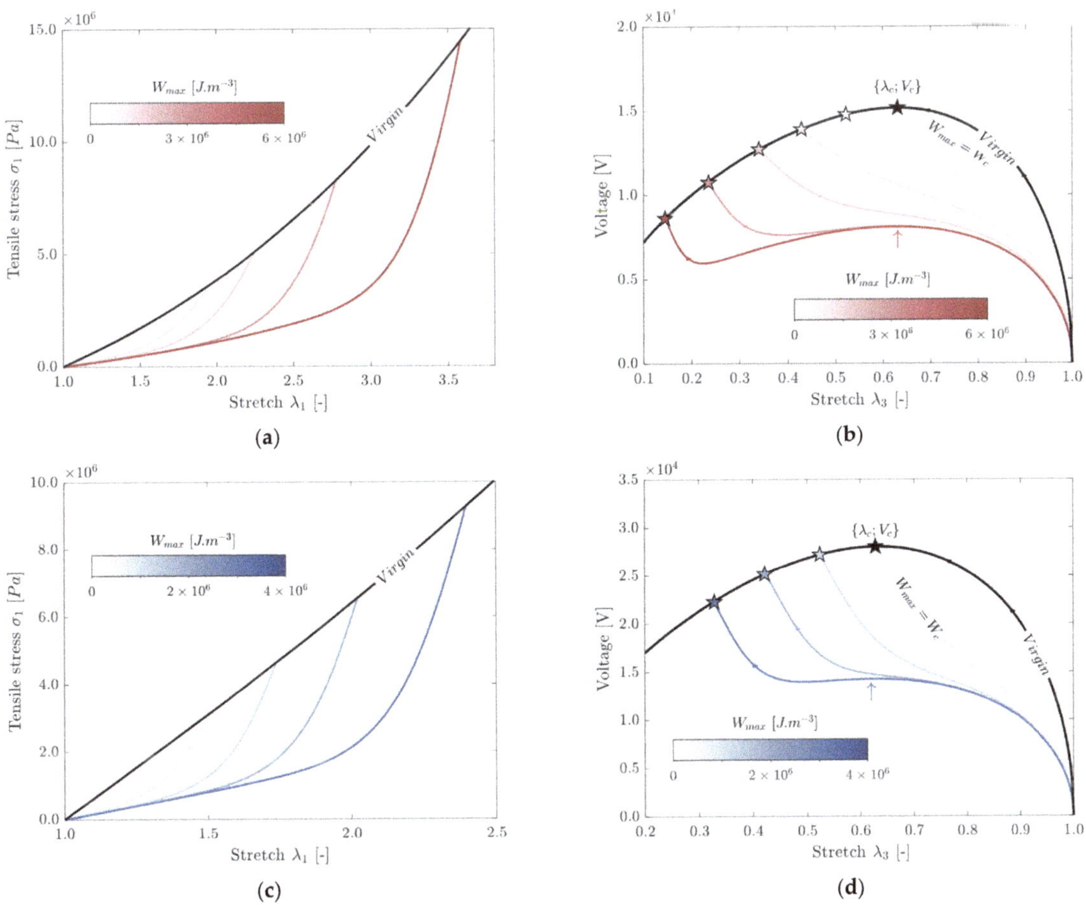

Figure 6. Uniaxial tensile stress derived from the Ogden–Roxburgh model (**left** column) and evolution of the critical instability (**right** column) for (**a**,**b**) the 50 ShA material; (**c**,**d**) the 70 ShA material. Initiation of the critical instability is represented with stars.

When considering a sample that has been stretched prior to the breakdown experiment, the Mullins effect alters the material's response to electrical stress, and the resultant voltage curve follows a different path before recovering one of the virgin materials. If the historical strain energy density W_{max} applied in the tensile experiment is below a threshold W_c, no consequence is predicted for the breakdown, as the irreversible critical point is identical to the point of the original virgin material. However, if W_{max} of a material has been raised above W_c in its loading history, its voltage curve will not pass through the instability point and will recover the curve of the virgin material in an unstable region (where thickness collapses without the need for additional voltage). This leads to breakdown voltages lower than V_c initiated at the locations of the colored stars, presented in Figure 6b,d.

Interestingly, for materials with significantly high values of historical SED, the voltage curve indicates a local maximum (upward arrow in Figure 6b,d). However, the decrease in the voltage is not monotonic beyond this local "pull-in instability"; thus, a further increase in voltage is required before recovering the permanently decreasing curve at the location of the stars (Figure 6b,d). A local instability does not necessarily imply an electrical breakdown [11,20]. However, once the curve of the virgin material is reached, the voltage becomes immediately unstable, resulting in a snap-through, which will ineluctably cause an electrical breakdown.

4. Results and Discussion

4.1. Mullins Effect

To confirm the theoretical predictions from the proposed model, the experimental DBS was measured for various values of maximum historical SED obtained from the tensile test experiment (Figure 7). The reported breakdown values for both model and experimental data were obtained from (1). For each individual mechanical loading case, the Weibull probability density function of the experimental electrical failures (2) is represented vertically. For the virgin sample and the first loading case of the 70 ShA elastomer, electrical breakdowns were not reached for all the testing locations due to the voltage limitation of the DC power supply. These corresponding data were treated as suspensions in the evaluation of the Weibull parameters and are represented by upward arrows in Figure 7b. For the virgin 50 ShA material, the mean of the experimental failure was measured at $\overline{E} = 140 \pm 6$ V·μm^{-1}, whereas for the 70 ShA material, a significantly higher DBS of $\overline{E} = 224 \pm 9$ V·μm^{-1} was obtained (uncertainties were determined from the two-sided 95% confidence bound interval). This increase in the DBS was anticipated by the EMI model introduced in Section 3.2 due to the substantial variation in mechanical properties between the two elastomers considered. For the 50 ShA material, the model predicts a breakdown value of 145 V·μm^{-1}, and that value is contained within the confidence bounds of the experimental results. For the 70 ShA material, a breakdown strength of 205 V·μm^{-1} is calculated for a virgin material, which is only 8% lower than the mean experimental value. The historical SED threshold under which no consequence is expected on the DBS has been verified experimentally as the mean DBS of the samples previously stretched to $\lambda_{max} = 1.3$ are very similar to the ones of the virgin materials ($\overline{E} = 135 \pm 6$ V·μm^{-1} for the 50 ShA and $\overline{E} = 210 \pm 7$ V·μm^{-1} for the 70 ShA). Above this threshold, the overall feature of the Mullins damage is also predicted accurately compared with experimental data obtained for the most severe loading cases. This finding confirms that breakdown is initiated at the location of the stars, and the local instability observed on the highly stretched samples is not causing a breakdown. The behavior observed here supports previous works [11] where it was found that pull-in instability does not necessarily imply an electrical failure. The Mullins damage is found to cause a plateau in the stretch–voltage curve (Figure 6b,d). This feature is interesting for dielectric elastomer actuators for which obtaining a large stretch amplitude from a low voltage variation is desirable from a transducer efficiency perspective. Although a preliminary stretch of the elastomer beyond its operating domain can be used to enhance its actuation response, further measurement would be required to conclude whether the plateau could be exploited or not. The instability phenomenon might lead to localized inhomogeneous deformation, which is known as creasing, cratering, snap-through, or wrinkling, and the typical distance between creases was found to be approximately the same as the thickness of the film [32]. In our experimental setup, the electrode covers a relatively wide surface (40 mm diameter), and the actuation region is hidden underneath the steel cylinder, preventing direct observation of the actuation mechanisms.

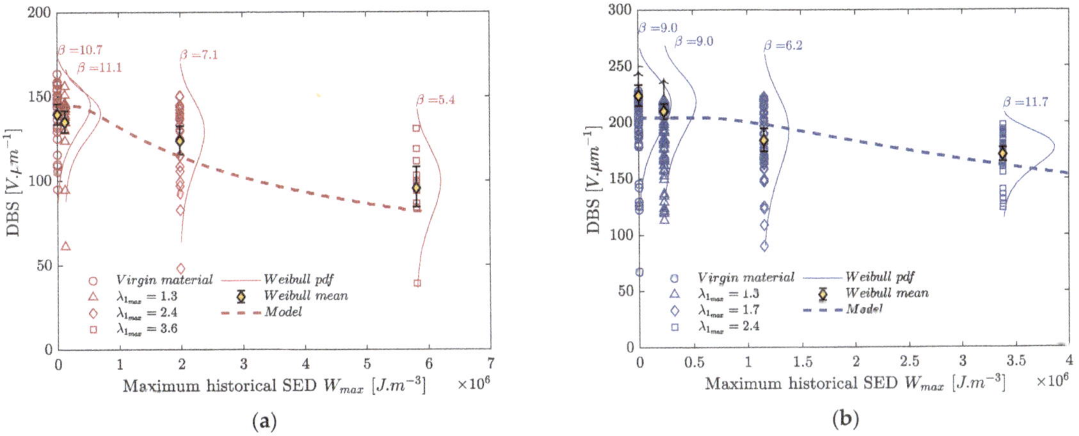

Figure 7. Comparison of experimental data to the EMI model: (**a**) 50 ShA material; (**b**) 70 ShA material.

In the Ogden–Roxburgh model, the damage parameter D is considered a scalar value indifferent to the direction of the principal strain. In the tensile experiment, the strain is applied along direction 1 (Figure 2), whereas the elastomer expands equibiaxially in the plane of the dielectric sheet when voltage increases. The Mullins effect-induced damage might preferably occur in the tensile direction, resulting in enhanced actuation along this axis. Further tests would be required to verify this anisotropy in the Mullins softening, with the possible outcome of tailoring the actuation response of dielectric elastomer actuators along a preferred orientation. This could be an alternative to the existing solutions that use orientated fibers to restrict deformation in one direction [33].

As the Mullins effect is found to reduce the breakdown strength significantly, it is worth discussing any potential mitigation to that degradation mechanism. The Mullins damage is attributed to complex interactions between polymeric chains and fillers. Therefore, it could be of interest to limit the silica content in PDMS composites while trying to stiffen the mechanical response using alternative strategies. Varying the crosslinking density of the elastomer network or changing the chemistry of the crosslinker could be considered in future research studies.

4.2. Mechanical Cycling

With the effect of cycle accumulation, the mechanical response is modified, leading to a softening of the elastomer and a reduction in the maximal stress reached in the mechanical cycle. Subsequent to the first cycle, minor changes are observed in the stress–stretch response between the loading and unloading phase, indicating that viscous losses can be considered negligible in proportion to the Mullins damage. Therefore, for the sake of clarity, it is chosen to display only the unloading curves in Figure 8, where the softening behavior is represented for the three stretch amplitudes evaluated. For the samples stretched to $\lambda_{1max} = 1.3$, the mechanical response is almost unaffected when the number of cycles increases. However, for the two other loading cases ($\lambda_{1max} = 1.7$ and $\lambda_{1max} = 2.4$), the softening is significant and culminates in about a 40% reduction in the peak tensile stress after 1000 cycles.

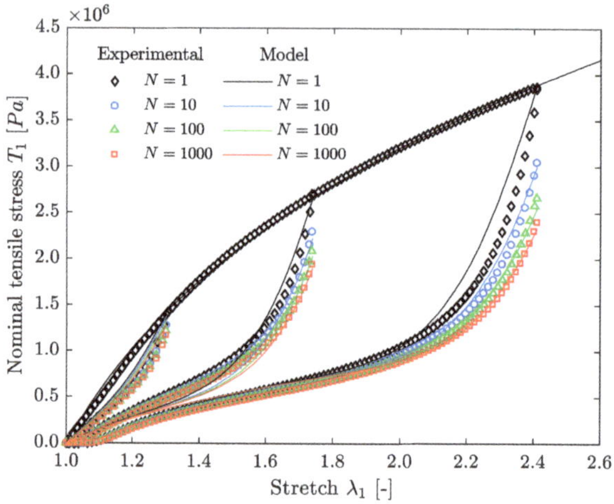

Figure 8. Softening of the 70 ShA material with cycling.

From a modeling perspective, this softening can be interpreted as an increase in the Mullins damage with the number of cycles. A modification of the Ogden–Roxburgh model is introduced where the value of W_{max} in (4) is replaced by a variable W^*, which increases with the number N of mechanical cycles (9). An exponential function (10) is chosen for the expression of W^* as it was found to represent the experimental cycle-induced softening accurately.

$$D = 1 - \frac{1}{r} erf\left[\frac{W^* - \widetilde{W}}{m + bW^*}\right] \tag{9}$$

$$W^* = W_{max} e^{A(1-\frac{1}{N^B})} \tag{10}$$

Parameters A = 0.24 and B = 0.25 were calculated, such as minimizing the error between the evolving experimental tensile stress and the stress derived from the modified Ogden–Roxburgh model (Figure 8). This approach constitutes a convenient way to model the experimental results but is only valid if the maximal stretch remains constant during the fatigue experiment. In the case of more complex random loadings, a different softening is expected, which would require the implementation of more sophisticated constitutive models.

Using this modified damage parameter in (8) allows us to predict the evolution of the EMI in the stretch–voltage diagram based on the number of cycle repetitions (Figure 9). For the lower stretch amplitude ($\lambda_{1max} = 1.3$), the location of the critical instability is found unaffected as the instability is identical to the snap-through of the virgin material, as represented by the black star in Figure 9. For the larger stretch amplitudes, only minor reductions in the breakdown voltage are expected after mechanical fatigue, as the location of the snap-through is marginally modified compared to a sample stretched only once ($N = 1$). In other words, the softening caused by the Mullins effect predominantly contributes to the location of the EMI, whereas the role of fatigue softening appears negligible.

To verify these model predictions, experimental breakdown tests were performed on unstretched film after a different number of mechanical cycles at three given stretch amplitudes (Figure 10). A good agreement was found between the EMI model and the mean of the Weibull distributions. It indicates that at this relatively low number of cycles, the breakdown strength is still driven by EMI. In this experimental setup, the number of cycles was limited to one thousand mechanical cycles because of the limited capacity of the tensile machine or premature failure of the samples in the clamping system. Extrapolating the result to long-term fatigue is not expected to affect the breakdown response further as the Mullins damage tends to stabilize towards an asymptotic endpoint as the number of cycles increases. However, for a large number of cycles, one could expect micro-cracks to propagate/nucleate, resulting in local electric weak points that cannot be captured from a global mechanical characterization. However, this type of flaw was not observed in the most severely loaded sample (λ_{max} = 2.4 and N = 1000 cycles) with microscope inspection under ×700 magnification, further fatigue experiments on longer duration can constitute future research work to better understand the long-term performance of dielectric elastomer generators.

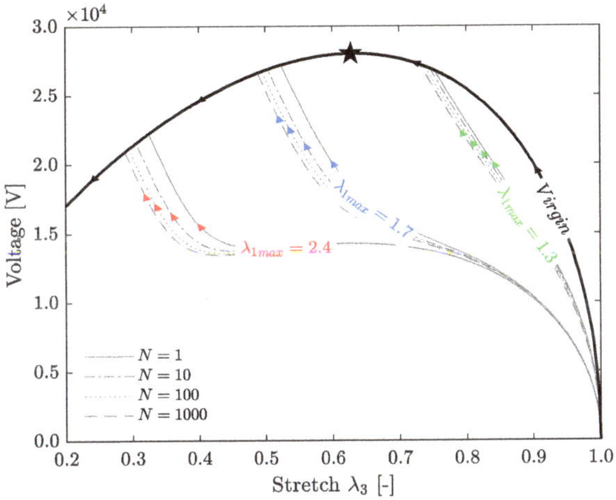

Figure 9. Evolution of the critical instability in the stretch–voltage diagram. The critical instability of a virgin material is represented with a black star.

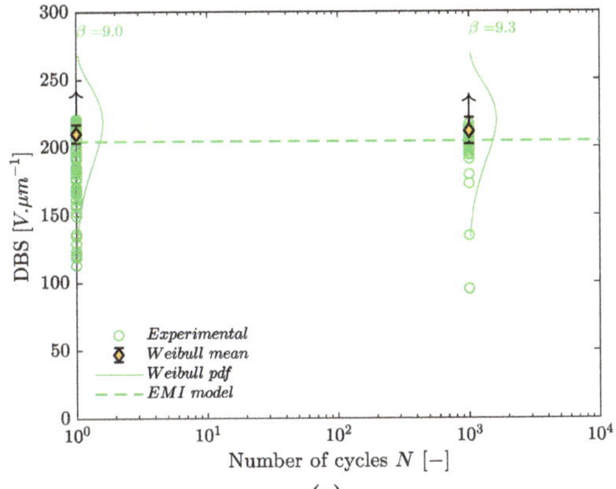

(a)

Figure 10. *Cont.*

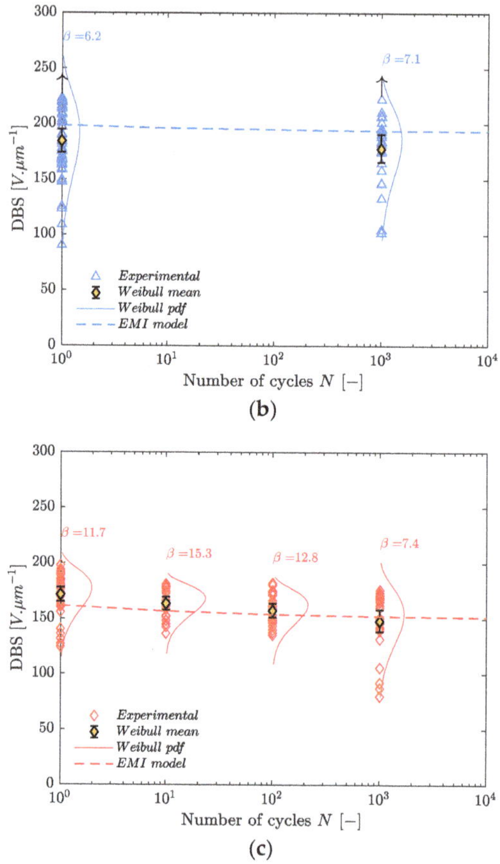

Figure 10. Evolution of the dielectric breakdown strength with mechanical cycling after a maximal stretch of: (**a**) $\lambda_1 = 1.3$; (**b**) $\lambda_1 = 1.7$; (**c**) $\lambda_1 = 2.4$.

4.3. Energy Density

The convertible energy W_e of DEGs is known to increase linearly with the relative permittivity ε_r and quadratically with an increasing applied electric field (11) [1].

$$W_e \propto \varepsilon_r {E_{BD}}^2 \tag{11}$$

Intense research activities are devoted to developing material formulations with higher relative permittivity [34], which is valuable for decreasing the operating voltage of dielectric actuators or enhancing their actuation stretch. However, it is worth discussing its beneficial contribution to DEGs. Inserting (8) into (11) shows that the convertible energy is actually independent of ε_r for a system in which the failure mechanism is driven by electro-mechanical instabilities (12). In other words, a rise in the relative permittivity is equally balanced by a reduction in the breakdown voltage in the overall energy balance. Although the underlying conclusion only holds if the relative permittivity remains unchanged with the accumulated mechanical damage, that property shall be considered carefully when attempting to develop new material formulations with enhanced energy densities.

$$W_e \propto -{\lambda_c}^2 D\mu\left({\lambda_c}^{\alpha} - {\lambda_c}^{-\frac{\alpha}{2}}\right) \tag{12}$$

Increasing the pre-stretch of the dielectric elastomer is a strategy that is often reported to prevent electro-mechanical instability and can be an option to overcome the limitation between permittivity and breakdown strength. However, in large-scale DEG, the failure mechanism is rarely driven by EMI, as the probability of defect-related breakdowns scales with the size of the transducer [14]. In such a case, increasing the permittivity might still be beneficial to increase the energy density of DEG.

5. Conclusions

The influence of multiple mechanical cycles on the DBS of PDMS elastomers has been investigated. The stretch amplitude of the first cycle was found to be of primary importance because the Mullins effect significantly changes the onset of electro-mechanical instability. Using stiffer PDMS elastomers is one approach for increasing the dielectric breakdown strength of a virgin material; however, this beneficial enhancement is partially counteracted by a higher softening resulting from the Mullins damage.

Subsequently, after this first loading, the breakdown strength remains nearly stable up to 1000 cycles. These experimental results are consistent with theoretical predictions obtained from a hyperelastic model that uses the damage accumulated along the fatigue experiment to derive the evolution of the electro-mechanical instability.

The outcome of this study helps to understand the relationship between dielectric performance and mechanical properties and contributes to gaining knowledge on the long-term reliability of DEGs. This constitutes one of the key milestones to reach before deploying these systems in commercial applications.

Author Contributions: Conceptualization, E.T., T.A., I.A.S. and P.H.F.M.; methodology, E.T., T.A., I.A.S. and P.H.F.M.; software, E.T.; validation, E.T., T.A., I.A.S. and P.H.F.M.; formal analysis, E.T., T.A., I.A.S. and P.H.F.M.; investigation, E.T., T.A., I.A.S. and P.H.F.M.; resources, T.A. and P.H.F.M.; data curation, E.T.; writing—original draft preparation, E.T.; writing—review and editing, E.T., T.A., I.A.S. and P.H.F.M.; visualization, E.T.; supervision, T.A. and I.A.S.; project administration, P.H.F.M. All authors have read and agreed to the published version of the manuscript.

Funding: This research was funded by SBM Offshore in the frame of the S3 Wave Energy Converter.

Data Availability Statement: Data presented in this study are available at the following DOI: https://doi.org/10.6084/m9.figshare.24334924.v2 (accessed on 1 November 2023).

Conflicts of Interest: SBM Offshore is funding a research project that aims at increasing the energy density of dielectric elastomer generators. The project is held by P.M., who acts as project coordinator and expert consultant. The University of Southampton is an academic partner for this project. The content of our study presents part of the work performed in the frame of this research project. E.T. is an employee of SBM Offshore and a PhD student from the University of Southampton. T.A. and I.S. are with the University of Southampton. P.M. acts as a project coordinator and expert consultant for the project. We hereby declare that the funder has no role in the study's design, in the collection, analysis, or interpretation of data, or the writing of the manuscript. The funder has reviewed and approved the manuscript without asking for modifications to the original version.

References

1. Moretti, G.; Rosset, S.; Vertechy, R.; Anderson, I.; Fontana, M. A review of dielectric elastomer generator systems. *Adv. Intell. Syst.* **2020**, *2*, 2000125. [CrossRef]
2. Kornbluh, R.D.; Eckerle, J.; McCoy, B. *A Scalable Solution to Harvest Kinetic Energy*; SPIE Newsroom: Bellingham, WA, USA, 2011.
3. Jean, P.; Wattez, A.; Ardoise, G.; Melis, C.; Van Kessel, R.; Fourmon, A.; Barrabino, E.; Heemskerk, J.; Queau, J.P. Standing wave tube electro active polymer wave energy converter. In Proceedings of the Electroactive Polymer Actuators and Devices (EAPAD), San Diego, CA, USA, 12–15 March 2012; SPIE: Bellingham, WA, USA, 2012; pp. 75–95.
4. Moretti, G.; Rosati, G.; Daniele, L.; Forehand, D.; Ingram, D.; Vertechy, R.; Fontana, M. Modelling and testing of a wave energy converter based on dielectric elastomer generators. *Proc. R. Soc. A* **2019**, *475*, 20180566. [CrossRef] [PubMed]
5. Vudayagiri, S.; Zakaria, S.; Yu, L.; Hassouneh, S.; Benslimane, M.; Skov, A. High breakdown-strength composites from liquid silicone rubbers. *Smart Mater. Struct.* **2014**, *23*, 105017. [CrossRef]

6. Kollosche, M.; Kofod, G. Electrical failure in blends of chemically identical, soft thermoplastic elastomers with different elastic stiffness. *Appl. Phys. Lett.* **2010**, *96*, 071904. [CrossRef]
7. Taine, E.; Andritsch, T.; Saeedi, I.; Morshuis, P. Effect of mechanical loading history on the electrical breakdown strength of dielectric elastomers. In Proceedings of the 2022 IEEE 4th International Conference on Dielectrics (ICD), Palermo, Italy, 3–7 July 2022; IEEE: Piscataway, NJ, USA, 2022; pp. 21–24.
8. Kofod, G.; Sommer-Larsen, P.; Kornbluh, R.D.; Pelrine, R. Actuation response of polyacrylate dielectric elastomers. *J. Intell. Mater. Syst. Struct.* **2003**, *14*, 787–793. [CrossRef]
9. Iannarelli, A.; Niasar, M.G.; Ross, R. The effects of static pre-stretching on the short and long-term reliability of dielectric elastomer actuators. *Smart Mater. Struct.* **2019**, *28*, 125014. [CrossRef]
10. Zakaria, S.; Morshuis, P.H.F.; Benslimane, M.Y.; Yu, L.; Skov, A.L. The electrical breakdown strength of pre-stretched elastomers, with and without sample volume conservation. *Smart Mater. Struct.* **2015**, *24*, 055009. [CrossRef]
11. Gatti, D.; Haus, H.; Matysek, M.; Frohnapfel, B.; Tropea, C.; Schlaak, H.F. The dielectric breakdown limit of silicone dielectric elastomer actuators. *Appl. Phys. Lett.* **2014**, *104*, 052905. [CrossRef]
12. Gerratt, A.P.; Bergbreiter, S. Dielectric breakdown of PDMS thin films. *J. Micromech. Microeng.* **2013**, *23*, 067001. [CrossRef]
13. Förster-Zügel, F.; Grotepaß, T.; Schlaak, H. Characterization of the dielectric breakdown field strength of PDMS thin films: Thickness dependence and electrode shape. In Proceedings of the Electroactive Polymer Actuators and Devices (EAPAD), San Diego, CA, USA, 9–12 March 2015; SPIE: Bellingham, WA, USA, 2015; p. 94300D.
14. Taine, E.; Andritsch, T.; Saeedi, I.A.; Morshuis, P.H.F. Size effect and electrical ageing of PDMS dielectric elastomer with competing failure modes. *Smart Mater. Struct.* **2023**, *32*, 105021. [CrossRef]
15. Diani, J.; Fayolle, B.; Gilormini, P. A review on the Mullins effect. *Eur. Polym. J.* **2009**, *45*, 601–612. [CrossRef]
16. Hanson, D.E.; Hawley, M.; Houlton, R.; Chitanvis, K.; Rae, P.; Bruce Orler, E.; Wrobleski, D.A. Stress softening in silica-filled polydimethylsiloxane provide insight into a mechanism for the Mullins effect. *Polymer* **2005**, *46*, 10989–10995. [CrossRef]
17. Clément, F.; Bakobza, L.; Monnerie, L. On the Mullins effect of in silica-filled polydimethylsiloxane networks. *Rubber Chem. Technol.* **2001**, *74*, 847–870. [CrossRef]
18. Georgousis, G.; Roumpos, K.; Kontou, E.; Kyritsis, A.; Pissis, P.; Koutsoumpis, S.; Mičušík, M.; Omastová, M. Strain and damage monitoring in SBR nanocomposites under cyclic loading. *Compos. Part B Eng.* **2017**, *131*, 50–61. [CrossRef]
19. Cantournet, S.; Desmorat, R.; Besson, J. Mullins effect and cyclic stress softening of filled elastomers by internal sliding and friction thermodynamics model. *Int. J. Solids Struct.* **2009**, *46*, 2255–2264. [CrossRef]
20. Koh, S.J.A.; Li, T.; Zhou, J.; Zhao, X.; Hong, W.; Zhu, J.; Suo, Z. Mechanisms of large actuation strain in dielectric elastomers. *J. Polym. Sci. Part B Polym. Phys.* **2011**, *49*, 504–515. [CrossRef]
21. Garton, C.G.; Stark, K.H. Electric strength of irradiated polythene. *Nature* **1995**, *176*, 1225–1226.
22. Zhao, X.; Suo, Z. Electromechanical instability in semicrystalline polymers. *Appl. Phys. Lett.* **2009**, *95*, 031904. [CrossRef]
23. Akbari, S.; Rosset, S.; Shea, H.R. Improved electromechanical behavior in castable dielectric elastomer actuators. *Appl. Phys. Lett.* **2013**, *102*, 071906. [CrossRef]
24. Zhang, M.; Denes, I.; Buchmeiser, M.R. Interplay between Mechanical Fatigue and Network Structure and Their Effects on Mechanical and Electrical Properties of Thin Silicone Films with Varying Stoichiometric Imbalance. *Macromol. Chem. Phys.* **2016**, *217*, 1558–1568. [CrossRef]
25. Vertechy, R.; Chen, Y.; Agostini, L.; Moretti, G.; Fontana, M.; Berselli, G. Fatigue life performances of silicone elastomer membranes for dielectric elastomer transducers: Preliminary results. In Proceedings of the Electroactive polymer actuators and devices (EAPAD), Denver, CO, USA, 4–7 March 2019; p. 27.
26. *ASTM-D149-20*; Standard Test Method for Dielectric Breakdown Voltage and Dielectric Strength of Solid Electrical Insulating Materials at Commercial Power Frequencies. ASTM: West Conshohocken, PA, USA, 2020.
27. Carpi, F.; Anderson, I.; Bauer, S.; Frediani, G.; Gallone, G.; Gei, M.; Graaf, C.; Jean-Mistral, C.; Kaal, W.; Kofod, G.; et al. Standards for dielectric elastomer transducers. *Smart Mater. Struct.* **2015**, *24*, 105025. [CrossRef]
28. Nelson, W.B. *Accelerated Testing: Statistical Models, Test Plans, and Data Analysis*; John Wiley & Sons: Hoboken, NJ, USA, 1990.
29. Ogden, R.W.; Roxburgh, D.G. A pseudo-elastic model for the Mullins effect in filled rubber. *Proc. R. Soc. Lond.* **1999**, *455*, 2861–2877. [CrossRef]
30. Hyouk Ryeol, C.; Kwangmok, J.; Nguyen Huu, C.; Minyoung, J.; Igmo, K.; Jachoon, K.; Joonho, L.; Jonghoon, L.; Jaedo, N.; Misuk, C.; et al. Effects of prestrain on behavior of dielectric elastomer actuator. In Proceedings of the Smart Structures and Materials 2005: Electroactive Polymer Actuators and Devices (EAPAD), San Diego, CA, USA, 7–10 March 2005; SPIE: Bellingham, WA, USA, 2005; pp. 283–291.
31. Tröls, A.; Kogler, A.; Baumgartner, R.; Kaltseis, R.; Keplinger, C.; Schwödiauer, R.; Graz, I.; Bauer, S. Stretch dependence of the electrical breakdown strength and dielectric constant of dielectric elastomers. *Smart Mater. Struct.* **2013**, *22*, 104012. [CrossRef]
32. Wang, Q.; Tahir, M.; Zang, J.; Zhao, X. Dynamic Electrostatic Lithography: Multiscale On-Demand Patterning on Large-Area Curved Surfaces. *Adv. Mater.* **2012**, *24*, 1947–1951. [CrossRef] [PubMed]

33. Lu, T.; Huang, J.; Jordi, C.; Kovacs, G.; Huang, R.; Clarke, D.R.; Suo, Z. Dielectric elastomer actuators under equal-biaxial forces, uniaxial forces, and uniaxial constraint of stiff fibers. *Soft Matter* **2012**, *8*, 6167–6173. [CrossRef]
34. Madsen, F.B.; Daugaard, A.E.; Hvilsted, S.; Skov, A.L. The Current State of Silicone-Based Dielectric Elastomer Transducers. *Macromol. Rapid Commun.* **2016**, *37*, 378–413. [CrossRef]

energies

MDPI

Article

Analysis of Uneven Distribution of Nodes Creating a Percolation Channel in Matrices with Translational Symmetry for Direct Current

Pawel Zukowski [1], Pawel Okal [2], Konrad Kierczynski [2,*], Przemyslaw Rogalski [2] and Vitalii Bondariev [2]

[1] Department of Economics, Vincent Pol University in Lublin, 2, Choiny Str., 20-816 Lublin, Poland; zukowski50pawel@gmail.com

[2] Department of Electrical Devices and High Voltage Technology, Faculty of Electrical Engineering and Computer Science, Lublin University of Technology, 38A, Nadbystrzycka Str., 20-618 Lublin, Poland; p.okal@pollub.pl (P.O.); p.rogalski@pollub.pl (P.R.); v.bondariev@pollub.pl (V.B.)

* Correspondence: k.kierczynski@pollub.pl; Tel.: +48-81-538-43-28

Abstract: In this study, the phenomenon of node percolation was tested using the Monte Carlo computer simulation method for square matrices with dimensions $L = 55$, 101 and 151. The number of samples for each matrix was 5×10^6. The spatial distributions of the coordinates of the nodes creating the percolation channel were determined, and maps of the density distribution of these nodes were created. It has been established that in matrices with finite dimensions, an edge phenomenon occurs, consisting of a decrease in the concentration of nodes creating a percolation channel as one approaches the edge of the matrix. As the matrix dimensions increase, the intensity of this phenomenon decreases. This expands the area in which values close to the maximum occur. The length distributions of the left and right clusters of non-conducting nodes were determined for the situation when the next randomly selected node connects them and thus reaches the percolation threshold. It was found that clusters whose dimensions are close to half of the matrix dimensions are most likely to occur. The research shows that both the values of the standard deviation of the percolation threshold and the intensity of the edge phenomenon are clearly related to the dimensions of the matrices and decrease as they increase.

Keywords: metrological approach; uncertainty of measurement; percolation phenomenon; percolation threshold; Monte Carlo method; computer simulation

Citation: Zukowski, P.; Okal, P.; Kierczynski, K.; Rogalski, P.; Bondariev, V. Analysis of Uneven Distribution of Nodes Creating a Percolation Channel in Matrices with Translational Symmetry for Direct Current. *Energies* **2023**, *16*, 7647. https://doi.org/10.3390/en16227647

Academic Editor: Mario Marchesoni

Received: 28 October 2023
Revised: 13 November 2023
Accepted: 15 November 2023
Published: 18 November 2023

1. Introduction

The study of the phenomenon known as percolation, derived from the Latin 'percolatio', is primarily aimed at studying the permeation of liquids and gases through porous substrates [1]. Percolation issues have ubiquitous applications in a variety of scientific and technological research fields, including chemistry [2], medicine [3], biology [4] and materials engineering [5]. In addition, the conceptual framework of percolation extends its utility to the determination of the propagation of electric current in entities characterized by disordered semiconductors [6].

The study of the percolation phenomenon is directed in two basic directions. One of these involves empirical research, during which the simplest way to observe a phenomenon is via the conduit of electric current [7]. Conversely, endeavors directed at the examination of gas or liquid flow have proved considerably more intricate to execute.

To observe the classical phenomenon of percolation in the context of electric current flow, certain preconditions must be met. The medium in which percolation occurs should consist of at least two phases. The first is referred to as the matrix, while the elements forming the second phase, referred to as the dispersed phase, are distributed throughout the matrix. The properties of these phases should differ significantly. Typically, the matrix is an insulating material, and the dispersed phase is conductive elements with macroscopic dimensions. As the concentration of the dispersed phase in the composite increases, its components form electrical connections via points of contact, gradually forming a complex network of structures known as clusters. When a certain critical concentration, known as the percolation threshold, is reached, a so-called infinite cluster connecting the electrodes, known as a percolation channel, is formed. There is a rapid change in the macroscopic electrical properties of the medium. In the case of metal–dielectric composites or nanocomposites, there is a sharp drop in resistivity and a change in the type of conductivity from dielectric to metallic [8]. Dielectric conductivity is characterized by the fact that the resistivity of the medium decreases with increasing temperature. However, in the case of metallic conductivity, resistivity increases with increasing temperature [9]. This is a type 2 phase transition.

As is known, the conventional limit between composites and nanocomposites is approximately 100 nm [10]. When at least one of the dimensions of the dispersed phase elements is below this limit, we are dealing with nanocomposites. Based on experimental studies, it was determined that the value of the percolation threshold in metal–dielectric nanocomposites depends on the material of the dispersed phase placed in matrices of various compositions. The percolation threshold can vary over a very wide range, starting from a few percent [11–14] to values close to 50% [15,16].

The phenomenon of percolation has been used in the description of various issues, such as the spread of epidemics [17–19], the reliability of computer networks [20] and the spread of fires [21,22], as well as in molecular biology [23], materials engineering [24,25] or the flow of electric current through conductive and non-conductive mixtures [26,27]. In recent decades, the percolation phenomenon has been successfully used to describe current conduction in metal–dielectric nanocomposites [28–30]. This is hopping conductivity based on the quantum mechanical phenomenon of electron tunneling [31–34]. Without taking this phenomenon into account, it is impossible to understand and explain the dependence of the current intensity in nanocomposites on the metallic phase content, current frequency and temperature.

The second line of research is theoretical research and modeling [35–37]. For a longer period, studies involving the simulation of percolation in two-dimensional networks with translational symmetry were mainly used to illustrate this phenomenon. Subsequent work in this direction amounted to more and more precise determination of the percolation threshold value [38–41]. Yes, it is interesting from the point of view of numerical calculations. However, such high accuracy is not needed, for example, in the analysis of experimental results.

In recent decades, a new field of materials has emerged, so-called 2D materials. These are two-dimensional materials with a thickness of one to several atoms. Such materials include graphene, MXenes and a number of others [42–55].

The structure of two-dimensional composites may contain atoms of impurities, as well as vacancies and internodes. This is evidenced, for example, by the observation of step conductivity caused by electron tunnelling in MXene-PCL nanocomposites [56]. As is known, tunnelling occurs when there are closely located potential wells with nanometer dimensions in the material [57]. This means that potential wells exist in MXene-PCL nanocomposites, created by inclusions with dimensions of one or several atoms. Positron annihilation studies [58] have shown that inclusions can be, for example, divacancy + an oxygen atom in the internode or vacancy + two oxygen atoms located in the internode.

The analysis of the distribution of such inclusions in 2D materials exactly coincides with the analysis of the percolation phenomenon in two-dimensional matrices with translational symmetry. Therefore, research on percolation in 2D lattice systems with translational symmetry, which was initially only model or illustrative, together with the development of technology for producing 2D materials, may discover both scientific, applied and practical features.

In this work, the percolation of nodes in square matrices was selected for analysis. This choice is because such matrices have high symmetry and are the least complicated systems compared to triangular, honeycomb and other even more complex matrices. The choice of the simplest and most symmetrical shape is since the properties of the percolation process in the case of complex networks may be masked by the geometric and analytical intricacies of complicated network models

Therefore, the aim of the work was:

- determining the value of the percolation threshold and the coordinates of the node interrupting the last percolation channel for each trial;
- development of maps and spatial distributions of nodes creating the percolation channel and the standard deviation of the percolation threshold;
- analysis of the probability of occurrence of clusters depending on their dimensions.

2. Research Method

The programming language Python was used to create the algorithm and simulation program for this study, with a detailed explanation provided in [59]. This work verified the correct operation of the program. Set theory served as the foundation for the calculations, as computers execute assignment processes significantly more quickly than computational processes. The dimension of the square network L was the sole variable. The 'Mersenne Twister' algorithm [60] used, a pseudo-random number generator, is often used in Monte Carlo simulations, as mentioned in [61,62]. A sequential selection of non-conducting nodes was carried out by the program. The initial node, randomly selected, constituted a set of a single element. Each subsequently chosen non-conducting node was then assessed to determine if it was part of any existing sets, that is, whether the selected node was within a one-unit distance vertically, horizontally or diagonally from any node in the preexisting set. Adding it to this set resulted in the formation of a cluster. In other cases, the node constituted a single-element set. An exception occurred when the selected node belonged to multiple clusters, when it was added to the merged clusters. Following the addition of a non-conducting node to an existing cluster, this cluster underwent a test to verify whether it contained points with coordinates $x = 1$ and $x = L$. This condition ascertained whether the cluster established a continuous connection between the right and left edges of the matrix. Meeting these criteria results in the formation of what is known as the 'infinite cluster' of non-conducting nodes and the attainment of the percolation threshold. During this step, the algorithm documented the coordinates of the initial randomly chosen node, the coordinates of the non-conducting node that disrupted the most recent percolation channel and the computed percolation threshold concentration and subsequently completed its operation. The algorithm was executed repeatedly until the specified number of samples, which was 5 million, was achieved. Simulations were conducted for matrices of sizes $L = 55$, 101 and 151.

3. Edge Phenomenon of Spatial Distribution of Nodes Forming a Percolation Channel

It was established that the percolation threshold values for matrices with dimensions $L = 51$, 101, 151 and sample numbers 5×10^4 and above do not depend on the matrix dimensions. In contrast, the values of the standard deviations decrease markedly with increasing matrix dimensions. In order to determine the sources of this phenomenon, maps of the spatial distribution of the nodes forming the percolation channel were made for matrices of dimensions 55, 101 and 151 for a sample number of 5×10^6. For this purpose, the coordinates of the non-conducting node that interrupts the last percolation channel were determined via Monte Carlo simulations in each sample. Figure 1 shows, as an example, a heat map [63] of the two-dimensional distribution of the nodes that form the percolation channel for a matrix of dimension 101.

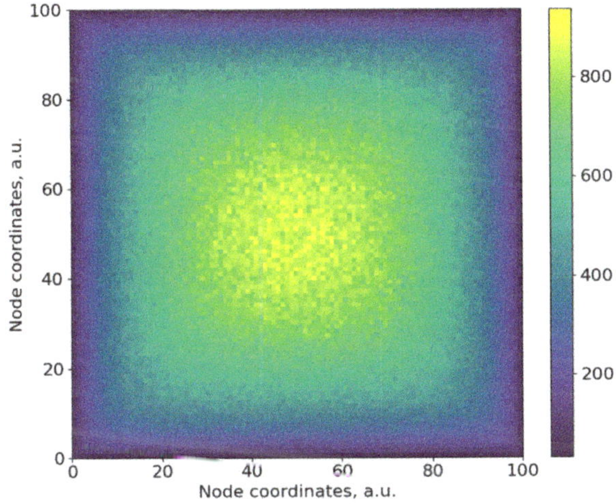

Figure 1. Two-dimensional heat map of the number of non-conductive nodes forming the percolation channel for matrix node coordinates of dimension $L = 101$ for 5×10^6 samples.

Figure 1 shows that there is an edge phenomenon in the matrices, which is that the nodes that form the percolation threshold are concentrated in the central part of the matrix. As each edge is approached, the number of these nodes decreases. In the central part, the number of nodes forming the percolation threshold is about an order of magnitude higher than the one near the edge.

In order to determine the sources of the decrease in the standard deviations of the percolation thresholds as the dimensions of the matrices increased, maps of the spatial distributions of the concentrations of the non-conducting nodes forming the percolation channel were made. Maps were made for matrices of dimensions 55, 101 and 151. To create the maps in a spreadsheet, after 5×10^6 samples had been performed, the coordinates of all matrix nodes were recorded along with the assigned numbers of non-conducting nodes forming the percolation channel. An ascending sorting of the results was then performed according to the number of nodes forming the percolation channel located in the matrix nodes. After this, ranges of numbers of non-percolating nodes were selected to divide the matrix into 10 zones. Table 1 shows the ranges of change in the number of nodes forming the percolation threshold in each zone and the colours of the nodes in the spatial maps, corresponding to each zone.

Figure 2 shows, as an example, the two-dimensional distribution of non-conductive nodes forming the percolation channel per node for a matrix with $L = 55$ divided into 10 zones.

Figure 2 perfectly illustrates the edge phenomenon, whereby the concentration of nodes forming the percolation channel decreases as the edge of the matrix is approached. Analysis of the spatial maps (Figure 2) has shown that, firstly, the distribution of nodes forming the percolation channel in the zones, closest to the centre of the matrix, is close to circular. Secondly, for nodes in further zones, there is a fourth-order axis of symmetry passing through the centre of the matrix perpendicular to it. This is characteristic of square matrices with $L \times L$ dimensions. Thirdly, the closer to the edge, the closer the zone shape is to square. Fourthly (Table 1), the highest density of nodes forming the percolation channel is near the centre of the matrix. The further away from the centre, the lower the content of these nodes. It is also evident from Figure 2 that a small proportion of the nodes, belonging to one zone, are located between the nodes of neighbouring zones. The shuffling of some of the nodes near the zone boundary is related to the fact that the boundary between zones is unitary and the distribution of percolation threshold values is a random distribution. For example, for a matrix with L = 55, the upper boundary of zone 2 is 999, while the lower boundary of zone 3 is 1000.

Table 1. Ranges of the number of non-conductive nodes forming the percolation channel per node of matrices with dimensions L = 55, 101, 151 in each zone, number of samples 5 × 10^6.

Zone Number	Range of Contents of the Nodes Forming the Percolation Channel			Zone Colour *
	L = 55	L = 101	L = 151	
1	599–0	169–0	69–0	
2	999–600	269–170	112–70	
3	1429–1000	349–270	152–113	
4	1729–1430	429–350	188–153	
5	1927–1730	499–430	222–189	
6	2104–1928	579–500	257–223	
7	2369–2105	649–580	289–258	
8	2519–2370	729–650	322–290	
9	2683–2520	799–730	354–323	
10	3019–2684	938–800	442–355	

* the colours in the table correspond to the colours in the Figure 2 and Figure 4.

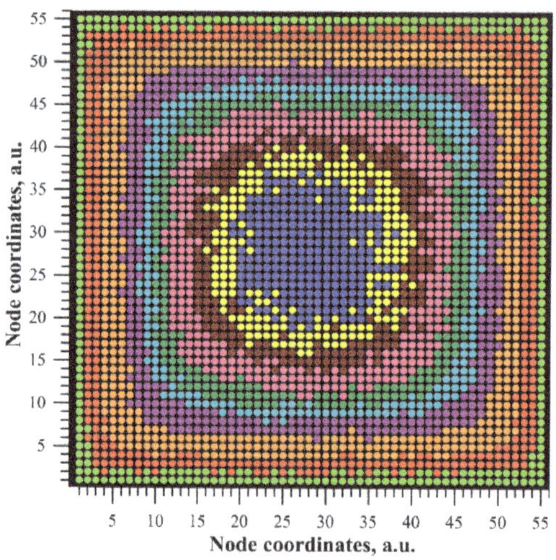

Figure 2. Spatial distribution of the number of non-conductive nodes forming a percolation channel per node of a matrix with L = 55, number of samples 5 × 10^6.

In order to describe the edge phenomenon numerically, Figure 3 shows the spatial distributions of the coordinates of the nodes forming the percolation channel. The distributions were determined along lines passing through the centres of the matrices perpendicular to the edges. The continuous lines in Figure 3 show the polynomial approximations of order six of the simulation results. The values of the coefficients of determination R^2 for the approximation waveforms are close to unity and are 0.9986, 0.9910 and 0.9863 for matrices of dimensions L = 55, 101 and 151, respectively. The closeness to unity of the R^2 coefficients indicates a good-quality approximation. Figure 3 shows that the highest concentration of nodes forming the percolation channel occurs near the centre of the matrix. As one moves away from the centre, the concentration of these nodes decreases. It is also evident from the figure that as the matrix dimensions increase, there is a slowdown in the rate of decrease in the concentration of nodes, forming the percolation threshold, towards the edges.

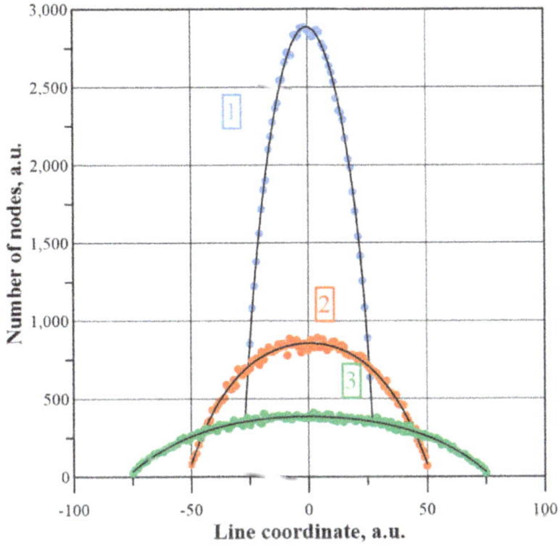

Figure 3. Coordinate distributions of the non-conducting nodes forming the percolation channel along lines passing through the centres of the matrices perpendicular to the sidewalls. 1—matrix $L = 55$, 2—101, 3—151, number of samples 5×10^6. Continuous lines—polynomial approximations of order six.

The intensity of this phenomenon decreases with increasing matrix dimensions. The edge phenomenon observed in this work for percolation in finite-dimensional matrices is very similar, from a qualitative point of view, to the laminar flow of a liquid through a pipe. There too, the highest flow velocity occurs at the centre of the pipe and the lowest at the inner surface [64].

4. Spatial Distributions of Standard Deviation

The figures shown in Table 1 are not useful for comparing standard deviation values. This is because each zone has a different number of nodes forming the percolation channel. Therefore, it proved impossible to compare the standard deviation values of the percolation threshold for different zones.

In order to compare the values of the standard deviations in the different zones, the matrix was divided into 10 zones, each containing an equal number of nodes forming the percolation channel of 5×10^5. To obtain this division, the coordinates of all the nodes of the matrix were recorded in a spreadsheet, together with the numbers of non-conductive nodes forming the percolation channel assigned to them. An ascending sorting of the results was then performed according to the number of non-conductive nodes forming the percolation channel located in the matrix nodes. A summation of the number of non-conductive nodes with the ascending result was then performed. In the zone numbered $1 \leq i \leq 10$, there are coordinates of nodes for which the sum of the number of nodes forming the percolation channel $\Sigma(i)$ is in the range

$$[(i-1) \cdot 5 \cdot 10^5 + 1] \leq \sum(i) \leq i \cdot 5 \cdot 10^5 \quad (1)$$

In this way, 10 zones were created, each containing 5×10^5 nodes, forming a percolation channel. Using a spreadsheet, three-dimensional spatial distributions of the nodes forming the percolation channel were plotted. Figure 4 shows, as an example, the three-dimensional spatial distribution of the nodes, forming the percolation channel for a matrix of size $L = 55$ for a sample number of 5×10^6. In the XY plane, the coordinates of the nodes of the square network are given. On the other hand, the values of the numbers of nodes forming the percolation channel per nodes of the square network were deposited along the vertical Z axis.

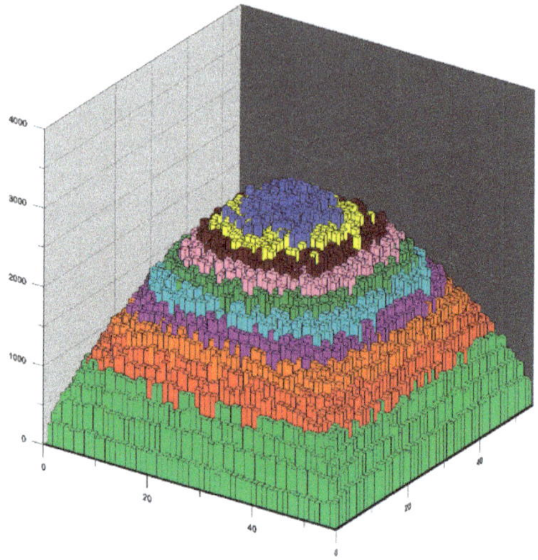

Figure 4. Three-dimensional spatial distribution of the nodes forming the percolation channel for a matrix of dimension $L = 55$ for a sample number of 5×10^6.

Figures 2–4 show that the highest concentration of nodes forming the percolation channel is found in the central part of the matrix. As one approaches the edge of the matrix, the concentration of nodes forming the percolation channe repeatedly decreases.

Based on the division into ten zones containing 5×10^5 nodes forming the percolation channel each, the mean values of the percolation thresholds and standard deviations were calculated, as shown in Table 2.

Table 2. Mean values of percolation thresholds and standard deviations for matrices with $L = 55$, 101 or 151 in each zone and zone colours. Number of samples in each zone: 5×10^5.

Zone Number, i, a.u.	$L = 55$		$L = 101$		$L = 151$	
	Average Value	Standard Deviation	Average Value	Standard Deviation	Average Value	Standard Deviation
1	0.59251	0.02485	0.59267	0.01616	0.59270	0.01208
2	0.59272	0.02467	0.59271	0.01600	0.59270	0.01195
3	0.59271	0.02456	0.59273	0.01600	0.59269	0.01194
4	0.59269	0.02457	0.59265	0.01597	0.59269	0.01192
5	0.59270	0.02450	0.59271	0.01590	0.59272	0.01192
6	0.59275	0.02450	0.59273	0.01590	0.59272	0.01190
7	0.59275	0.02448	0.59269	0.01590	0.59274	0.01188
8	0.59274	0.02440	0.59273	0.01587	0.59274	0.01185
9	0.59274	0.02438	0.59274	0.01587	0.59275	0.01186
10	0.59273	0.02435	0.59273	0.01583	0.59269	0.01186
Whole matrix	0.593	0.025	0.593	0.016	0.593	0.012

Figure 5 shows a histogram of the average values of the percolation threshold for the individual zones for matrices of 55, 101 and 151. The horizontal lines show the average values for the matrices. It can be seen from the figure that the lowest percolation threshold value occurs for zone 1 (outer) for a matrix with $L = 55$. The difference between the average value for the whole matrix and zone 1 is, however, very small, at around 0.00018. In this zone, as the matrix dimensions increase, the percolation threshold value approaches the average value for the whole matrix. In the following zones, random fluctuations in the percolation threshold value are observed. The difference between the smallest and largest results for zones 2 to 10 is less than 0.0001. This means that only for the matrix with $L = 55$ does the edge phenomenon have a slight effect on the percolation threshold value.

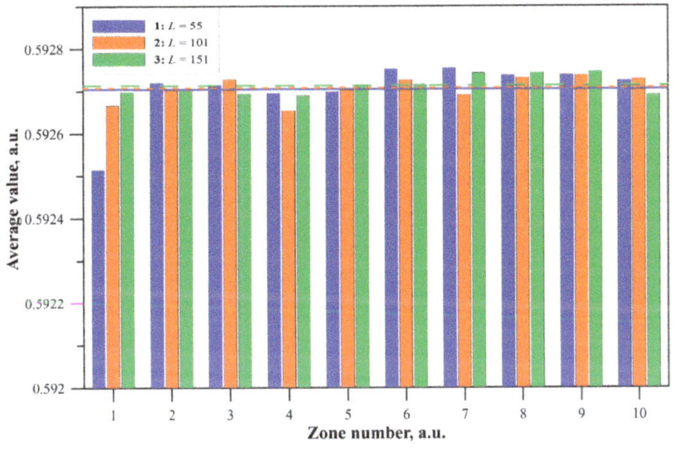

Figure 5. Histogram of percolation threshold values for zones 1 to 10 (Table 2) for matrices with L = 55—1, 101—2 and 151—3, number of samples in each zone 5×10^5. Horizontal lines—average values for matrices.

Figure 6 shows histograms of the standard deviation values for 10 zones containing 5×10^5 samples each and the average values for matrices of dimensions 55, 101 and 151. It can be seen from the figure that the edge phenomenon manifests itself in that the standard deviation values increase as the edge of the matrix is approached. The rate of change of the standard deviation in the zones decreases with increasing die dimensions. The differences in standard deviations between zone 1 and zone 10 are approximately 0.0005 for a die with $L = 51$, 0.00032 for $L = 101$ and 0.00021 for $L = 151$. Based on the analysis of the results for matrices with $L = 51$, 101 and 151, it can be concluded that increasing the die dimensions reduces the effect of the edge phenomenon on the spatial distributions of the standard deviation values.

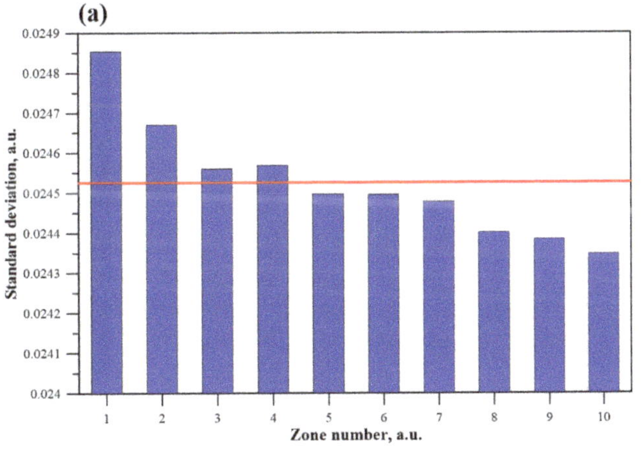

Figure 6. *Cont.*

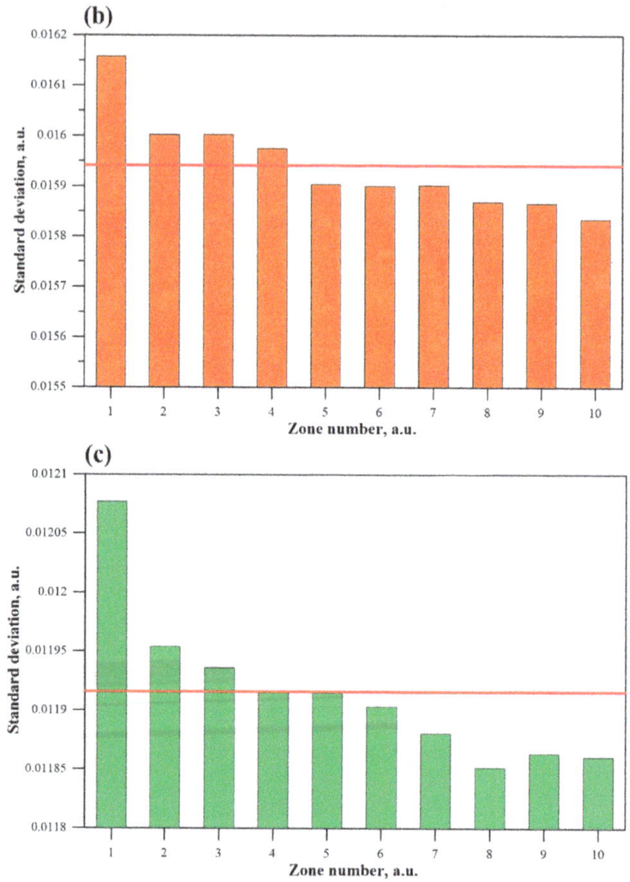

Figure 6. Histogram of the standard deviation values for zones 1 to 10, highlighted in Table 2., (**a**)—matrix with $L = 55$, (**b**)—matrix with $L = 101$, (**c**)—matrix with $L = 151$. Number of samples in each zone 5×10^5. Horizontal line (red colour)—mean value of the standard deviation for the matrix for the number of samples 5×10^6.

5. Analysis of the Size and Number of Non-Conductive Clusters as a Function of Matrix Dimensions

To further clarify the nature of the edge phenomenon, Figure 7, as an example, shows the dependence of the number of clusters consisting of non-conductive nodes on the number of non-conductive nodes inserted into a matrix with $L = 151$. The figure also shows the average number of non-conductive nodes in the clusters.

Figure 7 shows, as an example, the waveforms for 10 randomly selected samples. As can be seen from the figure, initially the clusters consist of single non-conductive nodes. As the number of introduced non-conductive nodes increases, a deviation from a linear relationship begins to occur, caused by an increase in the number of non-conductive nodes in the clusters. This is achieved by attaching the newly drawn non-conductive nodes to the existing ones. When a certain number of introduced non-conductive nodes is reached, the number of clusters reaches a maximum value. For a matrix with $L = 55$, this number is about 6×10^2, for $L = 101$—about 1.8×10^3 and for $L = 151$—about 4×10^3. With a further increase in the number of non-conductive nodes, the number of clusters starts to decrease. At the same time, the average number of non-conductive nodes per cluster increases rapidly. This is because newly introduced nodes connect neighbouring clusters to each other. The percolation threshold is reached when two adjacent clusters, one starting at the left edge of the matrix and the other starting at the right edge, are joined together. The sum of the lengths of these two clusters along the horizontal axis is $L-1$. Placing another non-conducting node between them results in a so-called infinite cluster, extending from one side edge of the matrix to the other. An infinite cluster composed of non-conductive nodes interrupts the possibility of current flow between the top and bottom edges of the matrix. The waveforms in Figure 7 end when the infinite cluster is formed and the percolation threshold is thus reached. It should be noted that a number of smaller clusters remain in the matrices after the percolation threshold is reached. Thus, in a matrix with L = 55, about 50 clusters remain, with an average number of non-conductive nodes in each cluster of about 25. In a matrix with L = 101, about 200 clusters remain, with an average number of nodes of about 25. In contrast, in a matrix with L = 151, the numbers are about 300 and about 30, respectively.

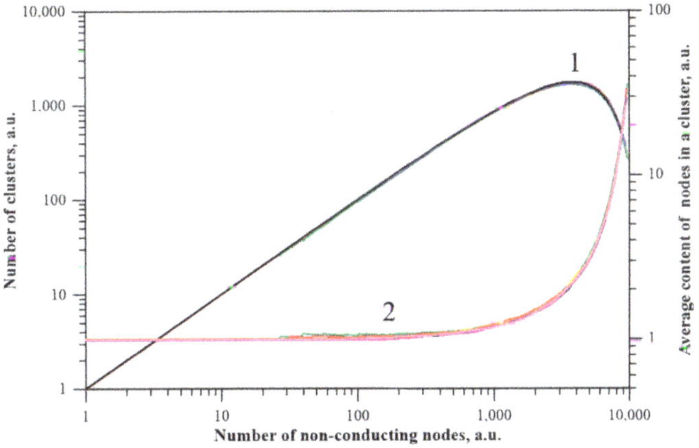

Figure 7. Dependence of the number of clusters consisting of non-conductive nodes—1 (left scale) and the average content of these nodes in a cluster—2 (right scale) on the number of non-conductive nodes for a matrix with L = 151.

Figure 8 shows the length distributions of the left and right clusters preceding the percolation threshold for matrices with L = 55, 101 and 151 for a sample number of 5×10^6. Drawing the next non-conductive node breaks the last percolation channel. Cluster lengths are measured along the horizontal axis: for the left cluster, from the left edge, while for the right cluster, from the right edge.

It can be seen from Figure 8 that the distributions of cluster lengths preceding the occurrence of the percolation threshold by a step for the left and right clusters are equal and symmetrical with respect to the centre of the matrix. This is due to the existence of a fourth-order square axis of symmetry in the matrix, perpendicular to the matrix and passing through its centre. It can be seen from Figure 8 that the larger one of the clusters is, the smaller the other should be. This is due to the fact that the sum of their lengths along the horizontal axis is $L-1$. The probability of there being pairs of clusters with different lengths, as can be seen from Figure 8, is lower than for clusters with similar dimensions. It should be noted that there is a non-zero probability of a cluster of zero length. This is the situation where one of the clusters has length $L-1$. Then the length of the other cluster is zero. This means that the percolation threshold will be reached when the next drawn node is on the side edge of the matrix. Its attachment to an already-existing cluster of length $L-1$ breaks the last percolation channel.

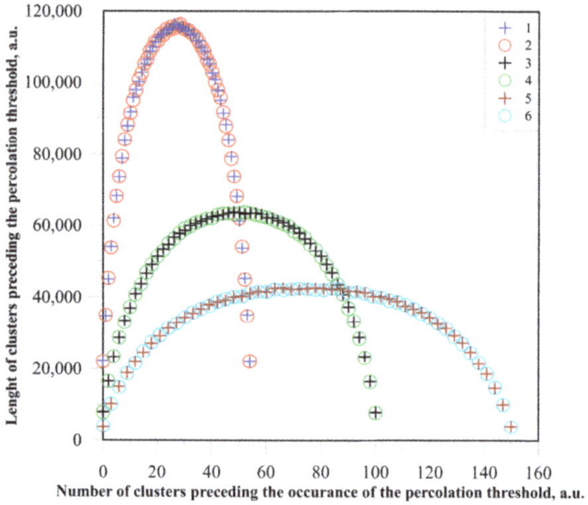

Figure 8. Length distributions of left and right clusters preceding the percolation threshold along the horizontal axis. 1, 3, 5—left clusters, distance calculated from the left edge of the matrix. 2, 4, 6—right clusters, distance calculated from the right edge. Matrices with L = 55—1, 2; 101—3, 4; 151—5, 6; number of samples 5×10^6.

It can be seen from Figure 8 that the left and right clusters are most likely to occur with dimensions that are close to half the dimensions of the matrix. The percolation channel will be broken when the next drawn non-conducting node is placed between these clusters. This means that the highest density of nodes forming the percolation channel is near the centre of the matrix. Similar considerations, given that there is a fourth-order axis of symmetry in square matrices, can be made for the vertical direction. In this direction, the probability of finding the node interrupting the last percolation channel will also be greatest near the centre of the matrix. This results in the highest probability of nodes forming a percolation channel being near the centre of the matrix.

It can be seen from Figures 3 and 8 that as the matrix dimensions increase, the intensity of the edge phenomenon decreases. As the matrix dimensions increase, there is also a decrease in the standard deviation values (Figure 5). This means that the values of the standard deviation and the intensity of the edge phenomenon are clearly related to the finite dimensions of the matrices and decrease with increasing matrix dimensions. It follows that as the dimensions of the matrix increase, the edge phenomenon will gradually disappear. This will result in a decrease in the value of the standard deviation. It should therefore be assumed that for matrices with large dimensions, the edge phenomenon will virtually disappear and the standard deviation values caused by it will tend towards zero.

6. Conclusions

At work, an in-depth analysis of the percolation phenomenon was conducted using computer simulations with the Monte Carlo method for square matrices of dimensions L = 55, 101 and 151. The number of samples for each matrix was 5×10^6. Spatial distributions of nodes breaking the last percolation channel were determined, and two- and three-dimensional density distribution maps of these nodes were created. Based on the spatial distributions of nodes breaking the last percolation channel, it was established that edge effects occur in matrices with finite dimensions. This phenomenon involves a decrease in the concentration of nodes breaking the last percolation channel as one approaches the edge of the matrix. Increasing the dimensions of the matrix slows down the tendency of reducing the number of nodes towards the edge. At the same time, the area with values close to the maximum expands. It was found that the intensity of edge effects and the values of the standard deviation of the percolation threshold decrease with an increase in the matrix dimensions.

The distributions of the lengths of left and right clusters were determined for the situation where the next randomly selected non-conductive node would be situated between them, connecting them and thereby causing the interruption of electrical current flow and reaching the percolation threshold. Cluster lengths were measured along the horizontal axis.

It was determined that the most probable occurrence is the presence of left and right clusters with dimensions close to half the dimensions of the matrix. A similar result, taking into account the fourth-order rotational symmetry in square matrices, can be obtained for the vertical direction. This means that the edge effect is unequivocally related to the probability of the occurrence of pairs of clusters, the connection of which by a non-conductive node breaks the last percolation channel. This probability is highest when the dimensions of each of these two clusters are close to half the dimensions of the matrix and decreases as one approaches the edge of the matrix.

Author Contributions: Conceptualization, P.Z., P.O. and K.K.; methodology, P.Z., P.O. and K.K.; software, P.O. and V.B.; validation, P.O. and K.K.; formal analysis, P.Z., P.O., K.K., P.R. and V.B.; investigation, P.O., K.K., P.R. and V.B.; resources, P.O. and K.K.; data curation, P.O. and V.B.; writing—original draft preparation, P.Z. and P.O.; writing—review and editing, K.K.; visualization, P.O., supervision, P.Z.; funding acquisition, P.O., K.K., P.R. and V.B. All authors have read and agreed to the published version of the manuscript.

Funding: The research was supported by the subsidy of the Ministry of Education and Science (Poland) for the Lublin University of Technology as funds allocated for scientific activities in the scientific discipline of Automation, Electronics, Electrical Engineering and Space Technologies—grants: FD-20/EE-2/701, FD-20/EE-2/702, FD-20/EE-2/705, FD-20/EE-2/707.

Data Availability Statement: Data are contained within the article.

Conflicts of Interest: The authors declare no conflict of interest.

References

1. Broadbent, S.R.; Hammersley, J.M. Percolation Processes. *Math. Proc. Camb. Philos. Soc.* **1957**, *53*, 629–641. [CrossRef]
2. Osetsky, Y.; Barashev, A.V.; Zhang, Y. Sluggish, Chemical Bias and Percolation Phenomena in Atomic Transport by Vacancy and Interstitial Diffusion in Ni Fe Alloys. *Curr. Opin. Solid. State Mater. Sci.* **2021**, *25*, 100961. [CrossRef]
3. Jiang, J.; Yu, X.; Lin, Y.; Guan, Y. PercolationDF: A Percolation-Based Medical Diagnosis Framework. *Math. Biosci. Eng.* **2022**, *19*, 5832–5849. [CrossRef] [PubMed]
4. Sahimi, M. Percolation in Biological Systems. In *Applied Mathematical Sciences (Switzerland)*; Springer: Berlin/Heidelberg, Germany, 2023; Volume 213, pp. 443–488.
5. Devpura, A.; Phelan, P.E.; Prasher, R.S. Percolation Theory Applied to the Analysis of Thermal Interface Materials in Flip-Chip Technology. In Proceedings of the ITHERM 2000. The Seventh Intersociety Conference on Thermal and Thermomechanical Phenomena in Electronic Systems (Cat. No.00CH37069), Las Vegas, NV, USA, 6 August 2002; Volume 1, pp. 21–28.
6. Evseev, V.A.; Konopleva, R.F.; Skal, A.S. Percolation in Semiconductors with Disordered Regions: Electrical Conductivity and Hall Coefficient. *Radiat. Eff.* **1982**, *66*, 167–172. [CrossRef]
7. Kirkpatrick, S. Percolation and Conduction. *Rev. Mod. Phys.* **1973**, *45*, 574–588. [CrossRef]
8. Tran, A.K.; Sapkota, A.; Wen, J.; Li, J.; Takei, M. Linear Relationship between Cytoplasm Resistance and Hemoglobin in Red Blood Cell Hemolysis by Electrical Impedance Spectroscopy & Eight-Parameter Equivalent Circuit. *Biosens. Bioelectron.* **2018**, *119*, 103–109. [CrossRef]
9. Santoso, S.; Beaty, H.W. *Standard Handbook for Electrical Engineers*; McGraw-Hill Education: London, UK, 2018; ISBN 1259642585.
10. *ISO–ISO/TS 80004-2:2015*; Nanotechnologies—Vocabulary—Part 2: Nano-Objects. ISO: Geneva, Switzerland, 2022. Available online: https://www.iso.org/standard/54440.html (accessed on 5 January 2022).
11. Du, A.-K.; Yang, K.-L.; Zhao, T.-H.; Wang, M.; Zeng, J.-B. Poly(Sodium 4-Styrenesulfonate) Wrapped Carbon Nanotube with Low Percolation Threshold in Poly(ε-Caprolactone) Nanocomposites. *Polym. Test.* **2016**, *51*, 40–48. [CrossRef]
12. Münstedt, H.; Starý, Z. Is Electrical Percolation in Carbon-Filled Polymers Reflected by Rheological Properties? *Polymers* **2016**, *98*, 51–60. [CrossRef]
13. Tu, Z.; Wang, J.; Yu, C.; Xiao, H.; Jiang, T.; Yang, Y.; Shi, D.; Mai, Y.-W.; Li, R.K.Y. A Facile Approach for Preparation of Polystyrene/Graphene Nanocomposites with Ultra-Low Percolation Threshold through an Electrostatic Assembly Process. *Compos. Sci. Technol.* **2016**, *134*, 49–56. [CrossRef]

14. Yang, K.; Huang, X.; Fang, L.; He, J.; Jiang, P. Fluoro-Polymer Functionalized Graphene for Flexible Ferroelectric Polymer-Based High-k Nanocomposites with Suppressed Dielectric Loss and Low Percolation Threshold. *Nanoscale* **2014**, *6*, 14740–14753. [CrossRef]
15. Koltunowicz, T.N.; Zukowski, P.; Boiko, O.; Saad, A.; Fedotova, J.A.; Fedotov, A.K.; Larkin, A.V.; Kasiuk, J. AC Hopping Conductance in Nanocomposite Films with Ferromagnetic Alloy Nanoparticles in a $PbZrTiO_3$ Matrix. *J. Electron. Mater.* **2015**, *44*, 2260–2268. [CrossRef]
16. Koltunowicz, T.N.; Zukowski, P.; Czarnacka, K.; Svito, I.; Fedotov, A.K. Percolation Phenomena in $Cu_x(SiO_y)_{100-x}$ Nanocomposite Films Produced by Ion Beam-Sputtering. *Acta Phys. Pol. A* **2015**, *128*, 908–912. [CrossRef]
17. Serrano, M.Á.; Boguñá, M. Percolation and Epidemic Thresholds in Clustered Networks. *Phys. Rev. Lett.* **2006**, *97*, 088701. [CrossRef]
18. Sander, L.M.; Warren, C.P.; Sokolov, I.M.; Simon, C.; Koopman, J. Percolation on Heterogeneous Networks as a Model for Epidemics. *Math. Biosci.* **2002**, *180*, 293–305. [CrossRef] [PubMed]
19. Moore, C.; Newman, M.E.J. Epidemics and Percolation in Small-World Networks. *Phys. Rev. E* **2000**, *61*, 5678–5682. [CrossRef] [PubMed]
20. Li, D.; Zhang, Q.; Zio, E.; Havlin, S.; Kang, R. Network Reliability Analysis Based on Percolation Theory. *Reliab. Eng. Syst. Saf.* **2015**, *142*, 556–562. [CrossRef]
21. Beer, T.; Enting, I.G. Fire Spread and Percolation Modelling. *Math. Comput. Model.* **1990**, *13*, 77–96. [CrossRef]
22. Duane, A.; Miranda, M.D.; Brotons, L. Forest Connectivity Percolation Thresholds for Fire Spread under Different Weather Conditions. *Ecol. Manag.* **2021**, *498*, 119558. [CrossRef]
23. Yang, M.; Bruck, H.A.; Kostov, Y.; Rasooly, A. Biological Semiconductor Based on Electrical Percolation. *Anal. Chem.* **2010**, *82*, 3567–3572. [CrossRef]
24. Katunin, A.; Krukiewicz, K. Electrical Percolation in Composites of Conducting Polymers and Dielectrics. *J. Polym. Eng.* **2015**, *35*, 731–741. [CrossRef]
25. Forero-Sandoval, I.Y.; Franco-Bacca, A.P.; Cervantes-Álvarez, F.; Gómez-Heredia, C.L.; Ramírez-Rincón, J.A.; Ordonez-Miranda, J.; Alvarado-Gil, J.J. Electrical and Thermal Percolation in Two-Phase Materials: A Perspective. *J. Appl. Phys.* **2022**, *131*, 230901. [CrossRef]
26. Otten, R.H.J.; Van Der Schoot, P. Continuum Percolation of Polydisperse Nanofillers. *Phys. Rev. Lett.* **2009**, *103*, 225704. [CrossRef] [PubMed]
27. Ukshe, A.; Glukhov, A.; Dobrovolsky, Y. Percolation Model for Conductivity of Composites with Segregation of Small Conductive Particles on the Grain Boundaries. *J. Mater. Sci.* **2020**, *55*, 6581–6587. [CrossRef]
28. Borisova, A.; Machulyansky, A.; Yakimenko, Y.; Bovtun, V.; Kempa, M.; Savinov, M. Broadband Dielectric and Conductivity Spectra of Dielectric—Metal Nanocomposites for Microwave Applications. In Proceedings of the 2013 IEEE XXXIII International Scientific Conference Electronics and Nanotechnology (ELNANO), Kiev, Ukraine, 16–19 April 2013; pp. 21–24.
29. Brouers, F.; Granovsky, A.; Sarychev, A.; Kalitsov, A. The Influence of Boundary Scattering on Transport Phenomena in Ferromagnetic Metal—Dielectric Nanocomposites. *Phys. A Stat. Mech. Its Appl.* **1997**, *241*, 284–288. [CrossRef]
30. Zukowski, P.; Koltunowicz, T.N.; Bondariev, V.; Fedotov, A.K.; Fedotova, J.A. Determining the Percolation Threshold for $(FeCoZr)_x(CaF_2)_{(100-x)}$ Nanocomposites Produced by Pure Argon Ion-Beam Sputtering. *J. Alloys Compd.* **2016**, *683*, 62–66. [CrossRef]
31. Żukowski, P.; Kołtunowicz, T.; Partyka, J.; Fedotova, Y.A.; Larkin, A.V. Hopping Conductivity of Metal-Dielectric Nanocomposites Produced by Means of Magnetron Sputtering with the Application of Oxygen and Argon Ions. *Vacuum* **2009**, *83*, S280–S283. [CrossRef]
32. Kołtunowicz, T.N.; Zhukowski, P.; Fedotova, V.V.; Saad, A.M.; Larkin, A.V.; Fedotov, A.K. The Features of Real Part of Admittance in the Nanocomposites $(Fe_{45}Co_{45}Zr_{10})_x(Al_2O_3)_{100-x}$ Manufactured by the Ion-Beam Sputtering Technique with Ar Ions. *Acta Phys. Pol. A* **2011**, *120*, 35–38. [CrossRef]
33. Koltunowicz, T.N. Dielectric Properties of $(CoFeZr)_x(PZT)_{100-x}$ Nanocomposites Produced with a Beam of Argon and Oxygen Ions. *Acta Phys. Pol. A* **2014**, *125*, 1412–1415. [CrossRef]
34. Kołtunowicz, T.N.; Bondariev, V.; Żukowski, P.; Fedotova, J.A.; Fedotov, A.K. AC Electrical Resonances in Nanocomposites with Partly Oxidized FeCoZr Grains Embedded in CaF_2 Ceramic Matrix—Effects of Annealing. *J. Alloys Compd.* **2020**, *819*, 153361. [CrossRef]
35. Webman, I.; Jortner, J.; Cohen, M.H. Numerical Simulation of Continuous Percolation Conductivity. *Phys. Rev. B* **1976**, *14*, 4737–4740. [CrossRef]
36. Qiao, R.; Catherine Brinson, L. Simulation of Interphase Percolation and Gradients in Polymer Nanocomposites. *Compos. Sci. Technol.* **2009**, *69*, 491–499. [CrossRef]
37. Charlaix, E. Percolation Threshold of a Random Array of Discs: A Numerical Simulation. *J. Phys. A Math. Gen.* **1986**, *19*, L533–L536. [CrossRef]
38. Jacobsen, J.L. High-Precision Percolation Thresholds and Potts-Model Critical Manifolds from Graph Polynomials. *J. Phys. A Math. Theor.* **2014**, *47*, 135001. [CrossRef]
39. Jacobsen, J.L. Critical Points of Potts and O(*N*) Models from Eigenvalue Identities in Periodic Temperley–Lieb Algebras. *J. Phys. A Math. Theor.* **2015**, *48*, 454003. [CrossRef]

40. Newman, M.E.J.; Ziff, R.M. Efficient Monte Carlo Algorithm and High-Precision Results for Percolation. *Phys. Rev. Lett.* **2000**, *85*, 4104–4107. [CrossRef] [PubMed]
41. de Oliveira, P.M.C.; Nóbrega, R.A.; Stauffer, D. Corrections to Finite Size Scaling in Percolation. *Braz. J. Phys.* **2003**, *33*, 616–618. [CrossRef]
42. Kim, S.; Gholamirad, F.; Yu, M.; Park, C.M.; Jang, A.; Jang, M.; Taheri-Qazvini, N.; Yoon, Y. Enhanced Adsorption Performance for Selected Pharmaceutical Compounds by Sonicated Ti_3C_2TX MXene. *Chem. Eng. J.* **2021**, *406*, 126789. [CrossRef]
43. Gogotsi, Y.; Anasori, B. The Rise of MXenes. *ACS Nano* **2019**, *13*, 8491–8494. [CrossRef]
44. Xu, Z. Fundamental Properties of Graphene. In *Graphene*; Elsevier: Amsterdam, The Netherlands, 2018; pp. 73–102.
45. Zhen, Z.; Zhu, H. Structure and Properties of Graphene. In *Graphene*; Elsevier: Amsterdam, The Netherlands, 2018; pp. 1–12.
46. Akhtar, M.; Anderson, G.; Zhao, R.; Alruqi, A.; Mroczkowska, J.E.; Sumanasekera, G.; Jasinski, J.B. Recent Advances in Synthesis, Properties, and Applications of Phosphorene. *NPJ 2d Mater. Appl.* **2017**, *1*, 5. [CrossRef]
47. Shahzad, F.; Alhabeb, M.; Hatter, C.B.; Anasori, B.; Man Hong, S.; Koo, C.M.; Gogotsi, Y. Electromagnetic Interference Shielding with 2D Transition Metal Carbides (MXenes). *Science* **2016**, *353*, 1137–1140. [CrossRef]
48. Bhimanapati, G.R.; Glavin, N.R.; Robinson, J.A. 2D Boron Nitride. In *Semiconductors and Semimetals*; Academic Press Inc.: Cambridge, MA, USA, 2016; Volume 95, pp. 101–147.
49. Li, X.; Zhu, H. Two-Dimensional MoS2: Properties, Preparation, and Applications. *J. Mater.* **2015**, *1*, 33–44. [CrossRef]
50. Ling, Z.; Ren, C.E.; Zhao, M.-Q.; Yang, J.; Giammarco, J.M.; Qiu, J.; Barsoum, M.W.; Gogotsi, Y. Flexible and Conductive MXene Films and Nanocomposites with High Capacitance. *Proc. Natl. Acad. Sci. USA* **2014**, *111*, 16676–16681. [CrossRef] [PubMed]
51. Mashtalir, O.; Naguib, M.; Mochalin, V.N.; Dall'Agnese, Y.; Heon, M.; Barsoum, M.W.; Gogotsi, Y. Intercalation and Delamination of Layered Carbides and Carbonitrides. *Nat. Commun.* **2013**, *4*, 1716. [CrossRef] [PubMed]
52. Naguib, M.; Mashtalir, O.; Carle, J.; Presser, V.; Lu, J.; Hultman, L.; Gogotsi, Y.; Barsoum, M.W. Two-Dimensional Transition Metal Carbides. *ACS Nano* **2012**, *6*, 1322–1331. [CrossRef]
53. Naguib, M.; Kurtoglu, M.; Presser, V.; Lu, J.; Niu, J.; Heon, M.; Hultman, L.; Gogotsi, Y.; Barsoum, M.W. Two-Dimensional Nanocrystals Produced by Exfoliation of Ti_3AlC_2. *Adv. Mater.* **2011**, *23*, 4248–4253. [CrossRef] [PubMed]
54. Berger, C.; Song, Z.; Li, T.; Li, X.; Ogbazghi, A.Y.; Feng, R.; Dai, Z.; Marchenkov, A.N.; Conrad, E.H.; First, P.N.; et al. Ultrathin Epitaxial Graphite: 2D Electron Gas Properties and a Route toward Graphene-Based Nanoelectronics. *J. Phys. Chem. B* **2004**, *108*, 19912–19916. [CrossRef]
55. Novoselov, K.S.; Geim, A.K.; Morozov, S.V.; Jiang, D.; Zhang, Y.; Dubonos, S.V.; Grigorieva, I.V.; Firsov, A.A. Electric Field Effect in Atomically Thin Carbon Films. *Science* **2004**, *306*, 666–669. [CrossRef]
56. Kołtunowicz, T.N.; Gałaszkiewicz, P.; Kierczyński, K.; Rogalski, P.; Okal, P.; Pogrebnjak, A.D.; Buranich, V.; Pogorielov, M.; Diedkova, K.; Zahorodna, V.; et al. Investigation of AC Electrical Properties of MXene-PCL Nanocomposites for Application in Small and Medium Power Generation. *Energies* **2021**, *14*, 7123. [CrossRef]
57. Mott, N.F. Electronic Process in Non-Crystalline Materials. *J. Non Cryst. Solids* **1968**, *1*, 1. [CrossRef]
58. Diedkova, K.; Pogrebnjak, A.D.; Kyrylenko, S.; Smyrnova, K.; Buranich, V.V.; Horodek, P.; Zukowski, P.; Koltunowicz, T.N.; Galaszkiewicz, P.; Makashina, K.; et al. Polycaprolactone-MXene Nanofibrous Scaffolds for Tissue Engineering. *ACS Appl. Mater. Interfaces* **2023**, *15*, 14033–14047. [CrossRef]
59. Zukowski, P.; Okal, P.; Kierczynski, K.; Rogalski, P.; Borucki, S.; Kunicki, M.; Koltunowicz, T.N. Investigations into the Influence of Matrix Dimensions and Number of Iterations on the Percolation Phenomenon for Direct Current. *Energies* **2023**, *16*, 7128. [CrossRef]
60. Matsumoto, M.; Nishimura, T. Mersenne Twister. *ACM Trans. Model. Comput. Simul.* **1998**, *8*, 3–30. [CrossRef]
61. Dean, P. A New Monte Carlo Method for Percolation Problems on a Lattice. *Math. Proc. Camb. Philos. Soc.* **1963**, *59*, 397–410. [CrossRef]
62. Noel, K. Analysis of Random Generators in Monte Carlo Simulation: Mersenne Twister and Sobol. *SSRN Electron. J.* **2016**. [CrossRef]
63. Wilkinson, L.; Friendly, M. The History of the Cluster Heat Map. *Am. Stat.* **2009**, *63*, 179–184. [CrossRef]
64. Tummers, M.J.; Schenker-van Rossum, M.C.; Delfos, R.; Twerda, A.; Westerweel, J. Turbulent flow and friction in a pipe with repeated rectangular ribs. *Exp. Fluids* **2023**, *64*, 160. [CrossRef]

energies

MDPI

Article

Monte Carlo Simulation of Percolation Phenomena for Direct Current in Large Square Matrices

Pawel Zukowski [1], Pawel Okal [2], Konrad Kierczynski [2,*], Przemyslaw Rogalski [2], Vitalii Bondariev [2] and Alexander D. Pogrebnjak [3]

1 Department of Economics, Vincent Pol University in Lublin, 2, Choiny Str., 20-816 Lublin, Poland; zukowski50pawel@gmail.com
2 Department of Electrical Devices and High Voltage Technology, Faculty of Electrical Engineering and Computer Science, Lublin University of Technology, 38A, Nadbystrzycka Str., 20-618 Lublin, Poland; p.okal@pollub.pl (P.O.); p.rogalski@pollub.pl (P.R.); v.bondariev@pollub.pl (V.B.)
3 Faculty of Electronics and Information Technology, Sumy State University, 2, Rymskogo-Korsakova Str., 40007 Sumy, Ukraine; alexp@i.ua
* Correspondence: k.kierczynski@pollub.pl; Tel.: +48-81-538-4328

Abstract: In this study, an in-depth analysis of the percolation phenomenon for square matrices with dimensions from $L = 50$ to 600 for a sample number of 5×10^4 was performed using Monte Carlo computer simulations. The percolation threshold value was defined as the number of conductive nodes remaining in the matrix before drawing the node interrupting the last percolation channel, in connection with the overall count of nodes within the matrix. The distributions of percolation threshold values were found to be normal distributions. The dependencies of the expected value (mean) of the percolation threshold and the standard deviation of the dimensions of the matrix were determined. It was established that the standard deviation decreased with the increase in matrix dimensions, ranging from 0.0262253 for a matrix with $L = 50$ to 0.0044160 for $L = 600$, which is almost six-fold lower. The mean value of the percolation threshold was practically constant and amounted to approximately 0.5927. The analysis involved not only the spatial distributions of nodes interrupting the percolation channels but also the overall patterns of node interruption in the matrix. The distributions revealed an edge phenomenon within the matrices, characterized by the maximum concentration of nodes interrupting the final percolation channel occurring at the center of the matrix. As they approached the edge of the matrix, their concentration decreased. It was established that increasing the dimensions of the matrix slowed down the rate of decrease in the number of nodes towards the edge. In doing so, the area in which values close to the maximum occurred was expanded. Based on the approximation of the experimental results, formulas were determined describing the spatial distributions of the nodes interrupting the last percolation channel and the values of the standard deviation from the matrix dimensions. The relationships obtained showed that with increasing matrix dimensions, the edge phenomenon should gradually disappear, and the percolation threshold standard deviation values caused by it will tend towards zero.

Keywords: percolation phenomenon; percolation threshold; uncertainty of measurement; metrological approach; computer simulation; Monte Carlo method

Citation: Zukowski, P.; Okal, P.; Kierczynski, K.; Rogalski, P.; Bondariev, V.; Pogrebnjak, A.D. Monte Carlo Simulation of Percolation Phenomena for Direct Current in Large Square Matrices. *Energies* **2023**, *16*, 8024. https://doi.org/10.3390/en16248024

Academic Editor: Hervé Morel

Received: 3 November 2023
Revised: 4 December 2023
Accepted: 7 December 2023
Published: 12 December 2023

1. Introduction

The phenomenon of percolation, research of which has been rapidly gaining momentum since the 1950s, is used in the description of various systems and phenomena [1–9]. It is a critical phenomenon describing phase transitions of the second type, the general idea of which is that in any medium made of two phases, elements of one phase that are independent due to random factors create a more complicated structure, which results in a change in the macroscopic properties of the entire medium because of a change in the concentration of the dispersed phase [10]. Thus, percolation theory encompasses models for both planar and three-dimensional random processes, along with the impact of their interplay. It serves to characterize systems featuring stochastic geometry and systems that are topologically disorganized. This is illustrated by the model of S.R. Broadbent and J.M. Hammersley, who, as precursors of the percolation theory, created the first stochastic model, which was an arbitrary network with a finite number of nodes [11]. Nodes in such a model are assigned two states, open or closed, and their number is marked as x. The probability p is the probability of their arrangement; therefore, there is a value of x for which the probability of percolation is non-zero, because the number of nodes creating the percolation channel is sufficient to create a continuous connection. This value is designated x_c and is called the percolation threshold [12,13]. Due to the adopted assumptions of the model, percolation phenomena are divided into two groups. If the states between nodes are assigned to the connections between them, then it is a bond percolation, while when the states are assigned to the entire nodes, it is a site percolation [14]. Due to the random nature of the phenomenon, the percolation threshold cannot be determined based on a single modeling.

Because in both cases the centers are discrete sets, their analysis is possible using computer simulation based on the Monte Carlo method [15]. Over an extended duration, investigations employing the simulation of the percolation phenomenon in two-dimensional networks exhibiting translational symmetry have been employed for the theoretical analysis of this phenomenon and, in certain instances, for its visual representation. Subsequent investigations in this realm have involved the increasingly precise determination of the percolation threshold value [16–21]. The record result for the uncertainty of determining the percolation threshold for square networks was obtained in publication [22], the value of which is approximately 10^{-8}. Nevertheless, such a high level of precision is unnecessary, for instance, in the analysis of experimental results, where an uncertainty of approximately 10^{-3} or 0.1% is sufficient.

In recent decades, a new field of materials has emerged, called 2D materials. These are two-dimensional materials with translational symmetry, consisting of flat lattices ranging in thickness from one to several atomic dimensions. Such materials are graphene, MXenes and a few others; see, for example, [23–38]. The examination of the dispersion of inclusions leading to conductivity through tunneling aligns precisely with the investigation of the percolation phenomenon in two-dimensional matrices exhibiting translational symmetry. Consequently, studying percolation in 2D lattice systems with translational symmetry, coupled with the exploration and advancement of technologies for fabricating 2D materials, could gain both applicative and practice attributes.

In the study cited as reference [39], researchers employed Monte Carlo computer simulations to investigate the percolation phenomenon in square matrices of different sizes. They analyzed a vast range of samples, varying from 5×10^4 to 5×10^6. Their findings revealed that the average percolation threshold remained nearly constant across different matrix sizes. However, a notable decrease, over two-fold, was observed in the standard deviation as the size of the matrix increased. In [40], it was established that in matrices with dimensions L = 55, 101 and 151, an edge phenomenon manifests, characterized by a reduction in the concentration of nodes forming a percolation channel as one approaches the edge of the matrix.

The aim of this work was an in-depth analysis of the parameters of the edge phenomenon for the percolation threshold in square matrices over a wide range of dimensions, from relatively small ones of L = 50 to large ones of L = 600, with the number of samples being 5×10^4 for each matrix. The residual goals of the work were:

- the determination of the two-dimensional spatial distributions of node coordinates interrupting the last percolation channel for square base matrices with dimensions from L = 50 to 600 for 5×10^4 samples;
- the determination of a formula describing the influence of matrix dimensions on the standard deviation values, based on the approximation of experimental results;

- the determination of the relationships describing the spatial distribution of nodes interrupting the last percolation channel depending on the dimensions of the matrix;
- the determination of the formula describing the maximum rate of reduction in the concentration of nodes interrupting the last percolation channel towards the edge of the matrix;
- the determination of the correlation between the intensity of the edge phenomenon for the distribution of nodes interrupting the last percolation channel and the value of the standard deviation of the percolation threshold depending on the dimensions of the matrix.

2. Research Method

An in-depth analysis of the parameters of the edge phenomenon for the percolation threshold of square matrices in a wide range of dimensions, which was the aim of this work, was performed using computer simulation. Both the algorithm itself and the program were described in [39]. Generally, the algorithm was based on set theory where nodes were described using Cartesian coordinates. Using a pseudorandom number generator, the program sequentially changed the states of the nodes while controlling and recording the evolution of cluster formation and the overall geometry of the simulated matrix. Moreover, when an infinite cluster was obtained, it recorded information about the percolation threshold value and the coordinates of the critical nodes that formed the percolation channel for further analysis. This process was repeated independently 5×10^4 times for matrices of size L = 50, 75, 100, 125, 150, 175, 200, 250, 300, 350, 400, 500 and 600.

3. Research and Analysis of Basic Percolation Parameters Depending on the Dimensions of the Matrix

3.1. Determination of the Parameters of the Probability Distribution of Percolation Thresholds Depending on the Dimensions of the Matrix

In this work, simulations were performed for matrices from L = 50 to L = 200 with a step of 25, from L = 200 to L = 400 with a step of 50, and for L = 500 and 600. In article [39], it was found that the standard deviation values of the percolation threshold for the range of simulation samples below 10^4 were unsatisfactory. Their stabilization occurred for sample numbers above 10^4. Therefore, for the purposes of the present work, the number of samples for each matrix was chosen to be 5×10^4. Based on simulations for each sample, the percolation threshold value and the coordinates of the node interrupting the final percolation channel were determined. The standard deviation value was also determined for each matrix. Based on the values obtained, histograms of the distribution of percolation threshold values were made and normal distributions were determined. These results are shown for selected matrices in Figures 1–3. Figure 4 shows the normal distributions for selected matrices with dimensions ranging from 50 to 600.

Figure 1 shows that there was an edge phenomenon in the matrices, such that the nodes that formed the percolation threshold were concentrated in the central part of the matrix. As each edge was approached, the number of these nodes decreased. In the central part, the number of nodes forming the percolation threshold was about an order of magnitude higher than that near the edge.

From an analysis of Figure 4 and Table 1, it is evident that the average percolation threshold value (indicated by the peak of the waveforms) remained relatively unchanged, considering the uncertainty range, regardless of the matrix size. Conversely, the standard deviation value (reflected in the breadth of the waveforms) showed a decreasing trend as the dimensions of the matrix increased. This resulted in an increase in the maximum value of the normal distributions.

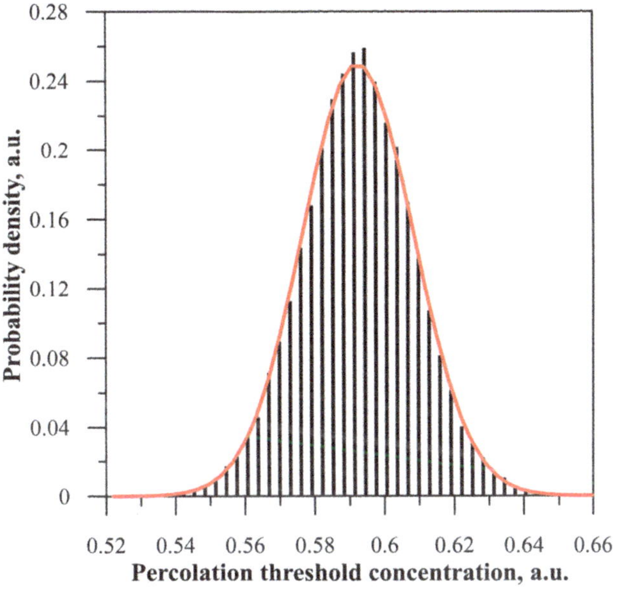

Figure 1. A graph showing the frequency distribution of percolation threshold values for a set of 5×10^4 samples in a matrix with dimensions $L = 100$, accompanied by a curve representing the normal distribution, depicted as a solid line.

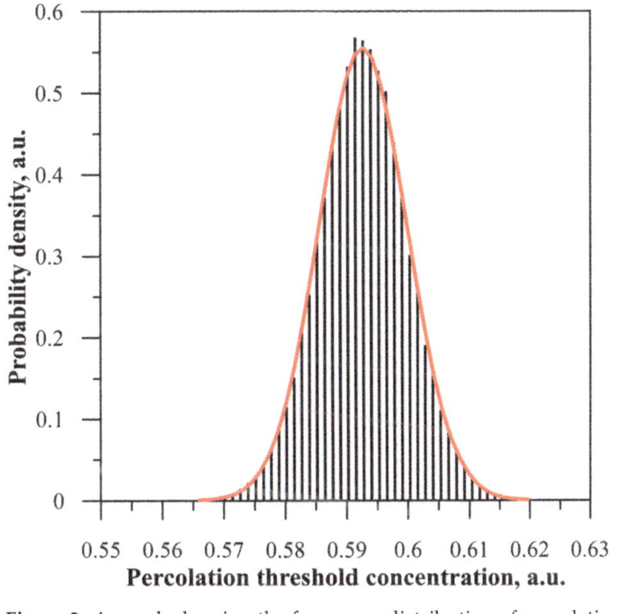

Figure 2. A graph showing the frequency distribution of percolation threshold values for a set of 5×10^4 samples in a matrix with dimensions $L = 300$, accompanied by a curve representing the normal distribution, depicted as a solid line.

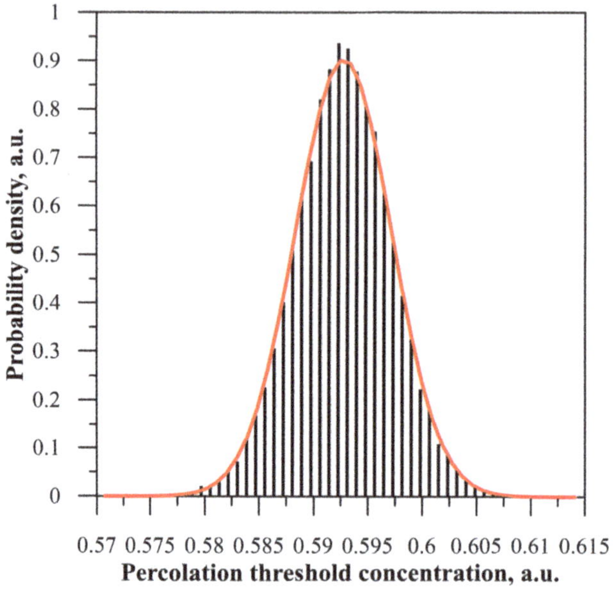

Figure 3. A graph showing the frequency distribution of percolation threshold values for a set of 5×10^4 samples in a matrix with dimensions L = 600, accompanied by a curve representing the normal distribution, depicted as a solid line.

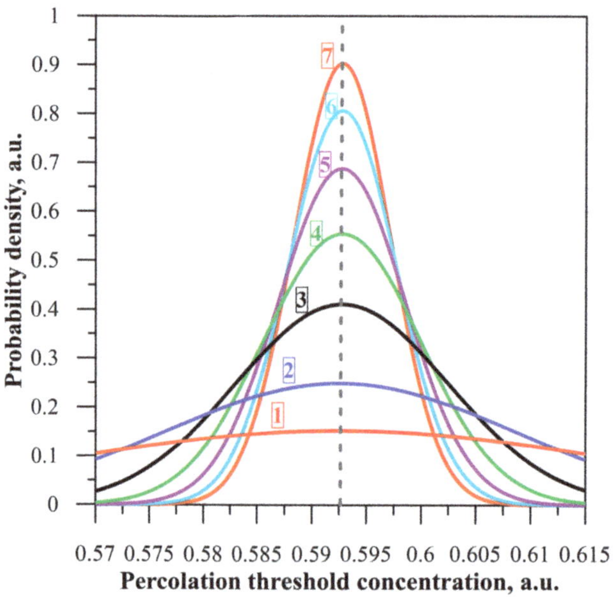

Figure 4. Normal distributions for L: 1—50, 2—100, 3—200, 4—300, 5—400, 6—500 and 7—600. Number of samples: 5×10^4.

Table 1. Dependence of percolation threshold mean value, standard deviation and coefficient of determination R^2 on matrix dimensions for 5×10^4 samples.

Matrix Dimensions L	Percolation Threshold Value, a.u.	Standard Deviation, a.u.	Coefficient of Determination R^2, a.u.
50	0.592187	0.0262253	0.99392
75	0.592663	0.0197870	0.99476
100	0.592443	0.0159748	0.99631
125	0.592767	0.0137418	0.99679
150	0.592616	0.0119424	0.99826
175	0.592718	0.0106823	0.99733
200	0.592726	0.0097261	0.99805
250	0.592751	0.0082911	0.99775
300	0.592732	0.0071884	0.99811
350	0.592717	0.0063989	0.99894
400	0.592713	0.0058001	0.99786
500	0.592726	0.0049406	0.99813
600	0.592738	0.0044160	0.99771

Using least-squares estimation, the R^2 coefficients of determination were determined for the normal distribution approximation of the percolation threshold values for matrices of different dimensions. The value of the R^2 coefficient for all matrices was greater than 0.993. This demonstrated the high accuracy of the approximation of the results using a normal distribution and that the percolation threshold values were random values. The results obtained are shown in Table 1. From the analysis of the dependence of the mean value of the percolation threshold on the matrix dimensions shown in Figure 5, it can be seen that for matrix dimensions of $L \leq 250$, the mean value varied from about 0.592187 to about 0.592767, which is about ±0.00058 or ±0.0979%. For values of $L > 250$ and above, the waveform became close to steady state and the variation decreased and ranged from 0.592712 to 0.592737, which is only ±0.000025 or ±0.0042%. This means that for $L > 250$, the fluctuations in the mean value of the percolation threshold decreased by more than 20-fold compared to the range with lower matrix sizes and were virtually negligible.

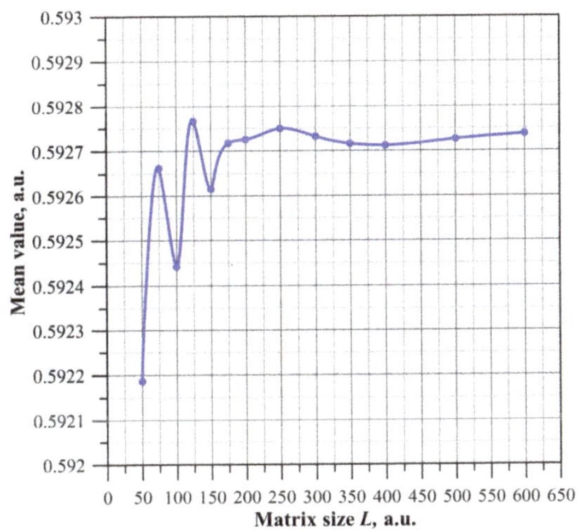

Figure 5. Dependence of mean value on matrix size for 5×10^4 samples.

Figure 6 illustrates how the value of the percolation threshold and its standard deviation varied with the size of the matrix. The data presented in Figure 6 clearly indicate that the value of the percolation threshold remained largely unaffected by changes in the dimensions of the matrix. In contrast, the standard deviation and the associated uncertainty of the percolation threshold decreased with increasing matrix dimensions.

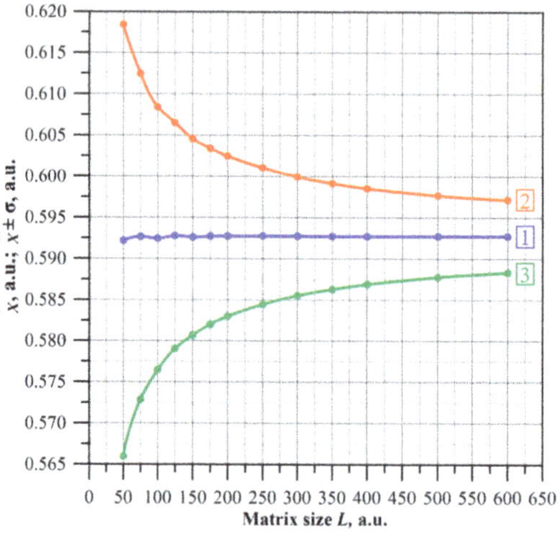

Figure 6. Dependence of percolation threshold mean value (1), mean + standard deviation (2) and mean − standard deviation (3) on matrix size L for 5×10^4 samples.

Figure 7 shows the dependence of the standard deviation on the matrix dimensions. From it, the dependence of the power approximation function on the matrix dimension L was obtained, given by the formula:

$$\sigma(L) \approx 0.045L^{-0.75}. \quad (1)$$

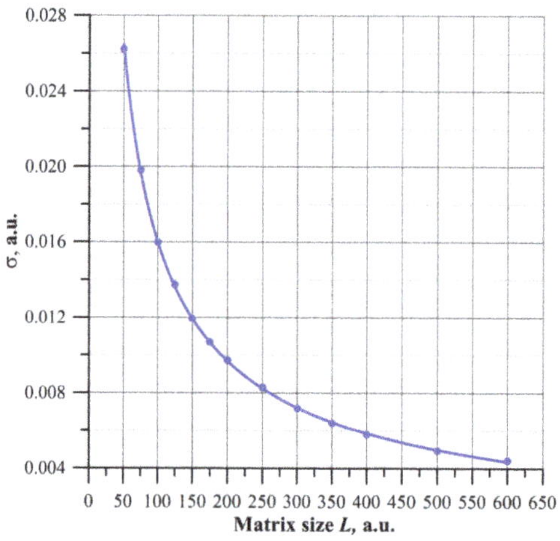

Figure 7. Dependence of the standard deviation on matrix size L for 5×10^4 samples.

The quality of the approximation, shown in Figure 7, is very good, as evidenced by the value of the coefficient of determination R^2 = 0.99982. This value is close to unity, and the difference between unity and R^2 is only 0.00018. It can be seen from Formula (1) that the standard deviation tends towards zero as the matrix dimension L increases.

3.2. The Edge Phenomenon of Node Coordinates Interrupting the Last Percolation Channel

Figure 6 demonstrates that the average percolation threshold value remained essentially constant, regardless of the size of the matrix. Conversely, as depicted in Figure 7, the standard deviation value showed a significant decrease as the dimensions of the matrix increased. Figure 7 shows that with an increase in matrix dimensions from 50 to 600, the standard deviation decreased by almost six-fold. In order to determine the cause of the decrease in standard deviation with increasing matrix dimensions, spatial distributions of the nodes interrupting the last percolation channel were developed.

For this purpose, the coordinates of the node that interrupted the last percolation channel were determined from a computer simulation for each successive sample. To achieve this, the computational simulations were employed to ascertain the coordinates of the node responsible for interrupting the final percolation channel in each consecutive sample. This node, also referred to as the node that formed the percolation threshold, caused the DC flow to disappear between the upper and lower edges of the matrix when it was removed.

A constant number of samples with increasing matrix dimensions resulted in a decreasing number of non-conducting nodes, interrupting the last percolation channel, at individual nodes of the square matrix. This made it impossible to compare the simulation results for matrices of different dimensions. In order to compare the density distributions of the nodes forming the percolation threshold, determined for matrices of different sizes, so-called containers were used. A container is an area containing a certain number of nodes of a network, in our case, a square network. As the dimensions of the matrix increase, the dimensions of the container increase proportionally to them. This causes the number of containers to remain constant in matrices of different dimensions. Therefore, as the matrix dimensions increase, the average number of nodes, interrupting the last percolation channel going into a container remains constant. This allows for a comparison of the spatial distributions of these nodes for matrices of different sizes. The symmetry of the square matrices was taken into account when determining the dimensions of the containers.

In this paper, the spatial distributions of the nodes forming the percolation threshold located in containers were determined for matrices with L dimensions of 50, 100, 200, 300, 400, 500 and 600. The distributions were made in the form of heat maps, often used for this purpose [x, y, z, n]. Figures 8–10 show, as examples, the distributions for matrices of dimensions 50, 200 and 500. The figures illustrate that the nodes constituting the percolation threshold were concentrated in the central part of the matrix. Approaching the edges, both top and side, the number of these nodes decreased. The edge phenomenon of the distribution of nodes forming the percolation threshold, observed in [40] for matrices of dimensions 55, 101 and 151, was present.

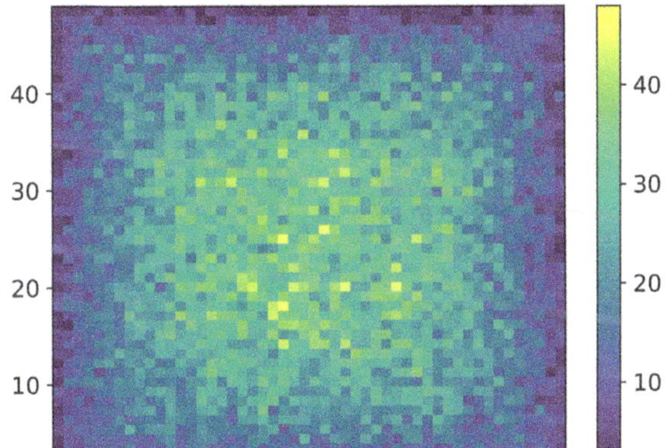

Figure 8. Two-dimensional heat map of the dependence of the number of non-conductive nodes interrupting the last percolation channel for the coordinates of the nodes of a matrix of dimensions $L = 50$ for 5×10^4 samples.

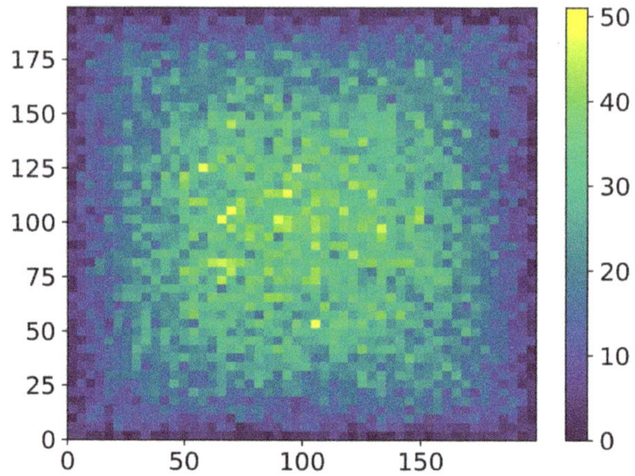

Figure 9. Two-dimensional heat map of the dependence of the number of non-conductive nodes interrupting the last percolation channel for the coordinates of the containers of a matrix of dimensions $L = 200$ for 5×10^4 samples.

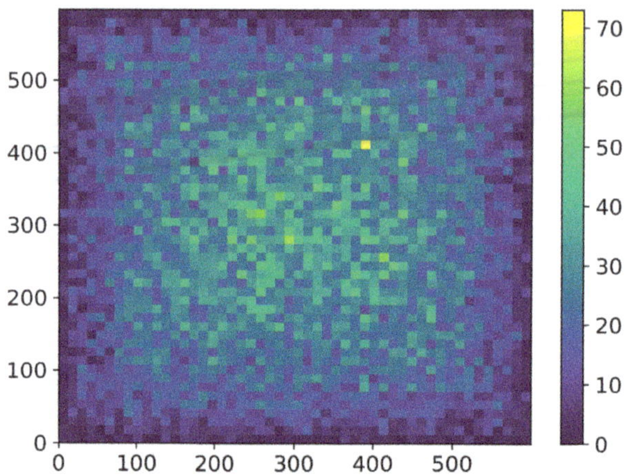

Figure 10. Two-dimensional heat map of the dependence of the number of non-conductive nodes interrupting the last percolation channel for the coordinates of the containers of a matrix of dimensions $L = 600$ for 5×10^4 samples.

In order to accurately analyze the distribution of nodes forming the percolation threshold, graphs of the dependence of the number of nodes forming the percolation threshold on container coordinates were developed. Figure 11 shows the simulation results as points and polynomial approximations as solid lines.

The computational and approximation results showed that for all matrices, the maximum was located in the center of the matrix, and that the values at the maximum for the different matrices were virtually the same. As the matrix edge was approached, there was a reduction in the concentration of nodes that interrupted the final percolation channel located within the confines of the containers. Increasing the size of the matrix slowed down the rate of decrease in the concentration of nodes interrupting the last percolation channel towards the edge. It can be seen from Figure 11 that as the matrix dimensions increased, the area in which the number of nodes forming the percolation threshold was close to the maximum value increased.

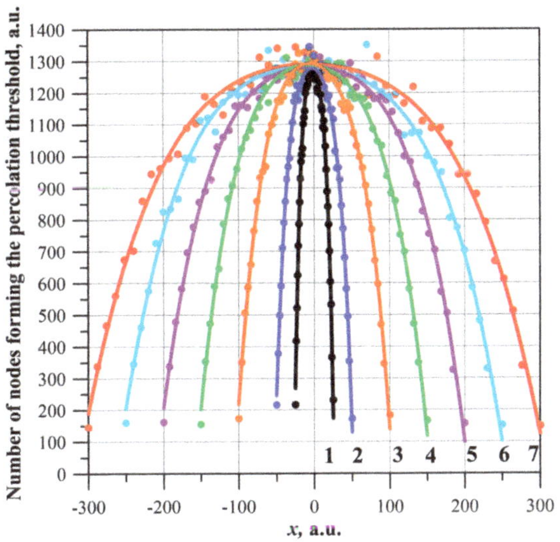

Figure 11. Number of nodes forming the percolation threshold in containers, distributed along the axes for matrix dimensions: 1—50, 2—100, 3—200, 4—300, 5—400, 6—500, 7—600 and polynomial approximations. Number of samples: 5×10^4.

The mean value of the R^2 determination coefficients for the polynomial approximations for all individuals was $R^2 \approx (0.992157 \pm 0.003208)$. This value indicated a good quality of fit of the approximating waveforms to the simulation results.

The rate of change of the distribution function of the number of nodes forming the percolation threshold was characterized by the derivative. Based on the approximating functions (Figure 11), the derivatives were calculated for the different dimensions of the matrix, as shown in Figure 12.

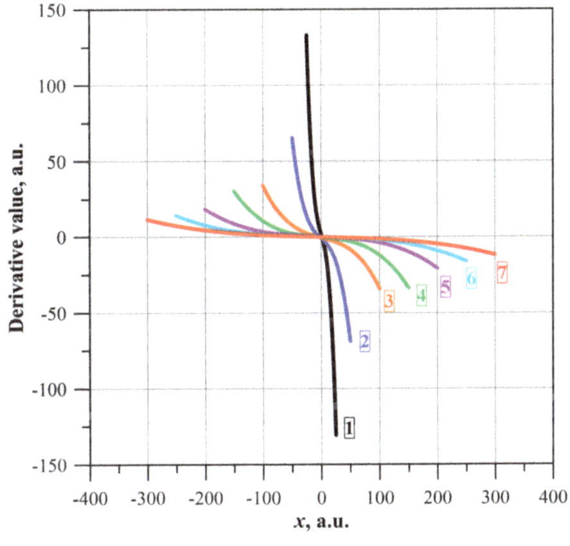

Figure 12. Plot of derivatives of the approximating function for the dimensions of the matrices: 1—50, 2—100, 3—200, 4—300, 5—400, 6—500 and 7—600.

It can be seen from Figure 12 that as the dimensions of the matrix increased, the value of the derivative decreases and becomes flatter. The largest value of each derivative is obtained at the side edge. Figure 13 shows the dependence of the modulus of the derivative value at the right-side edge on the dimensions of the matrix. The continuous line shows the waveform approximating this relationship.

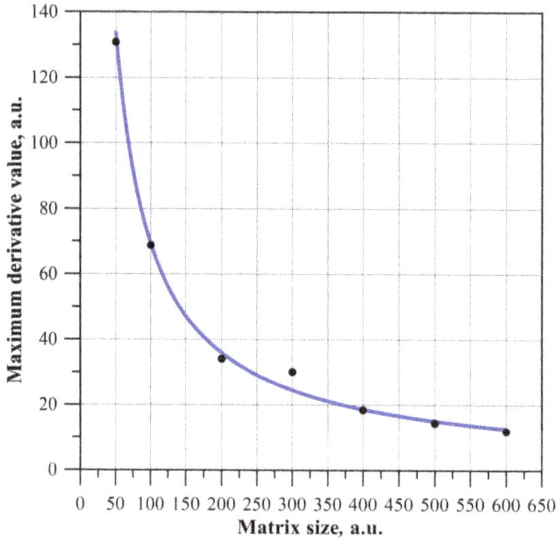

Figure 13. Dependence of the modulus of the maximum derived value on the matrix size *L*.

It can be seen from Figure 13 that the modulus of the maximum value of the derivatives decreased as the dimension *L* of the matrix increased.

The R^2 value of the approximating waveform from Figure 13 was close to unity and was 0.9863. This indicated good quality of the approximation. The equation of the approximating function was also obtained:

$$y \approx 5423L^{-1} \quad (2)$$

As can be seen from the figures above, the highest concentration of nodes interrupting the last percolation channel was near the center of the matrix. The further away from the center, the lower the content of nodes. As depicted in the aforementioned figures, the highest concentration of nodes interrupting the final percolation channel was situated in close proximity to the matrix's center. Moving away from the center resulted in a gradual decline in node concentration. This means that, for a finite-dimensional matrix, there was an edge phenomenon in that, as the edge of the matrix was approached, the concentration of nodes interrupting the last percolation channel decreased. As the dimensions of the matrix increase, there was a slowing down of the tendency of the concentration of nodes interrupting the last percolation channel to decrease towards the edge (Figures 11–13). In doing so, the area in which values that were close to the maximum occurred was expanded. This means that as the dimensions of the matrix increased, the edge phenomenon weakened. From Formula (2), it follows that for matrix dimensions going to infinity, the maximum value of the derivative should tend towards zero. A zero value of the derivative means that the spatial distribution of the number of nodes forming the percolation threshold becomes a constant value. An increase in the dimensions of the matrix also results in a decreasing value of the standard deviation of the percolation threshold, as in Formula (3). From Formula (3) for the standard deviation, it can be seen that this value can, with a good approximation, be described by the relation:

$$\sigma(L) \approx \sigma_0 L^{-0.75}. \quad (3)$$

On the other hand, the maximum value of the rate of reduction in the concentration of nodes, interrupting the last percolation (derivative) channel, is described by the formula:

$$K(L) \approx K_0 L^{-1} \quad (4)$$

A comparison of Formulas (3) and (4) shows that both parameters decrease as the dimensions of the square matrix L increases. This means that both the values of the standard deviation of the percolation threshold and the edge phenomenon are clearly related to the dimensions of the matrix. From Formulas (3) and (4), it can be seen that with increasing matrix dimensions, the edge phenomenon will gradually disappear, and the percolation threshold standard deviation values caused by it will tend towards zero.

4. Conclusions

In this paper, an in-depth analysis of the percolation phenomenon for square matrices with dimensions from $L = 50$ to 600 for 5×10^4 samples was performed using Monte Carlo computer simulations. In the study, the inverse logic of the simulation was used, which is that initially the matrix is completely filled with conducting nodes. The percolation threshold will be reached when the last percolation channel connecting the top and bottom edges of the matrix is broken. This will result in the disappearance of the DC flow between the upper and lower edges of the matrix. The value of the percolation threshold was defined as the quantity of nodes capable of conducting, remaining in the matrix before drawing the node interrupting the last percolation channel, related to the total number of nodes in the matrix.

On the basis of the simulations, it was determined that the distributions of percolation threshold values were normal distributions. Using this foundation, we explored how the expected (average) percolation threshold value and the standard deviation were influenced by variations in matrix dimensions. The study revealed that the average percolation threshold value remained largely unaltered by changes in matrix size.

It was found that the decrease in the standard deviation value of the percolation threshold observed in the study with increasing matrix dimensions is related to the occurrence of the edge phenomenon. The observed reduction in the standard deviation value of the percolation threshold, as the matrix dimensions increased, was associated with the presence of the edge phenomenon. The dependence of the standard deviation value on the matrix dimensions was determined, as well as the dependences describing the spatial distributions of the nodes interrupting the last percolation channel. It can be seen from the determined relationships that with increasing matrix dimensions, the edge phenomenon should gradually disappear and the percolation threshold standard deviation values caused by it will tend towards zero.

Author Contributions: Conceptualization, P.Z. and P.O.; methodology, P.Z., P.O., K.K. and A.D.P.; software, P.O., P.R. and V.B.; validation, P.O. and K.K.; formal analysis, P.Z., P.O., K.K., P.R., V.B. and A.D.P.; investigation, P.O., K.K., P.R. and V.B.; resources, P.O. and K.K.; data curation, P.O. and P.R.; writing—original draft preparation, P.Z. and P.O.; writing—review and editing, K.K.; visualization, P.O., supervision, P.Z.; funding acquisition, P.O., K.K., P.R. and V.B. All authors have read and agreed to the published version of the manuscript.

Funding: The research was supported by the subsidy of the Ministry of Education and Science (Poland) for the Lublin University of Technology as funds allocated for scientific activities in the scientific discipline of Automation, Electronics, Electrical Engineering and Space Technologies—grants: FD-20/EE-2/701, FD-20/EE-2/702, FD-20/EE-2/705, FD-20/EE-2/707.

Data Availability Statement: Data are contained within the article.

Conflicts of Interest: The authors declare no conflict of interest.

References

1. Osetsky, Y.; Barashev, A.V.; Zhang, Y. Sluggish, Chemical Bias and Percolation Phenomena in Atomic Transport by Vacancy and Interstitial Diffusion in Ni Fe Alloys. *Curr. Opin. Solid State Mater. Sci.* **2021**, *25*, 100961. [CrossRef]
2. Jiang, J.; Yu, X.; Lin, Y.; Guan, Y. PercolationDF: A Percolation-Based Medical Diagnosis Framework. *Math. Biosci. Eng.* **2022**, *19*, 5832–5849. [CrossRef]
3. Sahimi, M. Percolation in Biological Systems. In *Applied Mathematical Sciences (Switzerland)*; Springer: Berlin/Heidelberg, Germany, 2023; Volume 213, pp. 443–488.
4. Devpura, A.; Phelan, P.E.; Prasher, R.S. Percolation Theory Applied to the Analysis of Thermal Interface Materials in Flip-Chip Technology. In Proceedings of the Seventh Intersociety Conference on Thermal and Thermomechanical Phenomena in Electronic Systems—ITHERM 2000 (Cat. No.00CH37069), Las Vegas, NV, USA, 23–26 May 2000; Volume 1, pp. 21–28.
5. Evseev, V.A.; Konopleva, R.F.; Skal, A.S. Percolation in Semiconductors with Disordered Regions: Electrical Conductivity and Hall Coefficient. *Radiat. Eff.* **1982**, *66*, 167–172. [CrossRef]
6. Bychkov, E.; Tveryanovich, Y.; Vlasov, Y. *Ion Conductivity and Sensors*; Argonne National Lab: Argonne, IL, USA, 2004; pp. 103–168.
7. Rubie, D.C.; Nimmo, F.; Melosh, H.J. Formation of Earth's Core. In *Treatise on Geophysics*; Elsevier: Amsterdam, The Netherlands, 2007; pp. 51–90.
8. Peng, H.; Sun, X.; Weng, W.; Fang, X. Electronic Polymer Composite. In *Polymer Materials for Energy and Electronic Applications*; Academic Press: Cambridge, MA, USA, 2017; pp. 107–149. [CrossRef]
9. Sulaberidze, V.S.; Skorniakova, E.A. Phenomenon of the Percolation in Composite Materials Based on a Polymer Binder with a Dispersed Filler Phase. *IOP Conf. Ser. Mater. Sci. Eng.* **2020**, *919*, 022009. [CrossRef]
10. Essam, J.W. Percolation Theory. *Rep. Progress Phys.* **1980**, *43*, 833–912. [CrossRef]
11. Hammersley, J.M. A Generalization of McDiarmid's Theorem for Mixed Bernoulli Percolation. *Math. Proc. Camb. Philos. Soc.* **1980**, *88*, 167–170. [CrossRef]
12. Broadbent, S.R.; Hammersley, J.M. Percolation Processes. *Math. Proc. Camb. Philos. Soc.* **1957**, *53*, 629–641. [CrossRef]
13. Carreau, P.J.; Vergnes, B. Rheological Characterization of Fiber Suspensions and Nanocomposites. In *Rheology of Non-Spherical Particle Suspensions*; Elsevier: Amsterdam, The Netherlands, 2015; pp. 19–58.
14. Lebrecht, W.; Centres, P.M.; Ramirez-Pastor, A.J. Empirical Formula for Site and Bond Percolation Thresholds on Archimedean and 2-Uniform Lattices. *Phys. A Stat. Mech. Its Appl.* **2021**, *569*, 125802. [CrossRef]
15. Shlyakhtich, M.; Prudnikov, P. Monte Carlo Simulation of the Critical Behavior near the Percolation Threshold with Invaded Cluster Algorithm. *J. Phys. Conf. Ser.* **2021**, *1740*, 012009. [CrossRef]
16. Dean, P. A New Monte Carlo Method for Percolation Problems on a Lattice. *Math. Proc. Camb. Philos. Soc.* **1963**, *59*, 397–410. [CrossRef]
17. Dean, P.; Bird, N.F. Monte Carlo Estimates of Critical Percolation Probabilities. *Math. Proc. Camb. Philos. Soc.* **1967**, *63*, 477–479. [CrossRef]
18. Derrida, B.; Stauffer, D. Corrections to Scaling and Phenomenological Renormalization for 2-Dimensional Percolation and Lattice Animal Problems. *J. Phys.* **1985**, *46*, 1623–1630. [CrossRef]
19. De Oliveira, P.M.C.; Nóbrega, R.A.; Stauffer, D. Corrections to Finite Size Scaling in Percolation. *Braz. J. Phys.* **2003**, *33*, 616–618. [CrossRef]
20. Jacobsen, J.L. Critical Points of Potts and O(*N*) Models from Eigenvalue Identities in Periodic Temperley–Lieb Algebras. *J. Phys. A Math. Theor.* **2015**, *48*, 454003. [CrossRef]
21. Mertens, S. Exact Site-Percolation Probability on the Square Lattice. *J. Phys. A Math. Theor.* **2022**, *55*, 334002. [CrossRef]
22. Newman, M.E.J.; Ziff, R.M. Efficient Monte Carlo Algorithm and High-Precision Results for Percolation. *Phys. Rev. Lett.* **2000**, *85*, 4104–4107. [CrossRef]
23. Novoselov, K.S.; Geim, A.K.; Morozov, S.V.; Jiang, D.; Zhang, Y.; Dubonos, S.V.; Grigorieva, I.V.; Firsov, A.A. Electric Field Effect in Atomically Thin Carbon Films. *Science* **2004**, *306*, 666–669. [CrossRef] [PubMed]
24. Berger, C.; Song, Z.; Li, T.; Li, X.; Ogbazghi, A.Y.; Feng, R.; Dai, Z.; Marchenkov, A.N.; Conrad, E.H.; First, P.N.; et al. Ultrathin Epitaxial Graphite: 2D Electron Gas Properties and a Route toward Graphene-Based Nanoelectronics. *J. Phys. Chem. B* **2004**, *108*, 19912–19916. [CrossRef]
25. Naguib, M.; Kurtoglu, M.; Presser, V.; Lu, J.; Niu, J.; Heon, M.; Hultman, L.; Gogotsi, Y.; Barsoum, M.W. Two-Dimensional Nanocrystals Produced by Exfoliation of Ti_3AlC_2. *Adv. Mater.* **2011**, *23*, 4248–4253. [CrossRef]
26. Naguib, M.; Mashtalir, O.; Carle, J.; Presser, V.; Lu, J.; Hultman, L.; Gogotsi, Y.; Barsoum, M.W. Two-Dimensional Transition Metal Carbides. *ACS Nano* **2012**, *6*, 1322–1331. [CrossRef]
27. Mashtalir, O.; Naguib, M.; Mochalin, V.N.; Dall'Agnese, Y.; Heon, M.; Barsoum, M.W.; Gogotsi, Y. Intercalation and Delamination of Layered Carbides and Carbonitrides. *Nat. Commun.* **2013**, *4*, 1716. [CrossRef]
28. Ling, Z.; Ren, C.E.; Zhao, M.-Q.; Yang, J.; Giammarco, J.M.; Qiu, J.; Barsoum, M.W.; Gogotsi, Y. Flexible and Conductive MXene Films and Nanocomposites with High Capacitance. *Proc. Natl. Acad. Sci. USA* **2014**, *111*, 16676–16681. [CrossRef]
29. Li, X.; Zhu, H. Two-Dimensional MoS_2: Properties, Preparation, and Applications. *J. Materiomics* **2015**, *1*, 33–44. [CrossRef]
30. Bhimanapati, G.R.; Glavin, N.R.; Robinson, J.A. 2D Boron Nitride. In *Semiconductors and Semimetals*; Academic Press Inc.: Cambridge, MA, USA, 2016; Volume 95, pp. 101–147.

31. Shahzad, F.; Alhabeb, M.; Hatter, C.B.; Anasori, B.; Man Hong, S.; Koo, C.M.; Gogotsi, Y. Electromagnetic Interference Shielding with 2D Transition Metal Carbides (MXenes). *Science* **2016**, *353*, 1137–1140. [CrossRef] [PubMed]
32. Akhtar, M.; Anderson, G.; Zhao, R.; Alruqi, A.; Mroczkowska, J.E.; Sumanasekera, G.; Jasinski, J.B. Recent Advances in Synthesis, Properties, and Applications of Phosphorene. *NPJ 2D Mater. Appl.* **2017**, *1*, 5. [CrossRef]
33. Zhen, Z.; Zhu, H. Structure and Properties of Graphene. In *Graphene*; Elsevier: Amsterdam, The Netherlands, 2018; pp. 1–12.
34. Xu, Z. Fundamental Properties of Graphene. In *Graphene*; Elsevier: Amsterdam, The Netherlands, 2018; pp. 73–102.
35. Gogotsi, Y.; Anasori, B. The Rise of MXenes. *ACS Nano* **2019**, *13*, 8491–8494. [CrossRef] [PubMed]
36. Kim, S.; Gholamirad, F.; Yu, M.; Park, C.M.; Jang, A.; Jang, M.; Taheri-Qazvini, N.; Yoon, Y. Enhanced Adsorption Performance for Selected Pharmaceutical Compounds by Sonicated $Ti_3C_2T_X$ MXene. *Chem. Eng. J.* **2021**, *406*, 126789. [CrossRef]
37. Kołtunowicz, T.N.; Gałaszkiewicz, P.; Kierczyński, K.; Rogalski, P.; Okal, P.; Pogrebnjak, A.D.; Buranich, V.; Pogorielov, M.; Diedkova, K.; Zahorodna, V.; et al. Investigation of AC Electrical Properties of MXene-PCL Nanocomposites for Application in Small and Medium Power Generation. *Energies* **2021**, *14*, 7123. [CrossRef]
38. Diedkova, K.; Pogrebnjak, A.D.; Kyrylenko, S.; Smyrnova, K.; Buranich, V.V.; Horodek, P.; Zukowski, P.; Koltunowicz, T.N.; Galaszkiewicz, P.; Makashina, K.; et al. Polycaprolactone-MXene Nanofibrous Scaffolds for Tissue Engineering. *ACS Appl. Mater. Interfaces* **2023**, *15*, 14033–14047. [CrossRef]
39. Zukowski, P.; Okal, P.; Kierczynski, K.; Rogalski, P.; Borucki, S.; Kunicki, M.; Koltunowicz, T.N. Investigations into the Influence of Matrix Dimensions and Number of Iterations on the Percolation Phenomenon for Direct Current. *Energies* **2023**, *16*, 7128. [CrossRef]
40. Zukowski, P.; Okal, P.; Kierczynski, K.; Rogalski, P.; Bondariev, V. Analysis of uneven distribution of nodes creating a percolation channel in matrices with translational symmetry for direct current. *Energies* **2023**, *16*, 7647. [CrossRef]

energies

MDPI

Article

OLTC Fault detection Based on Acoustic Emission and Supported by Machine Learning

Andrzej Cichoń * and Michał Włodarz *

Department of Electric Power Engineering and Renewable Energy, Faculty of Electrical Engineering, Automatic Control and Informatics, Opole University of Technology, 45-758 Opole, Poland
* Correspondence: a.cichon@po.edu.pl (A.C.); michal.wlodarz@student.po.edu.pl (M.W.)

Abstract: Power transformers are an essential part of the power grid. They have a relatively low rate of failure, but removing the consequences is costly when it occurs. One of the elements of power transformers that are often the reason for shutting down the unit is the on-load tap changer (OLTC). Many methods have been developed to assess the technical condition of OLTCs. However, they require the transformer to be taken out of service for the duration of the diagnostics, or they do not enable precise diagnostics. Acoustic emission (AE) signals are widely used in industrial diagnostics. The generated signals are difficult to interpret for complex systems, so artificial intelligence tools are becoming more widely used to simplify the diagnostic process. This article presents the results of research on the possibility of creating an online OLTC diagnostics method based on AE signals. An extensive measurement database containing many frequently occurring OLTC defects was created for this research. A method of feature extraction from AE signals based on wavelet decomposition was developed. Several machine learning models were created to select the most effective one for classifying OLTC defects. The presented method achieved 96% efficiency in OLTC defect classification.

Keywords: on-load tap changer; acoustic emission; wavelet decomposition; diagnostic method

Citation: Cichoń, A.; Włodarz, M. OLTC Fault detection Based on Acoustic Emission and Supported by Machine Learning. *Energies* **2024**, *17*, 220. https://doi.org/10.3390/en17010220

Academic Editor: Tomasz Norbert Kołtunowicz

Received: 29 November 2023
Revised: 28 December 2023
Accepted: 29 December 2023
Published: 31 December 2023

1. Introduction

The main goal of the present research is to assess the possibility of detecting defects occurring in on-load tap changers (OLTCs). The analysis results of time courses of acoustic emission (AE) generated during switching processes using wavelet decomposition are presented in this paper. The creation of a measurement database containing a wide range of tested OLTC defects is also described.

OLTCs are one of the elements of transformer equipment. Their purpose is to allow the voltage levels on the power grid to change by varying the number of active coils in the transformer. The OLTC mechanism works under load conditions so that energy can be supplied to consumers continuously [1].

Power transformers are an essential part of the power grid and have a relatively low rate of failure. However, the cost of remedying the consequences of losses is high. This is why diagnostic tests are so necessary [2]. Increasing emphasis is being placed on condition assessment without shutting down transformers, and such an approach allows us to avoid the costs associated with power outages [3]. One of the elements most frequently causing transformer failures is OLTCs [4]. The reasons for their loss can be divided into three groups [5]:

- Oil faults;
- Mechanical faults;
- Electrical faults.

The research presented in this article focuses on the detection of mechanical damage. During switching, the dynamic forces acting on the OLTC contacts gradually deform the contacts. In uneven contact wear, the OLTC switches the individual phases non-simultaneously. The spring magazine stores mechanical energy from the electric drive and then releases it, ensuring quick switching. It consists of two springs mounted axially. If one of them is damaged, storing the appropriate amount of potential energy will not be possible, leading to the extension of the switching process.

There are many diagnostic methods to assess the technical condition of OLTCs. One of the most recognized and widely used methods is dynamic resistance measurement (DRM). This method is presented in [6]. In the research presented in this manuscript, this method is used to confirm the occurrence of a defect. The most significant disadvantage of DRM is the inability to perform diagnostics on a loaded OLTC.

A direction often chosen when developing OLTC diagnostic methods is the analysis of vibrations generated during switching. Vibration waveforms are challenging to interpret, and expert knowledge is required from the diagnostician to assess the technical condition. For this reason, research is being conducted to use machine learning (ML) tools to simplify the diagnostic process. To train the ML model, descriptors must be extracted from the signal. Many works have presented proposals for this process. In [7], the authors used FFT analysis for feature extraction. Another method of obtaining information from vibration waveforms using the wavelet transform is presented in [8]. An approach using features extracted from the time–frequency domain by a minimum-entropy deconvolution filter is shown in [9]. Feature extraction based on calculating statistical values in the time domain is shown in [10].

Another approach to obtaining data enabling the assessment of the technical condition of the OLTC is recording the AE signals generated during the switching operation. This method uses piezoelectric transducers with a larger bandwidth compared to vibration measurements. Partial discharges and electric arcs generate acoustic signals in higher frequency bands. Thus, combining OLTC diagnostics from a mechanical and electrical perspective is possible.

AE signals, similarly to vibroacoustic signals, are challenging to interpret. Thus, ML models are proposed to classify OLTC defects. Several data feature extraction methods have been presented in the literature. The usage of Hilbert transform was shown in [11]. An attempt to analyze OLTC AE signals in the time–frequency domain is presented in [12]. A feature extraction method joining spectral density, time and frequency measures, and wavelet coefficient energies is shown in [13].

This paper mainly investigates the possibility of OLTC mechanical defect detection using ML tools. The proposed diagnostic method uses wavelet decomposition to extract details from EA signals, and then numerical values are determined for each detail. The present research stands out from other studies by using an original database of waveforms of EA signals recorded for OLTCs with defects. A wide range of contact wear was modeled, which may contribute to the creation of a diagnostic system that allows for the assessment of not only whether the OLTC is functional or damaged but also the extent to which it is worn. Such a system will allow for the estimation of the life of the device and the planning of renovations more effectively. A measurement database for over 3500 OLTC switches was created. This allowed for effective training with large amounts of data and reliable validation of the effectiveness of defect classification. The literature proposes using various types of transducers, most often accelerometers. A universal broadband transducer was used during the measurements, allowing for simultaneous OLTC diagnostics for mechanical and electrical defects. The number of defects examined during this research was much higher than in the previously presented studies. The possibility of determining the degree of symmetrical and asymmetrical contact wear was examined. Additionally, a case of damage to the spring energy storage was modeled. A wide range of ML algorithms were used to classify defects and were compared in terms of the effectiveness of OLTC diagnostics. In order to achieve the best classification performance, Bayesian optimization was used.

2. Materials and Methods

Databases are used during the process of generating ML models. The more samples they contain, the more effective classifiers can be created. Currently, there is no open database containing OLTC measurements enabling the classification of defects with the level of detail desired by the authors. Therefore, it was decided to perform a series of measurements to provide research materials.

2.1. Experimental Setup

The conducted research used an actual OLTC model. The OLTC used consists of a selector shortened to six taps and a VEL-110-type power switch. The switch is shown in Figure 1. The device was placed in a tank filled with transformer oil, corresponding to the actual operating environment of the tested OLTC. The system has an electric drive, providing the mechanical energy necessary for switching.

Figure 1. Disassembled VEL-110 power switch.

To avoid the need to switch the OLTC manually, a PLC controller was used. It performed a series of switches, which significantly accelerated the measurements. The next step taken to minimize the need to interfere in the data acquisition process was to use the drive current measurement as a trigger for the measurement card.

During the measurements, the following defects that may occur in OLTCs were examined:

- Class 1: Fully functional OLTC;
- Class 2: Symmetrical contact wear by 1 mm;
- Class 3: Symmetrical contact wear by 2 mm;
- Class 4: Symmetrical contact wear by 3 mm;
- Class 5: Symmetrical contact wear by 4 mm;
- Class 6: Asymmetrical contact wear;
- Class 7: Broken spring;
- Class 8: Damaged contact surface.

Symmetrical contact wear occurs when the contacts for the three phases wear evenly (Classes 2–5). This manifests in the extension of the switching time, which causes a longer current flow through the additional resistance of the auxiliary circuits. This may lead to overheating of the resistors and, in the long run, to a complete loss of the OLTC's switching capacity. Several contact sets were used to model contact-wear defects, either symmetrical or asymmetrical. Each group was characterized by a varying degree of wear. Wear was simulated by mechanically machining the contacts. The view of the mounted, modified contacts is presented in Figure 2.

Figure 2. Modified OLTC contacts.

Asymmetrical wear of contacts (Class 6) results in the switching of individual phases, not simultaneously. The tests included asymmetrical switching of one phase. This defect was modeled using contacts with various degrees of wear.

The kinetic energy stored in the switch consists of two springs, one inside the other. A common cause of OLTC failure is damage to one of these springs. This defect was modeled by removing one of the springs (Class 7).

Class 8 covers contact damage by modifying their geometry without changing the effective contact thickness. This models extensive mechanical distortion, which often occurs in OLTCs. For this case, contacts with an increased contact area were used. The effective thickness of these elements was not changed.

2.2. Measurement Line

The measurement system used to record the AE signals generated by the OLTC during switching consisted of the following components:

- Piezoelectric transducer;
- Preamplifier;
- Amplifier;
- Data acquisition card;
- Laptop.

WD17AH was chosen as the acoustic wave transducer. This type of transducer was selected because it has a frequency response that matches the expected frequencies. Its additional advantage is high sensitivity. The data of the WD17AH piezoelectric transducer are shown in Table 1. The transducer was mounted to the OLTC tank using a magnetic holder.

Table 1. Technical data of the WD17AH transducer.

Frequency Band (kHz)	Peak Sensitivity, Ref V/(m/s)	Peak Sensitivity, Ref V/μbar
100–900	56 dB	−61 dB

The signal from the transducer was then amplified using a preamplifier with a gain of 20 dB. The 2/4/6 PREAMPLIFIER manufactured by Physical Acoustics Corporation was used. The next element of the measurement path was the Wide-Bandwidth AE Amplifier, also manufactured by Physical Acoustics Corporation.

An Acquitek CH3160 acquisition card was used to acquire the AE signals. The signals were sampled with a frequency of 350 kHz. The KEW 8146 Leakage Clap Sensor was also connected to the card, allowing for motor load current monitoring. The current signal was used as a trigger for the measuring card. When the current reached the threshold value, the measurement was triggered. AE signals were recorded using AcquiFlex v2.0 software dedicated to the card used. In Figure 3, the AE measurement line is shown.

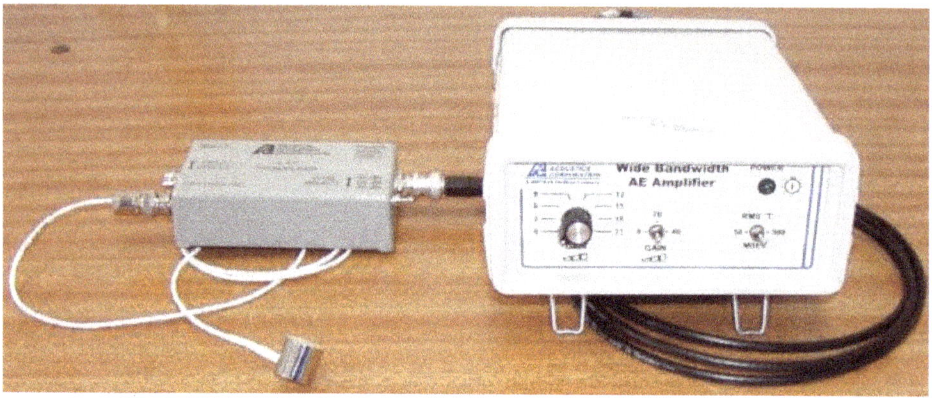

Figure 3. AE measurement line.

To enable verification of the occurrence of OLTC defects, the DRM method was used. This method is widely recognized for assessing the technical condition of OLTCs. An MT3 device was used to perform DRM. The device manufacturer provides OLTC.exe v2.10.510 software that allows for the control of MT3 and record waveforms. MT3 with a laptop is shown in Figure 4.

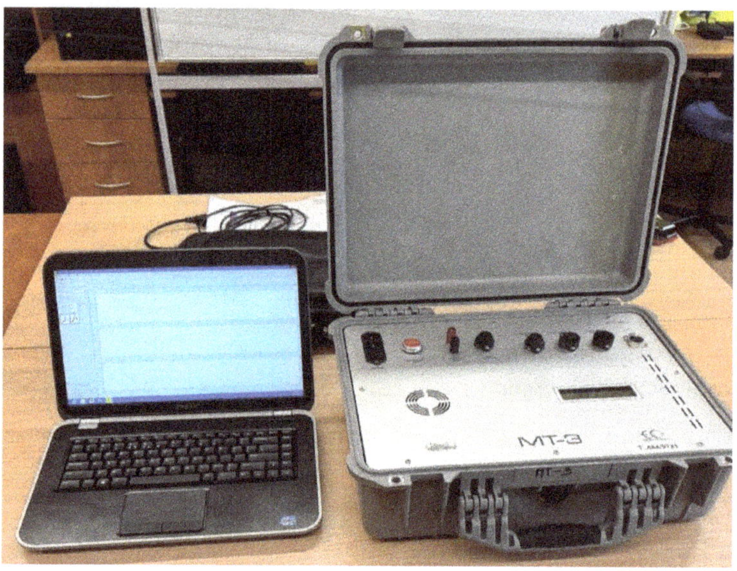

Figure 4. MT3 used for the DRM method.

Over 400 switches were made for each tested defect, during which measurements were taken. This allowed us to train ML tools with a significant amount of data and to check their effectiveness reliably. The general diagram of the measurement line is presented in Figure 5.

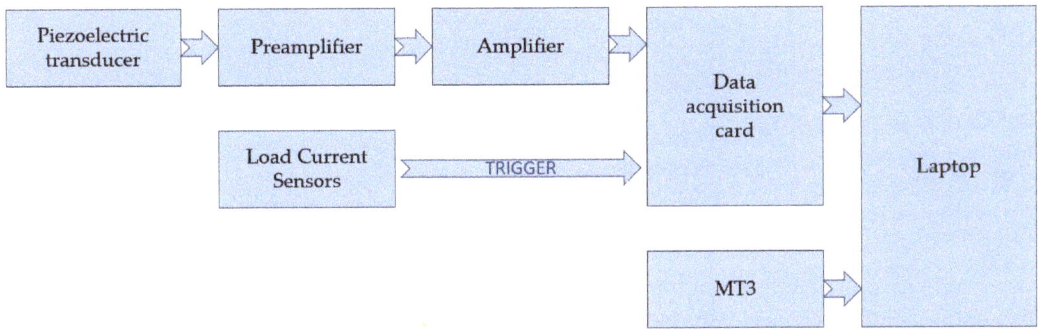

Figure 5. Measurement line diagram.

2.3. *Signal Processing and Classification*

Sample recording started when the drive started and ended when the switchover was completed. All signals were normalized by dividing them by the maximum value, so the amplitude of each signal was in the range <−1;1>. The recorded runs consisted of approximately 2 million samples. The fragment corresponding to the switch itself was much shorter. Therefore, to speed up further work, it was decided to cut out significant fragments of the signals. The algorithm searched for the maximum amplitude and cut out a chunk of 50,000 samples. An example of the original waveform and the detected fragment where the switching occurred are presented in Figure 6.

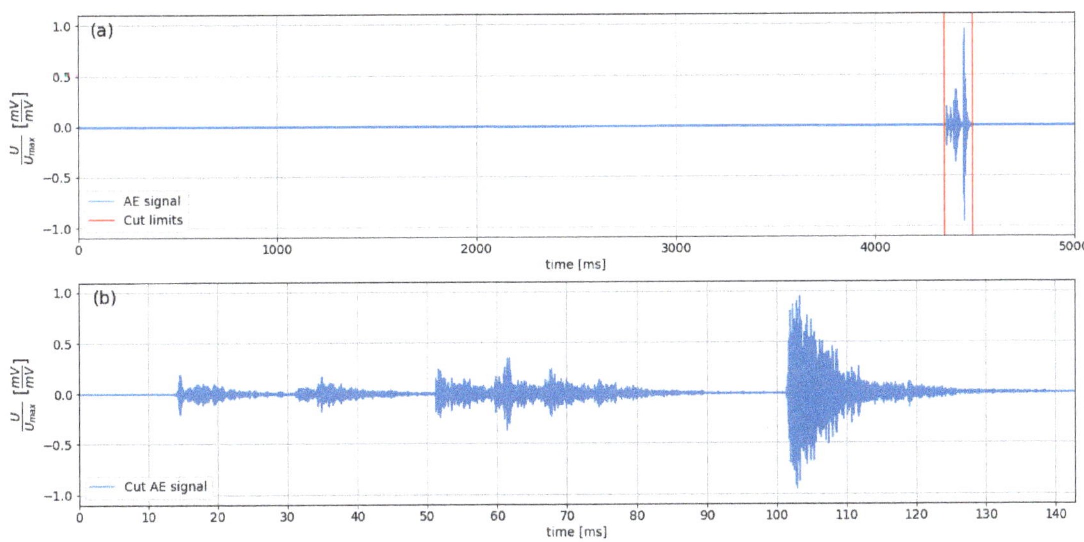

Figure 6. Cutting out signal fragments: (**a**) example original waveform with marked limits; (**b**) designated significant fragment.

The present research used discrete wavelet transform (DWT) to extract descriptors. DWT decomposes a waveform into several details. This is done by sliding a wavelet through the signal and multiplying it in each location. The results are the coefficients. The process is repeated for a larger-scale wavelet. An important issue when using DWT is the selection of the mother wavelet. During preliminary research, it was observed that the sym8 wavelet allows for achieving the best results in classifying OLTC defects. A similar mother wavelet was proposed in [14]. In the present study, the signal was decomposed into five details.

The following characteristic values were determined for each detail:

- Mean;
- Root mean squared;
- Form factor;
- Crest factor;
- Standard deviation;
- Skewness.

To achieve good results in ML training, the data must be normalized. A standard scaler was used for the data features. That is, the data features were shifted so that their mean was zero and scaled to achieve unit variance. The determined values were used as data for training ML models. The dataset was split in half into training and test datasets. The following ML models were selected:

- K-nearest neighbors (KNN);
- Decision tree (DT);
- Random forest (RF);
- Support-vector machine (SVM);
- Gradient boosting (GB);
- Adaptive boosting (AdaBoost).

KNN is a supervised learning classifier. Proximity is used to predict the class to which a given data point belongs. The basic assumption is that points belonging to the same classes are close to one another [15]. On the other hand, DT looks for the best series of tests to classify the data. This algorithm creates decision rules based on if-else statements. Each test is defined in a way that allows the data to be divided into classes as precisely as possible. Decision trees are very susceptible to overfitting [16]. This problem can be overcome by using RF. The RF classifier consists of a series of DTs. Each tree is trained on a different dataset. The classification result is the average output value of all trees [17].

SVM is a classifier whose training aims to determine a hyperplane separating examples belonging to two classes with a maximum margin. This algorithm is used to classify between two classes. When the issue of multi-class classification arises, the one-versus-rest approach is used. This approach involves training one classifier for each category. Each model predicts whether a data point belongs to the corresponding class [18]. Another technique used for classification in this research is GB, which generates a set of simpler ML models. Most often, these are DTs. In each iteration, a set of DTs are updated. They are fitted based on a negative gradient of the loss function [19]. A slightly different approach is used in AdaBoost classifiers. During model fitting, the significance of previously misclassified observations is boosted, and they are more likely to be used again to train the model to be more effective. As a weak learner for AdaBoost, DT was used [20].

Every classification model must be tuned to the currently examined dataset. Each model has different parameters, and their proper selection allows for a significant increase in classification efficiency. In the present research, the optimization of classifier parameters was performed using the Bayesian optimization method. This is a popular strategy for black-box function optimization. Initially, the optimizer checks the performance of the classifiers for random sets of parameters. Then, based on the results obtained, it approximates the objective function. The subsequent matched sets of parameters are determined at the maximum of the objective function. The algorithm continues for a given number of iterations.

3. Results

The first step in analyzing the results was to determine the actual existence of the modeled defects. The DRM was used for this purpose. Two sample current waveforms are presented in Figure 7. The presented results show the results for a fully functional OLTC (Figure 7a) and with asymmetry (Figure 7b). DRM results were analyzed for each defect.

A generally accepted way to interpret DRM results is to compare the times read from the waveforms. Figure 7 shows three characteristic times:

- T_1—from the moment of opening of the first primary contact to the closing of the second resistive contact;
- T_2—from the moment of closing the second resistive contact to the moment of opening the first resistive contact;
- T_C—total switching time.

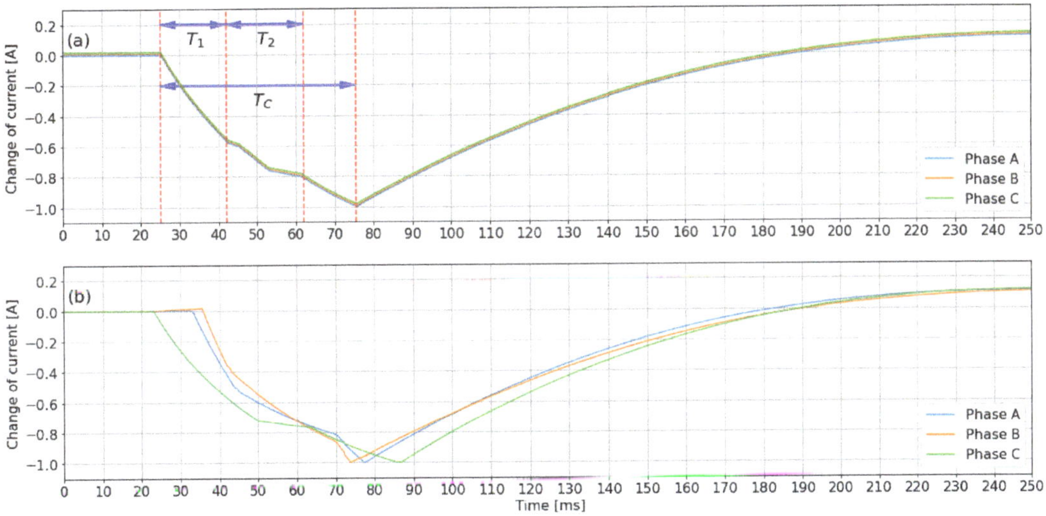

Figure 7. Example DRM results: (**a**) OLTC without defects (Class 1); (**b**) OLTC with asymmetry (Class 7).

The read characteristic times for all classes without asymmetry are presented in Table 2. The results presented are the average times read for ten randomly selected measurements from each class. Between Classes 1 and 5, a gradual increase in T_1 and a decrease in T_2 can be noticed. This is a characteristic symptom of contact wear. As expected, the damaged spring (Class 7) resulted in a significant extension of T_1, T_2, and T_C. For Class 8, there are no significant differences compared to Class 1. This is due to the lack of change in the contact thickness affecting the individual times.

Table 2. DRM characteristic times for classes without asymmetry.

Class	1	2	3	4	5	7	8
				Characteristic Times (ms)			
T_1	16	19	24	29	35	43	17
T_2	23	21	20	18	14	37	22
T_C	50	54	59	62	67	81	49

Checking the correctness of modeling the defect described in Class 6 is slightly different than in the case of the other classes. If OLTC switching asymmetry is checked, the characteristic times for individual phases should be compared. The times are not compared to those of other classes. The appearance of differences in the lengths of individual connection stages means the asymmetrical operation of the OLTC. The read times for Class 6 are presented in Table 3.

Table 3. DRM characteristic times and time differences for asymmetry (Class 6).

	Characteristic Times (ms)			Time Difference (ms)		
Phase	**A**	**B**	**C**	**A–B**	**B–C**	**C–A**
T_1	10	6	14	4	8	4
T_2	20	35	15	15	20	5
T_C	41	39	62	2	23	21

The interpretation of DRM measurements allowed for the detection of defects. It was found that the defect modeling methods used allowed for achieving the desired results. However, the DRM diagnostics did not allow for the diagnosis of mechanical deformation of the contact (Class 8).

Sample AE waveforms for each class are presented in Figure 8. Detailed interpretation of the results of the AE method is problematic because it is impossible to assign individual acoustic events to specific switching stages. The most visible defect is the asymmetrical operation of the OLTC. The area corresponding to contact switching (for Class 6: 45–90 ms) is characterized by more acoustic events with a lower amplitude. This is caused by non-simultaneous switching of the contacts of the individual phases. For defects without asymmetry, diagnosis is possible only by overlapping the recorded waveforms with the reference waveforms, but despite this procedure, the diagnostic results may be ambiguous. Therefore, we propose using ML to simplify OLTC diagnostics.

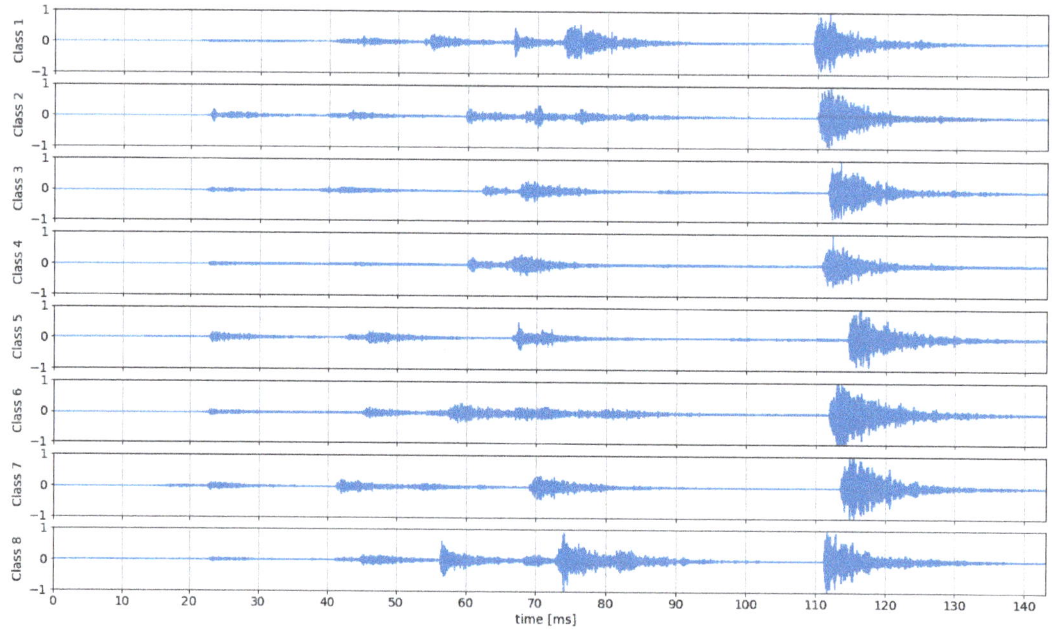

Figure 8. Example AE waveforms for the defects under research.

The method of feature extraction from AE waveforms proposed in this study is based on wavelet decomposition. Figure 9 shows the results of the decomposition of an example waveform. The statistical values calculated, 30 for each detail, constituted a database used to train the ML models.

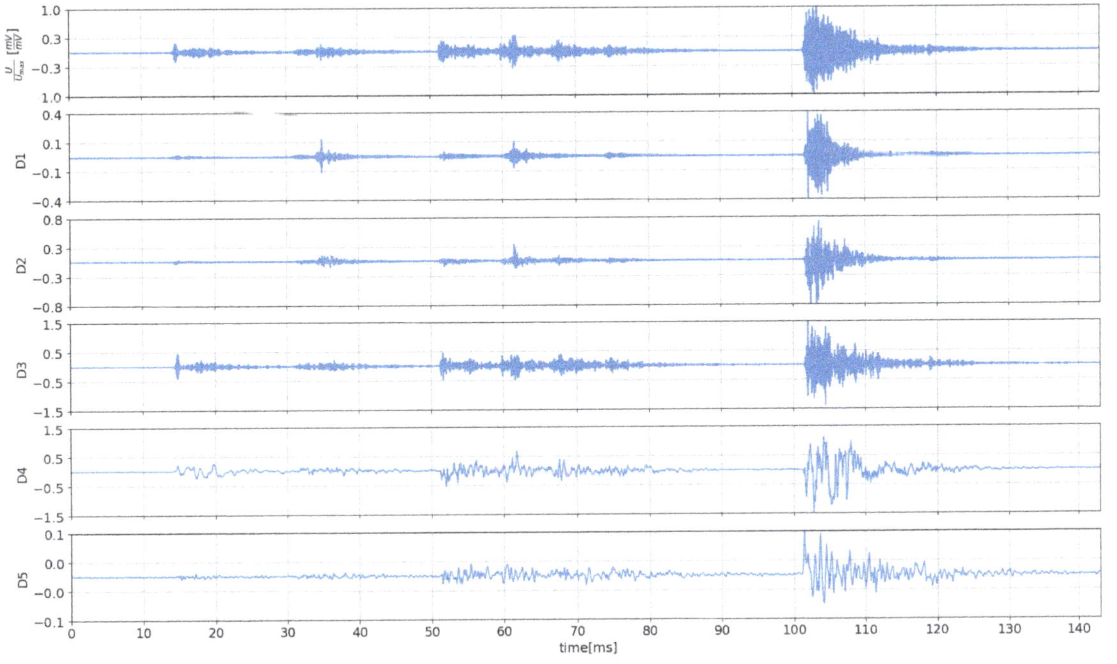

Figure 9. Original signal with its details created with the sym8 wavelet family.

The method used to extract the data features together with the selected method of optimizing the ML models allowed for achieving satisfactory results. The percentage accuracy results achieved using the individual algorithms are presented in Figure 10. Each model was trained five times. Each time, the training set was randomly selected from the entire set of data features. The presented results were achieved using half of the database as training data. The SVM algorithm performed the best, achieving an efficiency of 96%. The RF and GB algorithms also demonstrated effectiveness in classifying OLTC defects. Their efficacy was slightly worse than that of SVM. DT is susceptible to overfitting, especially in classification cases containing many classes. An example of such a situation is the presented OLTC defect classification. This is the reason for the relatively low effectiveness of DT in the proposed studies. Using many decision trees, the RF algorithm reduces the risk of overtraining. Using a weak classifier more advanced than, for example, tree stumps (one-level DT) made the model susceptible to overtraining. Additionally, the large number of classes in the training set increased the chance of this unfavorable situation. AdaBoost using DT turned out to be the worst classifier. The parameters of the most effective model are presented in Table 4.

Figure 11 presents the classification accuracy depending on the size of the training set. The dataset consists of features extracted from over 3500 tap changes. The smallest training set tested contained 1% of the total dataset. The rest of the data that were not used for training were used to test the model. Initially, one may notice a strong correlation between training dataset size and performance. More than 30% of the data were used as training data, and no significant increase in effectiveness was observed. The size of the created measurement database allows for effective training and validation of ML models.

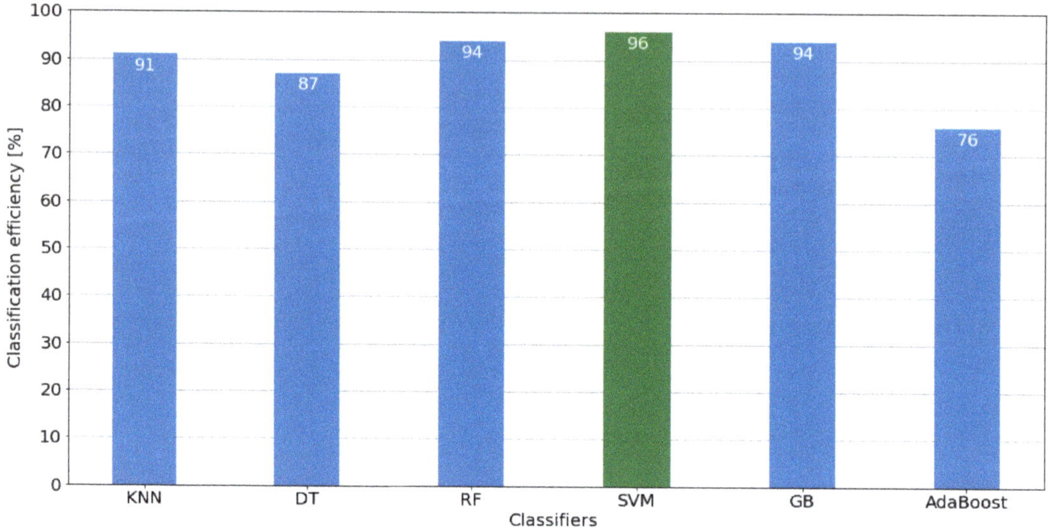

Figure 10. Classification accuracy for half of the dataset used as the training set. The efficiency of the best classifier is marked in green.

Table 4. SVM model parameters optimized using Bayesian optimization.

Kernel	Gamma	C
RBF	0.087	3.136

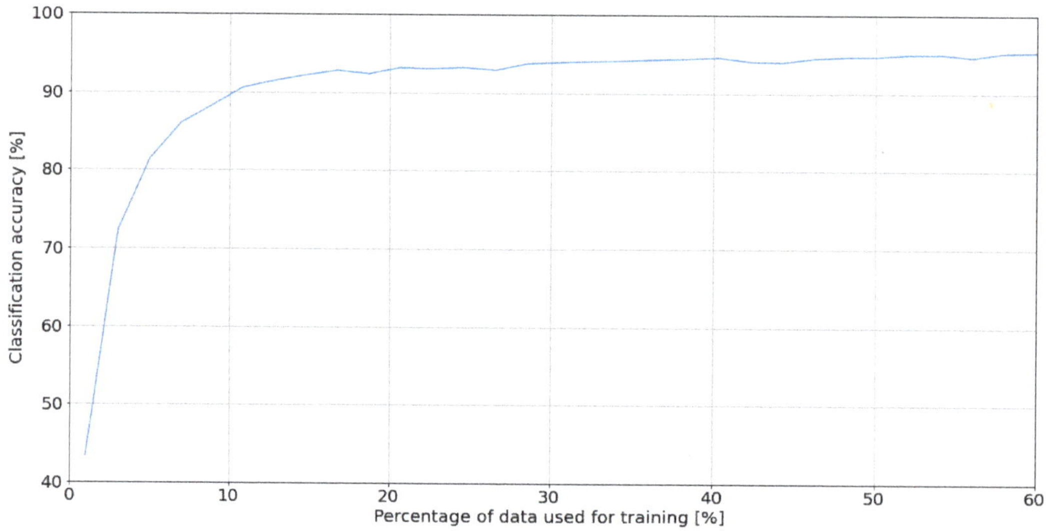

Figure 11. The efficiency of SVM classification for different sizes of the training set.

An analysis of the prediction effectiveness for individual classes of OLTC defects was performed for the best classifier. The results are shown in Figure 12. Every bar represents the percentage rate at which the corresponding class was predicted correctly. Each fault was diagnosed with a high efficiency of over 90%. The feature extraction method combined with SVM allows for determining the good technical condition of the OLTC, with 100%

compliance. It can be seen that the defect recognized with the greatest efficiency was Class 4, which corresponds to contact wear of 4 mm.

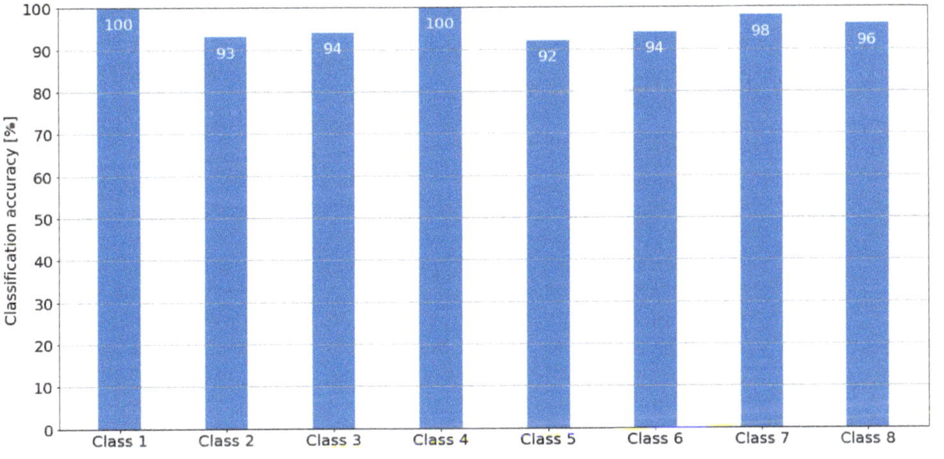

Figure 12. Classification accuracy of SVM for individual classes.

A confusion matrix allows for the determination of which defects are confused with one another. The confusion matrix for SVM, shown in Figure 13, was normalized over columns corresponding to accuracy in successful class prediction. The percentage values on the diagonal represent how many predictions for each class were made correctly. For the SVM model, the class with the least effective prediction was Class 2 (symmetrical contact wear by 1 mm). This was mainly confused with classes describing other degrees of contact wear.

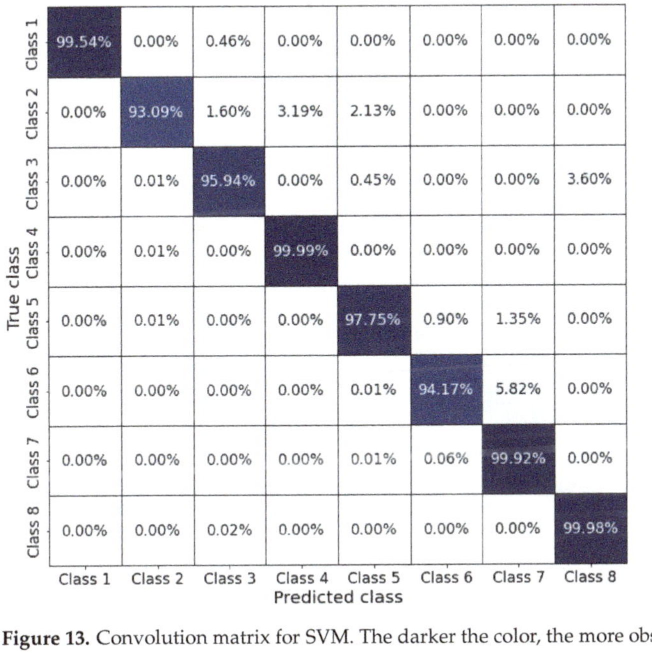

Figure 13. Convolution matrix for SVM. The darker the color, the more observations in a given area.

The most common classification error was incorrectly classifying a damaged spring (Class 7) as asymmetrically worn contacts (Class 6).

Generally speaking, each defect was classified with high efficiency. This is a satisfactory result considering the high level of detail used in dividing the causes of OLTC failure. Particular attention should be paid to the increased effectiveness of the proposed method in determining the absence of a defect. The SVM model produced no false positive predictions for Class 1.

4. Discussion

The research results presented in this article confirm the possibility of diagnosing OLTC faults using EA signals. The effectiveness in determining the OLTC defects was similar to the results presented in the literature [21–23]. However, our research used a measurement database containing a much larger number of defects occurring during OLTC operation. The possibility of diagnosing varying degrees of symmetrical contact wear, along with spring damage, contact deformation, and asymmetrical contact wear, was examined. The lack of false positive predictions of a properly functioning OLTC allows us to define the proposed tool as useful for maintaining OLTCs. Detailed fault diagnosis is expected to allow for the intelligent planning of transformer shutdowns for maintenance, reducing the risk of failure. The AE diagnostic method is a non-invasive method that will enable the creation of a system that monitors the OLTC status online, returning information about the OLTC status after each switchover. However, before introducing such a system, tests should be conducted on real OLTCs operating with transformers. The impact of noise on the fault classification performance should be investigated and, if necessary, a method should be developed to reduce noise in the signals.

The presented feature extraction method allowed us to achieve satisfactory results for all of the tested ML algorithms. The original measurement database created for this research contained large samples. This allowed for effective training of the ML models and reliable performance verification. It is expected that the proposed method of analyzing AE signals will allow for the effective diagnosis of various types of OLTC. For each type of switch, a database should be created storing AE measurements for devices with defects. It is proposed that such a database would contain similar amounts of data to that created for the purposes of the present research.

Future work will focus on determining additional descriptors to increase the effectiveness of diagnostic ML models. We also plan to test the efficacy of other artificial intelligence tools, such as neural networks. Convolutional neural networks are expected to enable further improvement of the diagnostic method using AE. Work will be carried out to determine the possibility of using the method described in this article to combine mechanical and electrical aspects of OLTC diagnostics.

The research presented here is part of broader research aimed at developing an online OLTC condition monitoring system.

Author Contributions: Conceptualization, A.C.; methodology, A.C.; software, M.W.; validation, A.C.; investigation, M.W.; data curation, M.W.; writing—original draft preparation, M.W.; visualization, M.W.; supervision, A.C.; project administration, A.C. All authors have read and agreed to the published version of the manuscript.

Funding: This research received no external funding.

Institutional Review Board Statement: Not applicable.

Informed Consent Statement: Not applicable.

Data Availability Statement: Data are contained within the article.

Conflicts of Interest: The authors declare no conflicts of interest.

References

1. Liu, X.; Aichhorn, A.; Liu, L.; Li, H. Coordinated Control of Distributed Energy Storage System with Tap Changer Transformers for Voltage Rise Mitigation under High Photovoltaic Penetration. *IEEE Trans. Smart Grid* **2012**, *3*, 897–906. [CrossRef]
2. Nadolny, Z. Design and Optimization of Power Transformer Diagnostics. *Energies* **2023**, *16*, 6466. [CrossRef]
3. Hong, K.; Pan, J.; Jin, M. Transformer Condition Monitoring Based on Load-Varied Vibration Response and Gru Neural Networks. *IEEE Access* **2020**, *8*, 178685–178694. [CrossRef]
4. Osmanbasic, E.; Skelo, G. Tap Changer Condition Assessment Using Dynamic Resistance Measurement. *Procedia Eng.* **2017**, *202*, 52–64. [CrossRef]
5. Ismail, F.B.; Mazwan, M.; Al-Faiz, H.; Marsadek, M.; Hasini, H.; Al-Bazi, A.; Yang Ghazali, Y.Z. An Offline and Online Approach to the OLTC Condition Monitoring: A Review. *Energies* **2022**, *15*, 200–500. [CrossRef]
6. Baran, P.; Varetsky, Y.; Kidyba, V.; Pryshliak, Y.; Sabadash, I. A DRM Technique for OLTC Testing by the Transformer Secondary Winding DC Voltage Curve. *Prz. Elektrotechniczny* **2022**, *98*, 17. [CrossRef]
7. Shang, R.; Peng, C.; Shao, P.; Fang, R. FFT-Based Equal-Integral-Bandwidth Feature Extraction of Vibration Signal of OLTC. *Math. Biosci. Eng.* **2021**, *18*, 1966–1980. [CrossRef] [PubMed]
8. Rivas, E.; Burgos, J.C.; García-Prada, J.C. Condition Assessment of Power OLTC by Vibration Analysis Using Wavelet Transform. *IEEE Trans. Power Deliv.* **2009**, *24*, 687–694. [CrossRef]
9. Kim, W.; Kim, S.; Jeong, J.; Kim, H.; Lee, H.; Youn, B.D. Digital Twin Approach for On-Load Tap Changers Using Data-Driven Dynamic Model Updating and Optimization-Based Operating Condition Estimation. *Mech. Syst. Signal Process* **2022**, *181*, 109471. [CrossRef]
10. Yan Yan, Y.Y.; Yan Yan, H.M.; Hongzhong Ma, D.S.; Dongdong Song, Y.F.; Yang Feng, D.D. OLTC Fault Diagnosis Method Based on Time Domain Analysis and Kernel Extreme Learning Machine. *J. Comput.* **2022**, 33, 91–106. [CrossRef]
11. Wotzka, D.; Cichon, A.; Manowski, P. Classification of OLTC Defects Based on AE Signals Measured by Two Different Transducers. In Proceedings of the 2019 20th International Scientific Conference on Electric Power Engineering, EPE 2019, Kouty nad Desnou, Czech Republic, 15–17 May 2019.
12. Raja Pagalavan, B.; Gayathri Monicka, J.; Somakumar, R. Fault Diagnosis of On-Load Tap-Changer Using Acoustic Emission Techniques. *Int. J. Sci. Technol. Res.* **2019**, *8*.
13. Wotzka, D.; Cichoń, A. Study on the Influence of Measuring Ae Sensor Type on the Effectiveness of Oltc Defect Classification. *Sensors* **2020**, *20*, 3095. [CrossRef] [PubMed]
14. Shanker, T.B.; Nagamani, H.N.; Punekar, G.S. Acoustic Emission Signal Analysis of On-Load Tap Changer (OLTC). In Proceedings of the 2013 IEEE 1st International Conference on Condition Assessment Techniques in Electrical Systems, CATCON, Kolkata, India, 6–8 December 2013.
15. Zhang, S. Challenges in KNN Classification. *IEEE Trans. Instrum. Meas.* **2021**, *71*, 1–14. [CrossRef]
16. Gabbay, F. *Computational Optimizations for Machine Learning*; MDPI: Basel, Switzerland, 2022; ISBN 978-3-0365-3187-8.
17. Ponomarenko, A.; Tatarintsev, S. Incorporating Financial Development Indicators into Early Warning Systems. *J. Econ. Asymmetries* **2023**, *27*, e00284. [CrossRef]
18. Zhu, M. Implementation of Support-Vector Machine Algorithm to Develop a Model for Electronic Commerce Energy Regulatory System. *Energy Rep.* **2021**, *7*, 2703–2710. [CrossRef]
19. Bentéjac, C.; Csörgő, A.; Martínez-Muñoz, G. A Comparative Analysis of Gradient Boosting Algorithms. *Artif. Intell. Rev.* **2021**, *54*, 1937–1967. [CrossRef]
20. Ghimire, D.; Lee, J. Geometric Feature-Based Facial Expression Recognition in Image Sequences Using Multi-Class AdaBoost and Support Vector Machines. *Sensors* **2013**, *13*, 7714–7734. [CrossRef] [PubMed]
21. Liu, J.; Wang, G.; Zhao, T.; Zhang, L. Fault Diagnosis of On-Load Tap-Changer Based on Variational Mode Decomposition and Relevance Vector Machine. *Energies* **2017**, *10*, 946. [CrossRef]
22. Li, S.; Dou, L.; Li, H.; Li, Z.; Kang, Y. An Innovative Electromechanical Joint Approach for Contact Pair Fault Diagnosis of Oil-Immersed On-Load Tap Changer. *Electronics* **2023**, *12*, 3573. [CrossRef]
23. Shang, R.; Peng, C.; Fang, R. A Segmented Preprocessing Method for the Vibration Signal of an On-Load Tap Changer. *Electronics* **2021**, *10*, 131. [CrossRef]

energies

MDPI

Article

Study on the Mechanism Effect of Bending Loads on the Decay-like Degradation of Composite Insulator GFRP Core Rod

Xiaoyu Yi [1,*], Likun Ding [2], Hongliang Liu [1], Jiaxin Zhang [1], Jie Liu [1], Zhaohui Li [3], Yuming Zhang [3], Ping Wang [3] and Jianghai Geng [3]

1 State Grid Hebei Energy Technology Service Co., Ltd., Shijiazhuang 050000, China; dyy_liuhl@he.sgcc.com.cn (H.L.); dyy_zhangjx@he.sgcc.com.cn (J.Z.); dyy_liuj@he.sgcc.com.cn (J.L.)
2 State Grid Hebei Electric Power Co., Ltd., Shijiazhuang 050000, China; dinglk@he.sgcc.com.cn
3 Hebei Key Laboratory of Power Transmission Equipment Security Defense, North China Electric Power University, Baoding 071003, China; 220222213161@ncepu.edu.cn (Z.L.); 120222101036@ncepu.edu.cn (Y.Z.); 51351571@ncepu.edu.cn (P.W.); gengjh@ncepu.edu.cn (J.G.)
* Correspondence: dyy_yixy@he.sgcc.com.cn

Abstract: This paper investigates the deterioration of, and the abnormal temperature rise in, the GFRP core rod material of compact V-string composite insulators subjected to prolonged alternating flexural loading under wind-induced stresses. The axial stress on the GFRP (Glass Fiber Reinforced Plastic) core rod, resulting from transverse wind loads, is a focal point of examination. By establishing a stress model and damage model, the paper simulates and computes the evolution of damage in the outer arc material of composite insulator core rods subjected to alternating flexural loads. Additionally, a multi-factor coupled aging platform is set up, integrating humidity, heat, and mechanical stress, to simulate the crazing deterioration process of composite insulators under alternating flexural loads. Experimental results reveal that during 400,000 alternating load cycles, the core rod underwent stages of surface damage, damage increasing, fatigue embrittlement, matrix hydrolysis, and fiber fracture. Simultaneously, the silicone rubber sheath on the outer side of the composite insulator's bending arc develops cracks over aging time, creating pathways for moisture ingress into the interface and core rod. The dielectric constant and dielectric loss factor of the aging region of the core rod increase to varying degrees compared to the non-aging part. Moreover, the degree of abnormal heating of the samples intensifies with the duration of aging experiments. These findings underscore the significance of understanding the aging and decay-like fracture process of compact line V-string composite insulators. They provide crucial insights for future research aimed at enhancing the material properties of composite insulator core rods.

Keywords: composite insulator; bending load; GFRP core rod; abnormal temperature rise

Citation: Yi, X.; Ding, L.; Liu, H.; Zhang, J.; Liu, J.; Li, Z.; Zhang, Y.; Wang, P.; Geng, J. Study on the Mechanism Effect of Bending Loads on the Decay-like Degradation of Composite Insulator GFRP Core Rod. *Energies* **2024**, *17*, 423. https://doi.org/10.3390/en17020423

Academic Editor: Tomasz Norbert Kołtunowicz

Received: 24 November 2023
Revised: 28 December 2023
Accepted: 11 January 2024
Published: 15 January 2024

1. Introduction

Composite insulators, a crucial component in electrical power systems, consist of silicone rubber sheaths, a fiberglass-reinforced plastic (FRP) core rod, and end fittings. These insulators play a vital role in bearing electrical and mechanical loads in transmission lines [1–4]. The fiberglass-reinforced epoxy resin (GFRP) core rod primarily serves the functions of electrical insulation and mechanical load support. At the same time, the external silicone rubber sheath is used to protect the core rod. It effectively guards against external moisture, dust, and chemical substances, preventing them from corroding the composite insulator core rod. In practical applications, composite insulators operate under diverse and challenging conditions experiencing various incidents that affect the secure and stable operation of transmission lines throughout their long-term service life. These incidents encompass flashovers under specific weather conditions, aging, and abnormal fractures of composite insulators [5–8]. Among these incidents, the anomalous fracture of composite insulators poses the most severe threat to the power system. Line faults resulting from such fractures cannot be promptly resolved within a short time frame, only by the replacement of new insulators to restore the power supply. Consequently, to ensure the secure and stable operation of the power grid, it becomes imperative to delve into the deterioration mechanisms of composite insulators.

Core rods are the main components of composite insulators that undertake mechanical loads. The deterioration mechanism of core rod materials is closely related to the reliability of composite insulator operation. Recent research by several scholars has delved into the artificial aging of composite insulator core rods. For instance, Yanfeng Gao et al. [9] conducted studies investigating core rod deterioration characteristics and designed corresponding simulation tests. Their research revealed that discharges in humid environments contributed to epoxy resin deterioration in the core rod. The microstructural features and epoxy resin deterioration observed in artificially aged samples closely resembled the actual core rod deterioration in practice. Leilei Zeng et al. [10] utilized a vacuum drying oven to conduct wet-heat aging tests on core rod specimens. They confirmed that the degree of core rod material deterioration deepened with increasing wet-heat exposure, as evidenced by scanning electron microscopy, Fourier transform infrared spectroscopy, and thermal analysis. Yongfei Zhao et al. [11] conducted wet-heat aging tests on short core rod samples at different temperatures using a constant-temperature water bath aging chamber. Their findings indicated that under wet-heat conditions, the core rod color darkened and the texture became softer. The surface temperature rise of degraded samples increased with higher wet-heat aging temperatures. The microstructural features and changes in epoxy resin content in artificially aged pieces exhibited good consistency with field-used core rods. Xinming Ma et al. [12] performed aging tests on core rod slices under different humidities and high-electric-field conditions, referencing the dielectric barrier discharge theory. After analyzing the aged samples through scanning electron microscopy, Fourier transform infrared spectroscopy, and X-ray energy-dispersive spectroscopy, they concluded that moisture accelerates the deterioration process of core rods. The studies mentioned above have primarily focused on analyzing the promoting effects of high electric fields and wet-heat environments on the deterioration of composite insulators.

In practical applications, composite materials often need to withstand static or dynamic mechanical loads, such as tension, compression, bending, or shear forces. These mechanical loads may cause microstructural changes in composite materials, thereby affecting their performance. Therefore, understanding the mechanism by which mechanical loads deteriorate composite materials is crucial [13–16]. According to incomplete statistics from a North China power grid company, between 2012 and 2020, they reported 683 cases of abnormal heating in 500 kV composite insulators, with 218 insulators likely having decay-like defects. Among these, 141 were V-string insulators mainly used in compact line configurations, accounting for 65% of the total [17]. The data above show that the composite insulators prone to decay-like defects are primarily compact V-string composite insulators. Therefore, there needs to be more research on the aging mechanisms specific to close V-string composite insulators, considering their operational environment and the characteristics of mechanical loads. The role of bending loads in the deterioration process of V-string composite insulators has yet to receive sufficient attention. The alternating mechanical load is one of the indispensable factors in the decay-like fracture process of composite insulators. There is no clear conclusion on the impact mechanism of mechanical loads on the cracking and deterioration process of composite insulators, as further in-depth research is needed.

In summary, previous research has certain limitations in the level of attention paid to the role of alternating loads in the core rod deterioration process. Furthermore, prior research has mostly used single-factor aging tests to age composite insulators which may differ from the actual aging process of composite insulators. This inspires us to design and conduct multi-factor coupled aging tests to simulate the deterioration mechanism of composite insulators under the interaction of multiple factors. Based on these observations, this paper employs simulation to analyze the microstructural damage evolution of composite insulator core rod materials under alternating bending loads. We establish a multi-factor composite insulator aging platform involving humidity, temperature, and mechanical stress, before going on to conducting artificial aging experiments on full-size composite insulators and studying the deterioration and abnormal temperature rise mechanisms under bending loads. The findings of this study are of significant importance for understanding the deterioration evolution process of V-string composite insulators and provide valuable insights for improving the material properties of composite insulator core rods which help to safeguard the secure operation of the power grid.

2. Microscopic Damage Evolution Simulation of GFRP Materials under Alternating Loads

2.1. Force Analysis of V-String Composite Insulators

When subjected to lateral wind acting parallel and perpendicular to the conductor, V-string composite insulators on compact lines experience axial pressure on the leeward insulator and axial tension on the windward insulator, assuming that the conductor connectors do not rotate. Consequently, the leeward insulator takes an arched bending state under the bending load, as Figure 1a depicts. In this bent state, the length of the line connecting the two ends of the composite insulator is shortened (compared to the normal state) by a distance known as the compression stroke ΔL, as shown in Figure 1b. The thick black line represents the V-string composite insulator in the normal state, while the thick red line represents the state after experiencing compression.

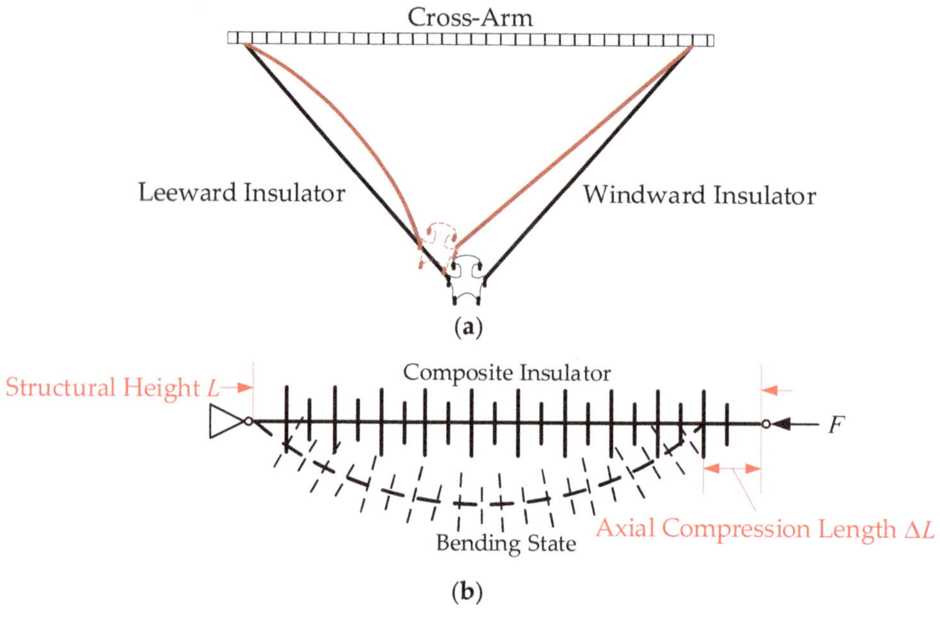

Figure 1. Bending state of v-string composite insulators under lateral wind loads. (**a**) Bending state, (**b**) compression stroke.

For quantitative analysis, the composite insulator with a structural height of L is divided into n segments, each with a length of $L_i = L/n$. It is assumed that the mass of the composite insulator is uniformly distributed, the material parameters are the same at all locations, and each cross-section has a standard circular shape with a diameter of D. After bending, the angular position of each segment of the V-string composite insulator under a lateral wind load is denoted as δ_i ($i = 1, 2, \ldots, n$). The mechanical model for the *ith* segment of the V-string composite insulator under a lateral wind load is depicted in Figure 2.

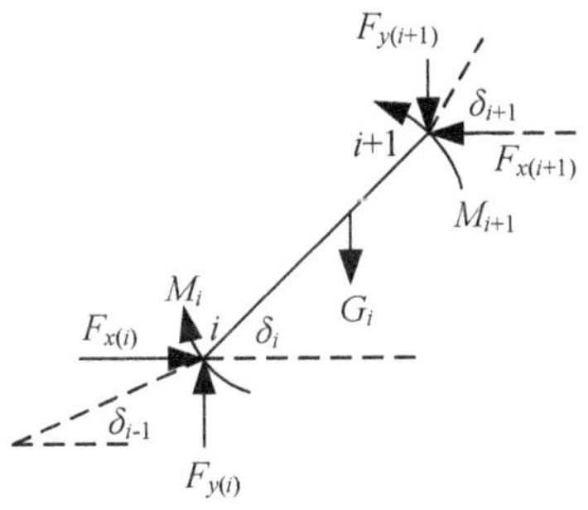

Figure 2. Mechanical model for the *ith* segment of the insulator.

From Figure 2, it can be observed that the condition for the *ith* segment to reach an equilibrium state is as follows:

$$\begin{cases} F_{x(i)} = F_{x(i+1)} \\ F_{y(i)} = F_{y(i+1)} + G_i \\ M_i + \frac{L}{2}\cos\delta_i(F_{y(i)} + F_{y(i+1)}) = M_{i+1} + \frac{L}{2}\sin\delta_i(F_{x(i)} + F_{x(i+1)}) \end{cases} \tag{1}$$

By employing the Newton–Raphson method to solve the above model, the maximum tensile and compressive stresses at any cross-section can be obtained as:

$$\begin{cases} \sigma^{+}_{\max(i)} = \frac{M_i D}{2I} - \frac{P_i}{A} \\ \sigma^{-}_{\max(i)} = -\frac{M_i D}{2I} - \frac{P_i}{A} \end{cases} \quad i = 1, 2, \ldots, n+1 \tag{2}$$

$$P_i = \begin{cases} F_{x(1)}\cos\delta_1 + F_{y(1)}\sin\delta_1 \quad i = 1 \\ F_{x(i)}\cos(\frac{\delta_{i-1}+\delta_i}{2}) + F_{y(i)}\sin(\frac{\delta_{i-1}+\delta_i}{2}) \quad i = 1, 2, \ldots, n \\ F_{x(n+1)}\cos\delta_{n+1} + F_{y(n+1)}\cos\delta_{n+1} \quad i = n+1 \end{cases} \tag{3}$$

If we consider one end of the composite insulator as the coordinate origin, the coordinates of any other point on the composite insulator can be expressed as:

$$\begin{cases} x_i = \sum\limits_{k}^{i-1} L_k\cos\delta_k \\ y_i = \sum\limits_{k}^{i-1} L_k\sin\delta_k \end{cases} \quad i = 1, 2, \ldots, n+1 \tag{4}$$

In the above equations, G_i represents the self-weight of each segment, $F_{x(i)}$ and $F_{y(i)}$ denote the horizontal and vertical forces, M_i and M_{i+1} are the bending moments at the two ends of the *ith* segment, D is the insulator cross-sectional diameter, I is the cross-sectional moment of inertia ($I = \pi D^4/64$), and A is the cross-sectional area.

The method outlined provides an analytical relationship between the axial compression and maximum stress in the composite insulator [18]. To calculate the critical bending load for the composite insulator, it can be viewed as an elongated strut with hinged supports at both ends. The critical bending load F in Figure 1b can be expressed as follows:

$$F = \frac{\pi^2 EI}{L^2} \tag{5}$$

$$I = \frac{\pi D^4}{64} \tag{6}$$

In the equation, E represents the elastic modulus of the composite insulator core rod and I is the moment of inertia in the rod's cross-section. D is the core rod diameter and L is the height of the composite insulator structure.

2.2. The 3D Model and Material Constitutive Model of GFRP Core Rod Material

For one typical I-string composite insulator, 30% to 50% of the year is influenced by low-level wind-induced vibrations. Characteristics of these wind-induced vibrations include high frequency and small amplitude; both need a significantly extended period to result in fatigue life. Usually, there are over 10 million cycles during operation and at least 30 million cycles in the experiment. Consequently, such low-level wind-induced vibrations have minimal impact on the mechanical performance of composite insulators. However, compact V-string composite insulators have a higher probability of experiencing secondary distance vibrations under the lateral wind. Moreover, the split wires can induce composite insulator bending motion through torsional and secondary distance vibrations [19,20]. The alternating bending loads associated with these vibrations have larger amplitudes and lower frequencies. Therefore, they can reach their fatigue life in a relatively short period of time (usually less than one million cycles). As a result, they significantly impact the mechanical performance of composite insulators. In summary, the alternating bending loads significantly contribute to the relatively higher occurrence of decay-like fracture in compact V-string composite insulators.

There is a significant difference in strength between the epoxy resin matrix and the fiberglass in GFRP core rod material. If the GFRP core rod material is regarded as a single-phase anisotropic homogeneous material, it will lead to distortion in the results when we calculate the properties of the fiber. The reason for this is that the fiber properties are replaced by the properties of the matrix that are more prone to damage, which does not accurately represent the material's behavior. Additionally, the damage and failure process at the fiber–matrix interface cannot be ignored. Therefore, in this section, the simulation and calculation of the damage evolution process of the material on the outer side of the bending arc of the GFRP core rod under alternating bending loads are performed using multiphase material finite element analysis and zero-thickness cohesive zone modeling.

A model is created using finite element analysis software, Abaqus 2021, with material parameters shown in Table 1. The model is a cuboid of GFRP material measuring 200 μm × 200 μm× 50 μm, extracted from the outer side of a core rod's bending circular arc. One face of the model corresponds to the outer surface of the core rod's bending circular arc, while the remaining faces are entirely contained within the core rod. The model includes 12 fibers with a diameter of 20 μm, as depicted in Figure 3.

Table 1. Material model parameters.

Material Phase	Parameter	Value
Fiberglass	Density	970 kg/m^3
	Modulus of elasticity	E_1 = 89,000 MPa, $E_2 = E_3$ = 8610 MPa
	Poisson's ratio	$\sigma_{12} = \sigma_{13}$ = 0.3, σ_{23} = 0.45 MPa
	Shear modulus	$G_{12} = G_{13}$ = 4160 MPa, G_{23} = 3000 MPa
Epoxy Matrix	Density	1200 kg/m^3
	Modulus of elasticity	3450 MPa
	Poisson's ratio	0.35
	Shear modulus	0
	Yield stress	74.7 MPa
Interface	Normal strength	1000 MPa
	Shear strength	385 MPa

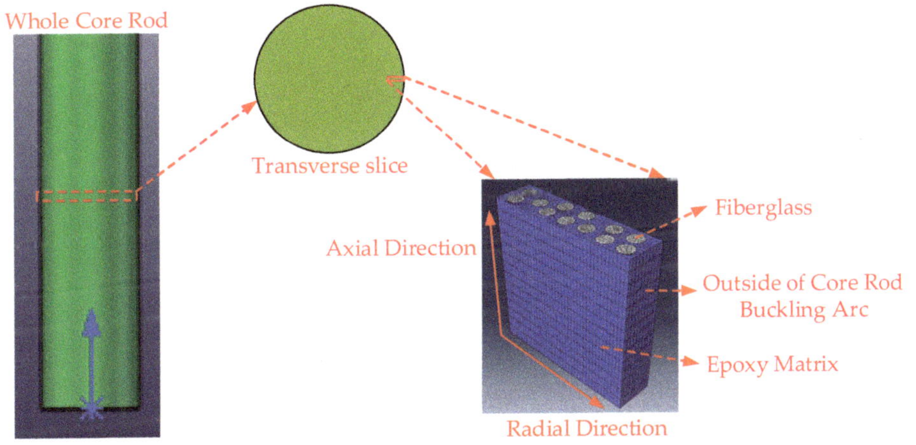

Figure 3. Three-dimensional polyphase model of GFRP core rod material.

Before conducting calculations, it is essential to define the material constitutive models for the matrix and fibers using the VUMAT (user subroutine for defining material constitutive relations) within the Abaqus software.

$$E = (1 - d_m)E_0 \tag{7}$$

$$d_m = \frac{\overline{u}^{pl}}{\overline{u}_f^{pl}} = L\frac{\varepsilon^{pl}}{\overline{u}_f^{pl}} \tag{8}$$

$$\overline{u}_f^{pl} = \frac{2G_f}{\sigma_y} \tag{9}$$

In the formula, E is the stiffness after the matrix damage evolution. E_0 is the stiffness before the damage and d_m varies from 0 to 1 as the stiffness decreases when the stiffness is 0, d_m = 1. $\overline{u}^{pl}$ stands for plastic deformation. $\overline{u}_f^{pl}$ is the plastic deformation during matrix damage failure, ε^{pl} represents the equivalent plastic strain, and L is the element length. G_f is the fracture energy per unit area and σ_y is the material yield strength.

First, the fiberglass is defined as transversely isotropic along the axial direction with no plastic deformation stage. When the minimum principal stress at a node exceeds the compressive strength limit or the maximum principal stress exceeds the tensile strength limit, it is considered fiber fracture failure. The corresponding elements will be removed. Second, the epoxy resin matrix is defined as an isotropic elastic–plastic material. According to the Mises yield criterion, matrix damage is initiated when the maximum principal stress in a matrix element exceeds the strength limit, resulting in a stiffness reduction, as expressed in Equation (7). At this point, the matrix elements experience damage failure and are deleted. After defining the material constitutive models, the fiber–matrix interface considered as a pure viscous interface with zero thickness and no material properties is assigned. This section employs cohesive interface elements based on the traction–separation law to simulate the fiber–matrix interface.

2.3. Damage and Deterioration Process of GFRP Core Rod Material under Alternating Loads

A composite insulator specimen with an insulating distance of 1000 mm, a core rod diameter of 18 mm, and a sheath thickness of 4 mm can be taken as an example. Experimental results show that its maximum axial compression displacement is 300 mm. By selecting 30% of the maximum axial compression displacement as the amplitude of the alternating bending load for the entire core rod, it can be proportionally scaled to a micro-scale model, resulting in an alternating load amplitude (L_m) of 48 μm. With an alternating load frequency set at 0.5 Hz, the number of cycles of alternating loads (bending and resetting as one cycle) for various load amplitudes is specified as 100,000 cycles, 200,000 cycles, 300,000 cycles, and 400,000 cycles. The deterioration status of the GFRP core rod material under these conditions is illustrated in Figure 4.

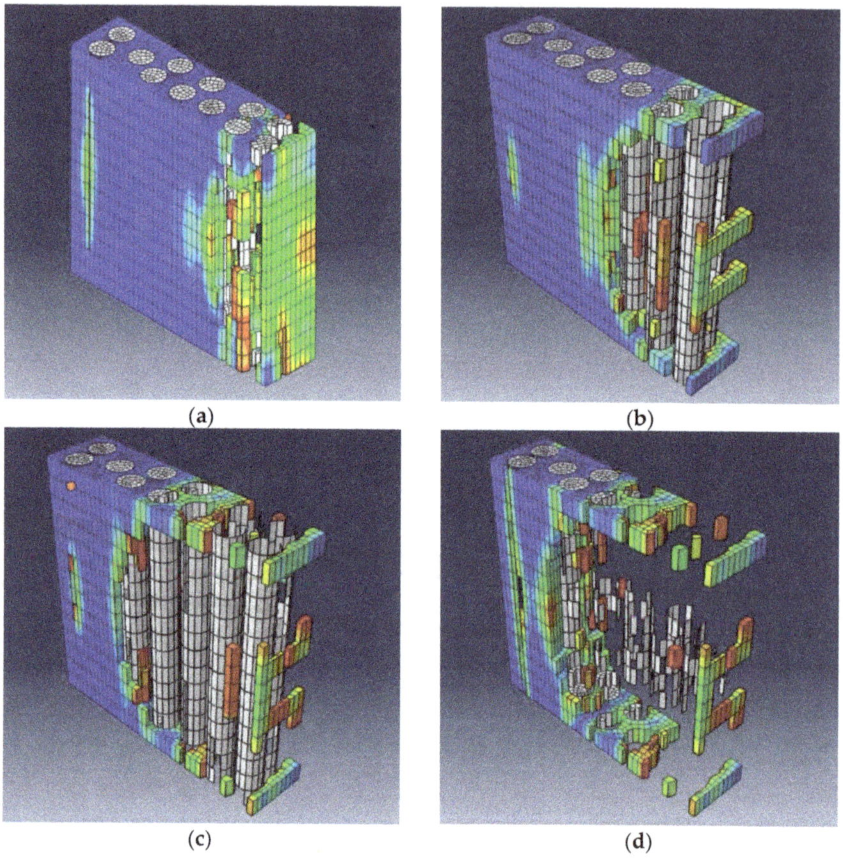

Figure 4. Deterioration process of GFRP core rod material under bending loads; (**a**) N = 1 × 10^5; (**b**) N = 2 × 10^5; (**c**) N = 3 × 10^5; (**d**) N = 4 × 10^5.

The blue color in Figure 4 represents the initial color of the epoxy resin matrix unit. The initial color of the fiberglass unit is white. As fatigue intensifies, the color of the epoxy resin unit gradually develops from blue to light green, green, yellow, orange, light red and finally red. Figure 4 provides a visual representation of the damage evolution process of GFRP core rod materials under the influence of alternating bending loads. As the number of cycles of alternating loads increases, the GFRP core rod material gradually experiences fatigue and damage. Fatigue is initiated on the outer surface of the core rod's bending arc and propagates toward the core rod's interior. Simultaneously, with an increase in the amplitude of the alternating load, the fatigue rate accelerates, leading to a greater extent of damage. After subjecting the core rod to 100,000 cycles of alternating loads, the epoxy resin matrix on the outer surface of the core rod experiences significant stress. When an element matrix surpasses the maximum principal stress, the element matrix is deleted, indicating damage and detachment in that portion of the epoxy matrix. The detached matrix debris accumulates at the core rod–sheath interface. After applying alternating loads until 200,000 cycles, the glass fibers undergo fracture damage once they lose the protection of the epoxy matrix. The broken fiber debris detaches from the core rod's surface, accumulating at the core rod–sheath interface. Upon reaching 400,000 cycles of alternating loads, the material on the core rod's surface, with a thickness of approximately 100 μm, experiences varying degrees of damage. The decay-like composite insulators retrieved from sites also exhibit a significant accumulation of white cotton-like residues at the local core rod–sheath interface, as depicted in Figure 5. This phenomenon further confirms the rationality of the simulation results.

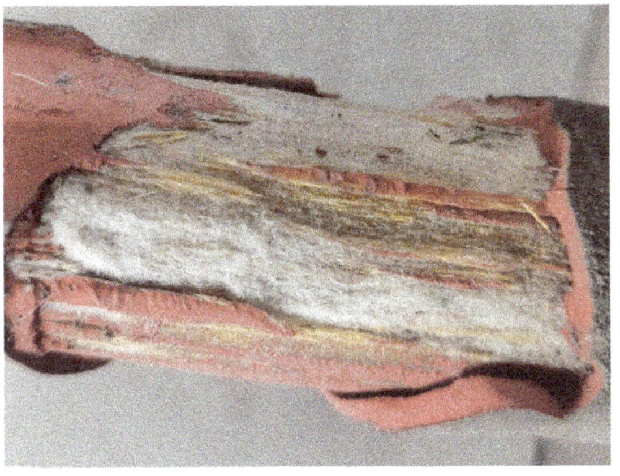

Figure 5. Debris at the core rod–sheath interface of a brittle composite insulator.

In summary, the outer surface of the composite insulator core rod's bending arc under alternating loads experiences fatigue damage. As the number of cycles of alternating loads increases, the damaged area gradually expands and propagates toward the core rod's interior. The mixture of debris generated in the damaged area, comprising epoxy matrix and fibers, accumulates at the core rod–sheath interface. These observations have been corroborated in actual instances of decay-like composite insulators [21].

3. Multi-Factor Aging Test of Composite Insulators under Wet-Hot and Alternating Loads

3.1. Sample Selection

The composite insulator samples elected for this experiment were provided by Xiangyang State Grid Composite Insulator Co., Ltd. (Xiangyang, China). The specific parameters of the specimens are shown in Table 2.

Table 2. Parameters of composite insulator samples.

Parameter	Value
Rated mechanical tensile load	70 kN
Insulation distance	1000 mm
Min nominal creepage distance	3150 mm
Number of umbrella skirts (large/small)	25(9/16)
Core rod diameter	18 mm
Sheath thickness	4 mm
Maximum axial compression	300 mm

3.2. Experimental Setup

An artificially accelerated aging test platform with multiple factors, including humidity, temperature, and mechanical load, was constructed to simulate the actual operating conditions of V-string composite insulators. The main components of this test platform include a temperature and humidity control module and an alternating load application module. The alternating load application module was primarily utilized to mimic the unique mechanical characteristics of compact line V-string composite insulators. It accomplishes this by employing a servo actuator cylinder to exert reciprocating compressive and reset motions on the composite insulator, thereby applying alternating flexural loads.

Referring to the environmental conditions in areas where composite insulators are prone to deterioration in China, the artificially accelerated aging test of composite insulators was set with an ambient temperature of 20–25 °C and a relative humidity of 75–80% RH. According to DL/T 1580-2016 [22], GB/T 34937-2017 [23], and GB/T 13096-2008 [24], combined with the operating condition on-site of composite insulators, it was determined that the axial compression amount of the alternating bending load applied in the test is 90 mm and the frequency is 0.5 Hz. The composite insulator subjected to bending loads is shown in Figure 6.

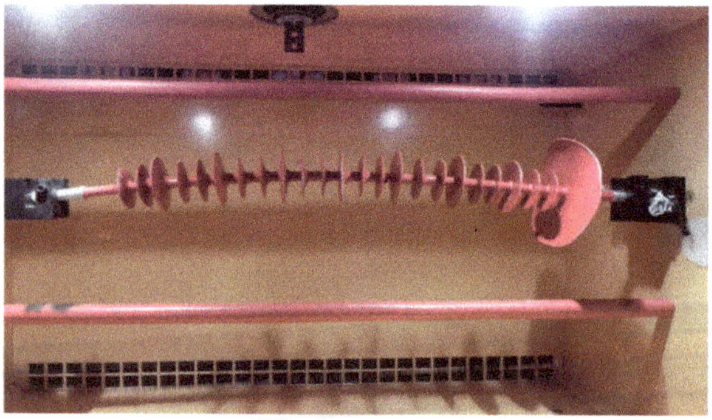

Figure 6. Schematic diagram of the composite insulator bearing a bending load.

3.3. *Testing and Characterization*

3.3.1. Scanning Electron Microscopy Testing

Microscopic morphology analysis was conducted on surface material samples of composite insulator core rods at various aging stages using the SEM 230 scanning electron microscope produced by Alfa Chemistry Company in New York, NY, USA. The purpose was to investigate the changes in surface morphology at the point of buckling deterioration in GFRP core rods.

3.3.2. Thermogravimetric Analysis

The PerkinElmer STA6000 thermal gravimetric analyzer produced by PerkinElmer, Waltham, MA, USA, was employed to examine the thermal decomposition characteristics of surface material samples from composite insulator core rods. The test covered a temperature range of 30 °C to 800 °C, featuring a temperature rise rate of 10 °C/min, and utilized nitrogen as a protective gas.

3.3.3. Fourier Transform Infrared Spectroscopy Analysis

The IRTracer-100 Fourier transform infrared spectrometer produced by Shimadzu Corporation in Kyoto, Japan, conducted attenuation total reflection scanning tests on surface material samples from composite insulator core rods at various aging stages. The distinction in functional group content before and after aging was determined by analyzing the distinct absorbance peaks of functional groups on the characteristic infrared light peaks.

3.3.4. Charged Temperature Rise Experiment

Different composite insulator samples subected to varying alternating load cycles were tested for electrical temperature rise. To make the heating phenomenon more pronounced, the composite insulator samples were placed in a constant temperature and humidity chamber at 20 °C and 75% RH for 150 h before the test to ensure saturated moisture absorption [25,26]. Afterward, a 64 kV, 50 Hz AC voltage was applied to the high-voltage end fittings of the composite insulator while grounding the low-voltage end fittings. One hour after applying the AC voltage, the temperature rise images of the composite insulator were recorded using a FLIR T1040 handheld thermal imaging camera produced by Teledyne FLIR in Wilsonville, OR, USA.

3.3.5. Moisture Absorption Test

The most severely deteriorated section of the core rod from the composite insulator sample was used to prepare test specimens. The specific preparation steps are as follows. First, we cut a 40 mm long segment from the location with the maximum flexural deflection, which is the red-dashed rectangle area in Figure 7. Secondly, we prepared this segment into 20 pieces of specimens with a thickness of 2 mm and a diameter of 18 mm, as shown in Figure 7. After that, we used 100-grit sandpaper to uniformly sand the cross-section of the prepared samples. Continually, after grinding, we placed the pieces in a constant temperature and humidity chamber at 50 °C for 72 h. Finally, we measured each slice's dielectric loss factor and the saturated moisture absorption rate after drying.

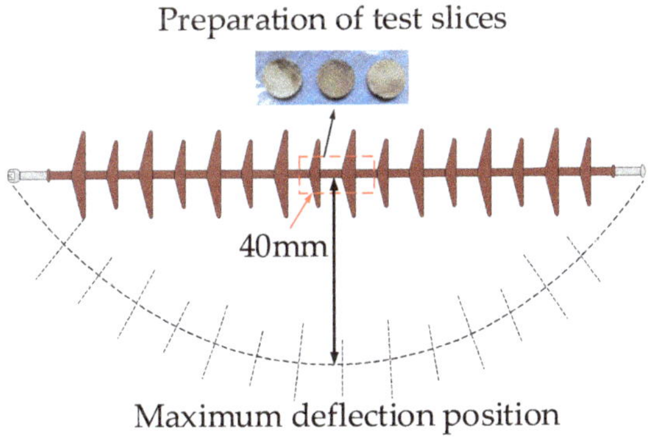

Figure 7. Schematic diagram of the sampling area and preparation for test pieces.

We conducted moisture absorption and weight gain tests on core rod materials in accordance with ASTM D5229/D5229M-20 [27]. The FA224C electronic analytical balance produced by the Shanghai Lichen Instrument Technology Co., Ltd. in Shanghai, China was selected for testing, with an accuracy of 0.1 mg. Each experiment was set up with 3 repeated samples. Samples were regularly taken from deionized water for weighing, and then quickly immersed back into ionized water to continue the moisture absorption test. The weighing results were taken as the average of the 3 samples, and the process was repeated until the weight of the samples no longer changed, that is, when they reached the saturation and moisture absorption level. The formula for calculating the moisture absorption rate M_t of the sample is as follows [28]:

$$M_t = \frac{W_t - W_0}{W_0} \times 100\% \tag{10}$$

In the formula, W_0 is the initial weight of the sample and W_t is the weight of the sample at time t. The average of the saturation moisture absorption rates measured on 20 test pieces of each core rod was taken to characterize the saturation moisture absorption rate of the corresponding brittle composite insulator test piece at a temperature of 20 °C.

3.3.6. Dielectric Loss Factor Test

The dielectric loss factor of the test piece was measured using the YG9100 fully automatic anti-interference precision dielectric loss tester produced by Shanghai Yanggao Electric Appliance Co., Ltd., Shanghai, China. Before the test, the upper and lower cross-sections of the test piece were uniformly pasted with tin foil. This ensured full contact between the cross-section and the measuring electrode, avoiding discharge interference from air gaps. The test frequency was 50 Hz. We then took the average value of the dielectric loss factor measured on 20 test pieces of each core rod to characterize the dielectric loss factor of the corresponding degraded composite insulator test piece.

3.4. *Experiment Result*

3.4.1. Appearance Inspection

Before conducting the test, pre-treatment of the specimens was performed. The sample was visually inspected to ensure they were free from any visible damage. Subsequently, the specimens' surfaces were cleaned using deionized water. After cleaning, the sample was placed in a constant temperature and humidity chamber at 50 °C for 72 h to ensure thorough drying. After drying, the sample was transferred to the wet–thermal–mechanical multifactor composite insulator artificial accelerated aging test platform to initiate the aging test. Throughout the testing process, observations were made to record any changes in the appearance of the specimens. Furthermore, the protective sheath was removed for a comparative analysis of the core rod's morphology.

As shown in Figure 8, after 100,000 cycles of alternating load applications, the external appearance of the composite insulator exhibited no significant changes. However, after 200,000 cycles, cracks appeared in the insulator sheath. When the number of alternating load cycles reached 400,000, the insulator sheath exhibitd noticeable and extensive cracking in multiple areas.

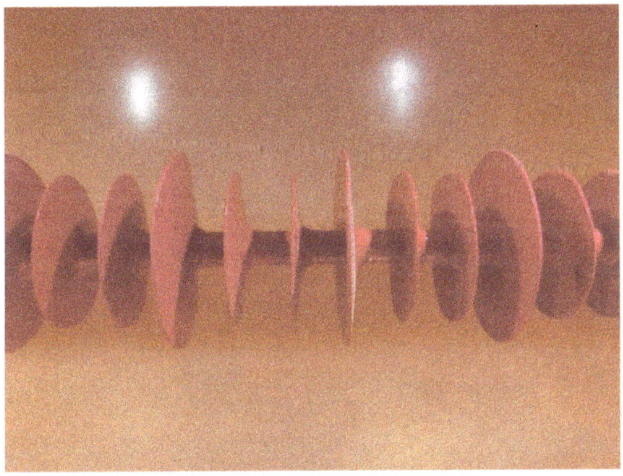

(**a**)

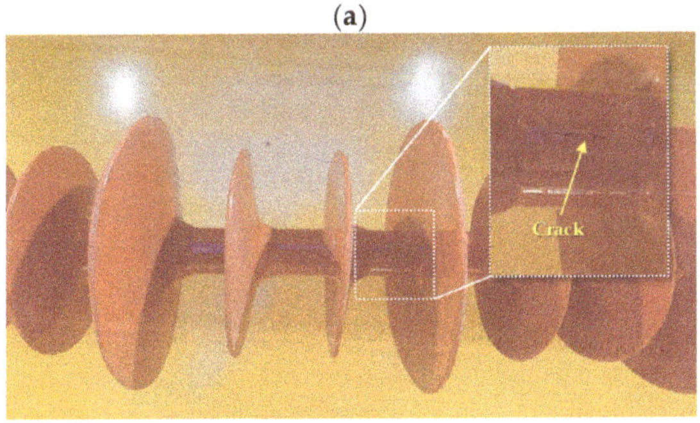

(**b**)

Figure 8. *Cont.*

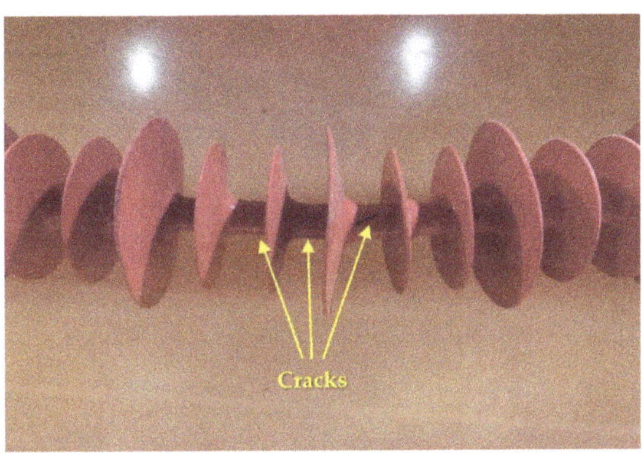

(**c**)

Figure 8. Appearance of composite insulators tested after applying different number of times of alternating loads: (**a**) 100,000 times, (**b**) 200,000 times, and (**c**) 300,000 times.

To further examine the deteriorated appearance of the core rod, a strip process was performed on the aged composite insulator sheath. The morphology of the core rod at the location with the maximum flexural deflection in the stripped composite insulator is depicted in Figure 9.

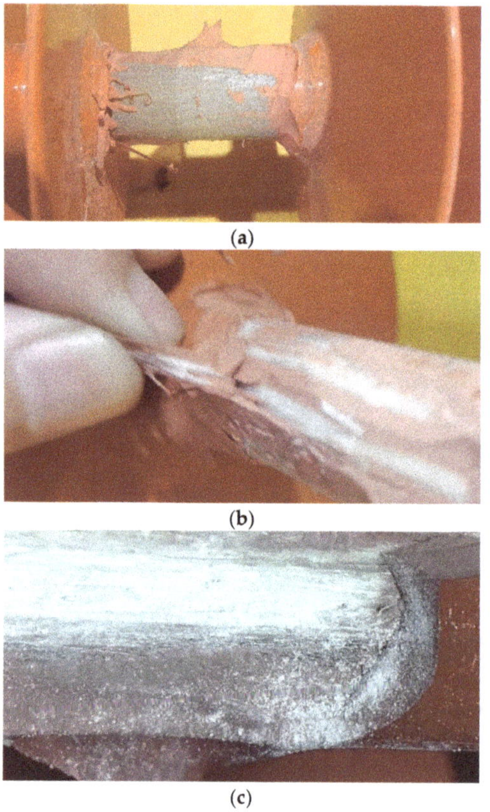

Figure 9. Appearance of composite insulator core rod tested after applying different number of times of alternating loads: (**a**) 100,000 times, (**b**) 200,000 times, and (**c**) 300,000 times.

Figure 9 shows that with an increasing number of cycles of alternating load applications, the outer part of the core rod's curved arc gradually deteriorated. After 100,000 cycles of alternating load applications, the adhesion between the composite insulator sheath and the core rod remained in good condition. The core rod's surface slightly whitened, with no matrix fragments spilling out. After 200,000 cycles, partial delamination of the insulator sheath occurred, and there were some core rod fragments at the interface between the sheath and the core rod. When the number of alternating load cycles reached 400,000, it was found that a significant amount of white matrix powder spilled out when the composite insulator sheath was stripped. The core rod showed signs of hydrolysis, and its color changed from pale green to light yellow.

3.4.2. Microscopic Morphology Analysis

Figure 10 shows the microstructure of the core rod under four different fatigue states observed using a scanning electron microscope. Comparative analysis reveals that the extent of deterioration on the outer side of the core rod's curved arc surface deepened with an increasing number of alternating bending load application cycles. After 100,000 cycles of alternating load applications, the glass fibers on the core rod surface remained intact without any signs of breakage. The epoxy resin also adhered well to the fibers, with no exposed glass fibers. However, there were instances of interface cracking between some fibers and the epoxy matrix, as shown in the red-dashed area in Figure 10. After 200,000 cycles of alternating loads, the core rod's surface exhibited increased fatigue damage at locations with more significant flexural deflection. Still, no apparent hydrolysis suggests that moisture did not significantly penetrate the core rod area. After 300,000 cycles of alternating loads, the core rod's surface epoxy matrix was severely damaged, and signs of deterioration were present, indicating that moisture had penetrated the interior of the composite insulator sheath in significant amounts and the matrix had started hydrolysis. After 400,000 cycles of alternating loads, the epoxy resin matrix material had undergone extensive hydrolysis at the severely deteriorated areas of the core rod's surface. Some epoxy resin matrix fragments adhered to the fiber surface. Some fibers lost the epoxy resin matrix's protection and were fractured [29,30].

3.4.3. Thermogravimetric Analysis Results

By heating and measuring the weight of samples at different temperatures, the epoxy resin content in the core rod could be analyzed, and the degree of deterioration of artificially aged core rods could be quantitatively analyzed. The thermogravimetric test results at the buckling point of composite insulator core rods before and after aging are shown in Figure 11. The epoxy resin decomposition temperature was reached when the temperature rises to 320 °C. When the temperature reached 430 °C, the epoxy resin was basmati-factorially decomposed, and the remaining masses of the samples before and after aging accounted for 81.63% and 89.95%, respectively. After the multi factor aging test, the epoxy resin in the core rod of the test sample degraded and the content decreased.

3.4.4. Fourier Transform Infrared Spectroscopy Analysis Results

Through Fourier transform infrared spectroscopy analysis, it can be seen that there is no significant difference in characteristic peaks between the core rod after multiple factors aging and the unaged core rod. However, there was a difference in intensity between the same characteristic peak. As shown in Figure 12, the strength of C=O and Si−O in the core rod material had decreased to a certain extent, while the strength of O−H had increased, indicating that as aging progresses, the continuous invasion of moisture into the core rod in a high humidity environment leads to the continuous deterioration of the composite material.

From the above physical and chemical analysis results, it can be seen that as aging progresses, water continuously invades the composite insulator core rod, the epoxy resin matrix continuously hydrolyzes, and the degree of deterioration also intensifies.

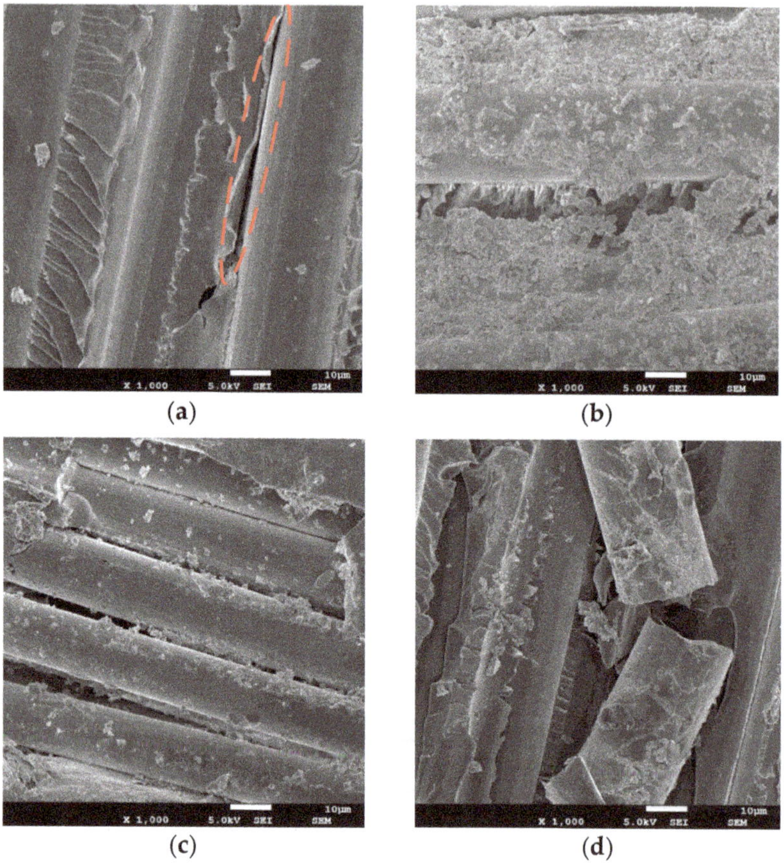

Figure 10. Microscopic morphology of core rods with different fatigue times; (**a**) N = 1 × 10^5; (**b**) N = 2 × 10^5; (**c**) N = 3 × 10^5; (**d**) N = 4 × 10^5.

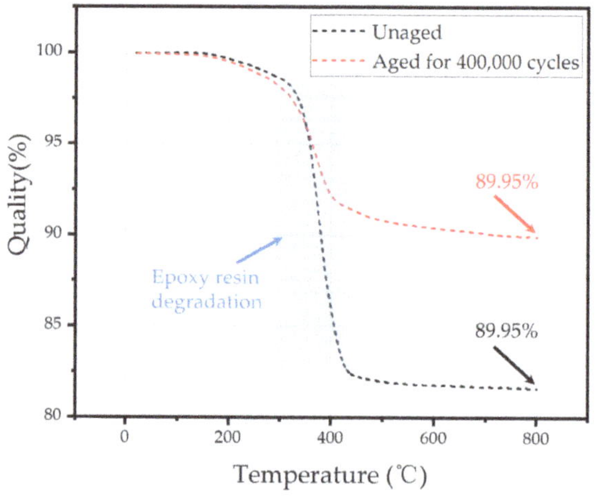

Figure 11. Thermogravimetric analysis results of composite insulator core rods before and after aging.

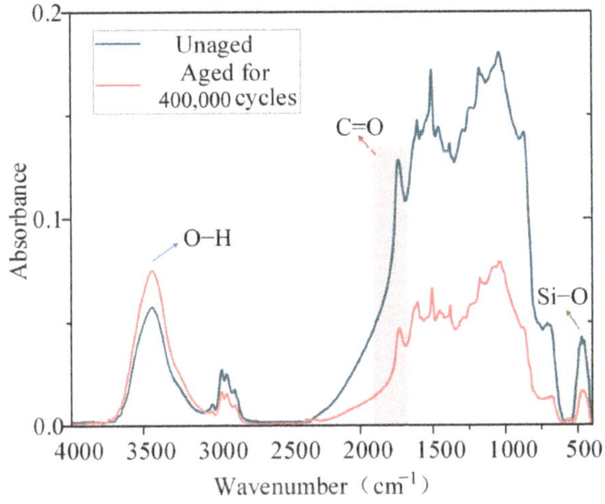

Figure 12. Infrared spectroscopic analysis results of composite insulator core rods before and after aging.

3.4.5. Artificial Aging Test Sample Charged Temperature Rise Test

The infrared images of the abnormal temperature rise of composite insulators after applying different alternating loads are shown in Figure 13.

Number of applied cyclic loads	100,000	200,000	300,000	400,000
Infrared image of composite insulator	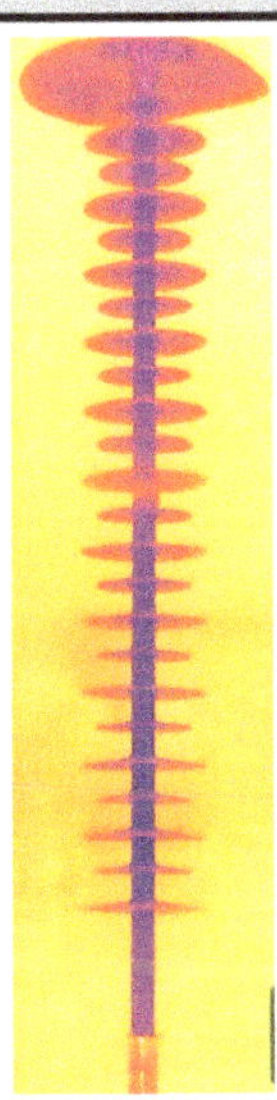	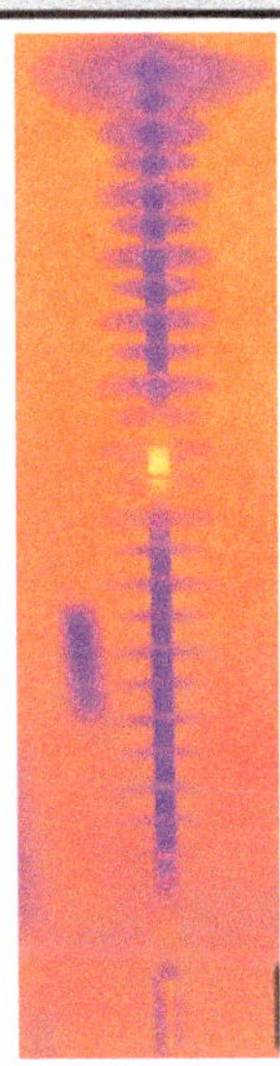	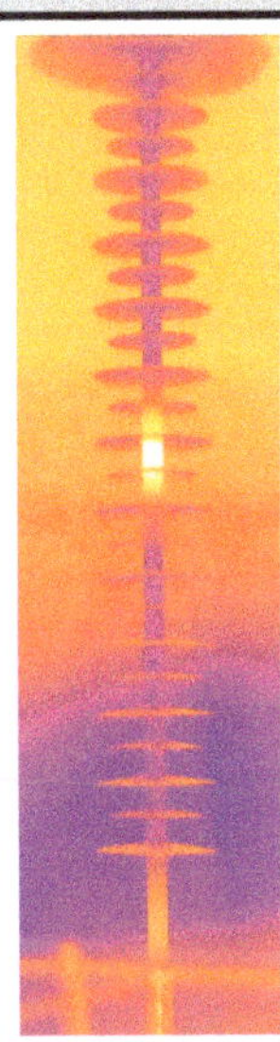	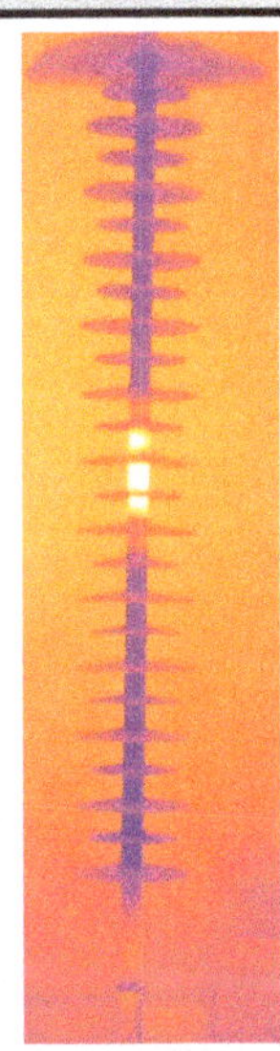
Maximum temperature rise	1.6K	4.2K	9.3K	16.8K

Figure 13. Infrared heating images of composite insulators after being electrified under different cycles of alternating loads.

Figure 13 shows that as the number of alternating load cycles increases, the deterioration of the composite insulator samples progressively worsens—both the temperature rise range and the degree of temperature rise increase. When 100,000 cycles of alternating loads were applied, there was a slight temperature rise in the central region of the composite insulator, with a maximum temperature rise value of 1.6 K. Continuing to exert alternating load cycles up to 200,000, the temperature rise area of the composite insulator expanded in a small range, with the maximum temperature rise reaching 4.2 K. When the number of alternating load cycles reached 300,000, the temperature rise area of the composite insulator further expanded, and the maximum temperature rise value increased to 9.3 K. At this point, the heating phenomenon became more pronounced. Finally, when 400,000 alternating load cycles were applied, the maximum temperature rise amplitude of the composite insulator reached 16.8 K. The temperature rise area and the maximum temperature rise value were primarily located at the maximum flexural deflection of the composite insulator, indicating that this location experiences more severe deterioration.

3.4.6. Analysis of Moisture Absorption and Dielectric Loss Characteristics

The test specimens had uniform tin foil coverage on their top and bottom surfaces. It is assumed that external moisture only permeates from the sides of the samples. This assumption aligns with the moisture ingress observed during the temperature rise test. We then weighed and recorded the initial weight of each test piece after drying using an electronic balance. Each test piece was then placed in a constant temperature and humidity box to absorb moisture at a temperature of 20 °C and a humidity rate of 75%. We then took it out and weighed it immediately every 6 h, and calculated the moisture absorption rate of the test piece, totaling 480 h.

The results of the dielectric loss factor and saturation moisture absorption tests for each sample are presented in Figure 14. The figure shows that both the dielectric loss factor and saturation moisture absorption of the test samples exhibited a trend of initially slow increase, followed by a rapid rise, and then a subsequent slow growth with an increase in the number of times of alternating load applications. The test results indicate that the samples extracted from intact insulators had a dielectric loss factor and saturation moisture absorption rate of 3.45% and 0.04%, respectively. However, when subjected to 400,000 alternating load applications, the deteriorated core rod's dielectric loss factor and saturation moisture absorption rate reached 30.68% and 0.6%, respectively, significantly increasing compared to the new samples.

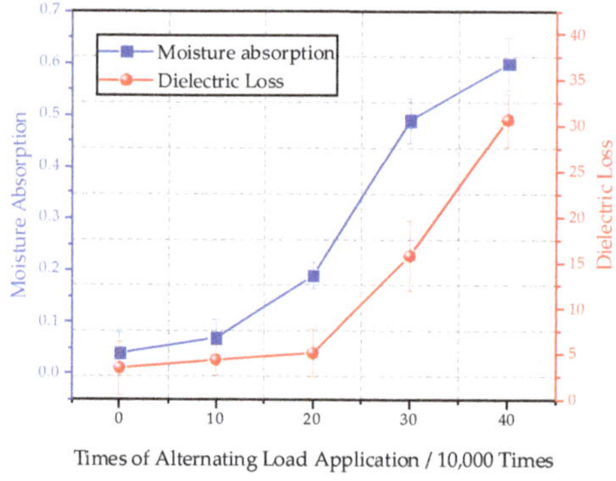

Figure 14. Moisture absorption and dielectric loss characteristics of composite insulator thin specimens.

4. Analysis and Discussion

The abnormal heating phenomenon in aged composite insulators is mainly attributed to dielectric losses. These losses are induced by the polarization effect of the dielectric material when exposed to the power frequency alternating electric field [31]. The equivalent electrical circuit model for the core rod material is illustrated in Figure 15.

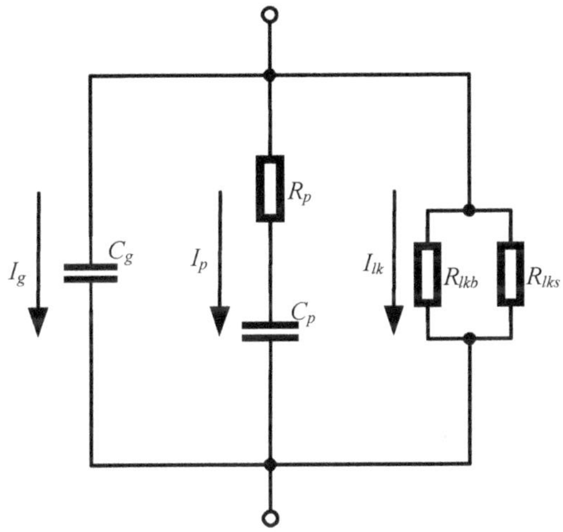

Figure 15. Dielectric equivalent circuit diagram.

In the diagram, C_g represents the capacitance equivalent to lossless polarization, while R_p and C_p represent the equivalent resistance and capacitance formed by lossy polarization, respectively. R_{lk} is the leakage resistance, which can be further divided into bulk leakage resistance and surface leakage resistance. I_{lk} represents the leakage current. For aged composite insulators, the decrease in insulation resistance in the deteriorated section of the core rod increases the leakage current, contributing relatively slightly to the temperature rise. Under the influence of high electric field strength in AC voltage, the polarization loss in the decay-like part of the composite insulator under AC power frequency is the leading cause of its abnormal temperature rise.

Furthermore, the power loss per unit volume of the dielectric can be expressed as:

$$p = EJ_r = E^2\omega\varepsilon \tan\delta \tag{11}$$

In Equation (11), E represents the distribution of the electric field intensity in the dielectric. J_r is the total active current density within the dielectric. ω is the angular frequency of the power frequency sinusoidal electric field. ε is the dielectric constant and $tan\delta$ is the loss angle tangent of the dielectric. This equation shows that the dielectric power loss is closely related to the magnitude of the electric field intensity and the dielectric loss factor. Furthermore, the scanning electron microscopy testing of slices of composite insulator core rods at different aging stages and physicochemical analysis showed that the surface epoxy matrix of the aged core rod material gradually deteriorated. The epoxy resin hydrolyzed continuously, resulting in numerous voids and microcracks inside. The emergence of these micro-defects not only led to distorted and abnormally increased electric fields in that region but also provided pathways for accelerated moisture ingress. An increase in the saturation moisture absorption rate macroscopically represents this. Additionally, water molecules are polar molecules. When they exist in the dielectric, they interact with the dielectric molecules, leading to the polarization of dielectric molecules and, consequently, an increase in dielectric material polarization loss. Coupled with the increase in dielectric loss in the core rod material during the aging process, macroscopically represented by the rise in the dielectric loss factor, the abnormal heating of the composite insulator is further promoted.

5. Conclusions

This article establishes a force model and a damage model to simulate and calculate the evolution of the damage process of the material outside the bending arc of the composite insulator core rod under alternating bending loads. A wet thermal mechanical multi-factor coupled artificially accelerated aging platform for composite insulators is established, and deterioration simulation tests for V-string composite insulators are carried out. The main conclusions are as follows:

(1) Compact V-string composite insulators were found to be susceptible to the influence of alternating flexural loads. A detailed examination of the damage evolution in GFRP core rod material revealed that fatigue damage was initiated from the outer surface of the core rod's bending arc due to alternating flexural loads. It propagated towards the core rod's interior, beginning with epoxy resin matrix damage on its surface, subsequently leading to fiber breakage.
(2) The establishment of a multi-factor composite insulator artificially aging platform, combining moisture, heat, and mechanical stress, allowed for accelerated aging tests. Microscopic morphology observation and physicochemical analysis indicates that while subjecting the test specimen to 400,000 alternating load cycles, the core rod underwent stages of surface damage, an increase damage, fatigue embrittlement, matrix hydrolysis, and fiber fracture. The deterioration primarily occurred on the outer side of the composite insulator's bending arc.
(3) Electric temperature rise tests were conducted on composite insulator specimens subjected to different numbers of alternating load cycles. It was observed that the deterioration of composite insulator specimens gradually intensified with an increasing number of alternating load cycles. Both the temperature rise range and the degree increased, with the temperature rise region and maximum temperature rise primarily occurring at the point where the bending deflection of the composite insulator was at its maximum, indicating that this position was the most severely deteriorated part.
(4) Both the dielectric loss factor and saturation moisture absorption rate of the test samples exhibited an initially slow increase, followed by rapid growth and then a gradual increase with an increasing number of alternating load cycles. Test results showed that the dielectric loss factor and saturation moisture absorption rate for slices from the new insulator were 3.45% and 0.04%, respectively. However, after 400,000 alternating load cycles, the deteriorated core rod's dielectric loss factor and saturation moisture absorption rate significantly increased to 30.68% and 0.6%, respectively, in contrast to the new samples.
(5) As the times of alternating flexural loads increased, the depth of core rod deterioration continuously deepened. The microcracks and pores formed inside the core rod distort the electric field and provided pathways for accelerated moisture ingress. This, in turn, increased dielectric material polarization loss, causing abnormal temperature to rise in the deteriorated section of the composite insulator.

Author Contributions: Data curation, L.D., H.L. and J.G.; formal analysis, X.Y. and P.W.; funding acquisition, X.Y.; investigation, Z.L.; methodology, X.Y., Y.Z. and P.W.; project administration, L.D. and J.L.; resources, H.L.; software, J.Z., Z.L. and Y.Z.; supervision, J.L. and P.W.; validation, J.Z. and J.G.; visualization, Z.L.; writing—original draft, X.Y., L.D. and Y.Z.; writing—review and editing, H.L., J.Z., J.L. and J.G. All authors have read and agreed to the published version of the manuscript.

Funding: This research was funded by the State Grid Corporation of China Technology Project, Grant Number TSS2022-11, and the APC was funded by State Grid Corporation of China.

Data Availability Statement: Dataset available on request from the authors.

Conflicts of Interest: Authors Xiaoyu Yi, Hongliang Liu, Jiaxin Zhang, Jie Liu were employed by the company State Grid Hebei Energy Technology Service Co., Ltd. Author Likun Ding was employed by the company State Grid Hebei Electric Power Co., Ltd. The remaining authors declare that the research was conducted in the absence of any commercial or financial relationships that could be construed as a potential conflict of interest.

References

1. Liang, X.; Gao, Y.; Wang, J.; Li, S. Rapid Development of Silicone Rubber Composite Insulator in China. *High Volt. Eng.* **2016**, *42*, 2888–2896. (In Chinese)
2. Yang, L.; Yu, J.; Guo, Y.; Chen, S.; Tan, K.L.; Li, S. An Electrode-Grounded Droplet-Based Electricity Generator (EG-DEG) for Liquid Motion Monitoring. *Adv. Funct. Mater.* **2023**, *33*, 2302147. [CrossRef]
3. Liu, Z. Present Situation and Prospects of Applying Composite Insulators to UHF Transmission Lines in China. *Power Syst. Technol.* **2006**, *30*, 1–7. (In Chinese)
4. Guan, Z.; Peng, G.; Wang, L. Application and Key Technical Study of Composite Insulators. *High Volt. Eng.* **2011**, *37*, 513–519. (In Chinese)
5. Dai, J. Study of Brittle Fracture of High Voltage Composite Insulators. Master's Thesis, Tsinghua University, Beijing, China, 2006. (In Chinese).
6. Liang, X.; Huang, L.; Wang, S. Composite insulator flashovers caused by bird dropping and unknown reasons. *Power Syst. Technol.* **2001**, *25*, 13–16. (In Chinese)
7. Zhang, Y.; Hou, S.; Geng, J.; Gong, Y.; Zhong, Z. Diagnosis Method of Decay-like Composite Insulators in a High-Humidity Environment Based on Characteristic Coefficient of Temperature Rise Gradient. *Polymers* **2023**, *15*, 2715. [CrossRef] [PubMed]
8. Wang, J.; Liang, X.; Gao, Y. Failure analysis of decay-like fracture of composite insulator. *IEEE Trans. Dielectr. Electr. Insul.* **2014**, *21*, 2503–2511. [CrossRef]
9. Gao, Y.; Liang, X. Study on Decay-Like Fracture of Composite Insulator, Part II: Experimental Simulation and Preventive Method Discussion of Decay-Like Fracture. *Proc. CSEE* **2016**, *36*, 5070–5077+5132. (In Chinese)
10. Zeng, L. Study on Abnormal Heating and Hygrothermal Aging Characteristics of Composite Insulators. Master's Thesis, South China University of Technology, Guangzhou, China, 2019. (In Chinese).
11. Zhao, Y. Study on the Deterioration Characteristics of the Core of Composite Insulators under Hygrothermal Environment. Master's Thesis, North China Electric Power University, Beijing, China, 2018. (In Chinese).
12. Ma, X. Research on the Influence of Moisture Immersion on the Internal Deterioration Process of Composite Insulator. Master's Thesis, Dalian University of Technology, Dalian, China, 2022. (In Chinese).
13. Noonan, M.; Obande, W.; Ray, D. Simulated End-of-Life Reuse of Composites from Marine Applications Using Thermal Reshaping of Seawater-Aged, Glass Fibre-Reinforced Acrylic Materials. *Compos. Part B Eng.* **2024**, *270*, 111118. [CrossRef]
14. Stupar, S.; Vuksanović, M.M.; Mijin, D.Ž.; Milanović, B.C.; Joksimović, V.J.; Barudžija, T.S.; Knežević, M.R. Multispectral Electromagnetic Shielding and Mechanical Properties of Carbon Fabrics Reinforced by Silver Deposition. *Mater. Chem. Phys.* **2022**, *289*, 126495. [CrossRef]
15. Stupar, S.; Vuksanović, M.M.; Mijin, D.Ž.; Bučko, M.; Joksimović, V.J.; Barudžija, T.S.; Tanić, M.N. Functional Nano-Silver Decorated Textiles for Wearable Electronics and Electromagnetic Interference Shielding. *Mater. Today Commun.* **2023**, *34*, 105312. [CrossRef]
16. Nicolalde, J.F.; Yaselga, J.; Martínez-Gómez, J. Selection of a Sustainable Structural Beam Material for Rural Housing in Latin América by Multicriteria Decision Methods Means. *Appl. Sci.* **2022**, *12*, 1393. [CrossRef]
17. Chen, Y.; Xue, W.; Liu, Y. Influence of Silicone Rubber Content of Shed and Sheath on Composite Insulator Lifespan. *High Volt. Eng.* **2022**, *48*, 736–752. (In Chinese)
18. Le, B.; Hou, L.; Wang, L.; Zhu, P. Study on loading characteristics of V type composite insulator used in 330 kV compact line. *Proc. CSEE* **2005**, *25*, 91–95. (In Chinese)
19. Diana, G.; Cheli, F.; Manenti, A.; Nicolini, P.; Tavano, F. Oscillation of Bundle Conductors in Overhead Lines due to Turbulent Wind. *IEEE Trans. Power Deliv.* **1990**, *5*, 1910–1922. [CrossRef]
20. Zhu, K.; Liu, B. Calculation of Torsional Stiffness of Bundle Conductors in Overhead Transmission Lines. *Power Syst. Technol.* **2010**, *34*, 210–214. (In Chinese)
21. Zhang, Z.; Pang, G.; Lu, M.; Gao, C.; Jiang, X. Research on Silicone Rubber Sheds of Decay-Like Fractured Composite Insulators Based on Hardness, Hydrophobicity, NMR, and FTIR. *Polymers* **2022**, *14*, 3424. [CrossRef] [PubMed]
22. *DL/T 1580-2016*; Technical Specification for Core of Long Rod Composite Insulators for a. c. or d. c. Systems. China National Energy Administration: Beijing, China, 2016.
23. *GB/T 34937-2017*; Insulators for Overhead Lines—Composite Suspension and Tension Insulators for d. c. Systems with a Nominal Voltage Greater than 1500 V—Definitions, Test Methods and Acceptance Criteria. China National Standardization Administration: Beijing, China, 2017.
24. *GB/T 13096-2008*; Test Method for Mechanical Properties of Pultruded Glass Fiber Reinforced Plastic Rods. China National Standardization Administration: Beijing, China, 2008.
25. Zhang, Z.; Huang, Q.; Geng, J.; Liu, Q.; Zhang, S. Defect Identification of Composite Insulator Based on Infrared Image. *Polymers* **2022**, *14*, 2620. [CrossRef]
26. Gao, S.; Liu, Y.; Li, L. A Comparative Study of Abnormal Heating Composite Insulators. *Polymers* **2023**, *15*, 2772. [CrossRef]
27. *ASTM D5229/D5229M-20*; Standard Test Method for Moisture Absorption Properties and Equilibrium Conditioning of Polymer Matrix Composite Materials. ASTM International: West Conshohocken, PA, USA, 2020.
28. Li, X.; Zhang, Y.; Chen, L.; Fu, X.; Geng, J.; Liu, Y.; Gong, Y.; Zhang, S. Study on the Ageing Characteristics of Silicone Rubber for Composite Insulators under Multi-Factor Coupling Effects. *Coatings* **2023**, *13*, 1668. [CrossRef]

29. Wang, Q.; Bao, W.; Gao, Y.; Liu, S.; Liu, S.; Zuo, Z.; Wu, C.; Liang, X. Influence of Surface Discharge on Resin Degradation in Decay-like Fracture of Composite Insulators. *Polymers* **2023**, *15*, 790. [CrossRef] [PubMed]
30. Yuan, Z.; Wang, C.; Jin, L.; Tu, Y.; Zhang, Y.; An, Z.; Zhao, Y. A Modified Langmuir Model for Moisture Diffusion in UGFRE of Composite Insulator Considering the Composite Degradation. *Polymers* **2022**, *14*, 2922. [CrossRef] [PubMed]
31. Zhong, Z.; Chen, Y.; Liu, Y.; Hou, S.; Geng, J. Study on the influence mechanism of ambient humidity on the temperature rise of decay-like composite insulators. *High Volt.* **2022**, *7*, 916–924. [CrossRef]

energies MDPI

Article

Insulation Resistance Degradation Models of Extruded Power Cables under Thermal Ageing

Xufei Ge, Fulin Fan *, Martin J. Given and Brian G. Stewart *

Institute for Energy and Environment, University of Strathclyde, Glasgow G1 1XW, UK; xufei.ge@strath.ac.uk (X.G.); m.given@strath.ac.uk (M.J.G.)
* Correspondence: f.fan@strath.ac.uk (F.F.); brian.stewart.100@strath.ac.uk (B.G.S.)

Abstract: Insulation resistance (IR) is an essential metric indicating insulation conditions of extruded power cables. To deliver reliable IR simulation as a reference for practical cable inspection, in this paper, four IR degradation models for cross-linked polyethylene-insulated cables under thermal ageing are presented. In addition, the influences of methodologies and temperature profiles on IR simulation are evaluated. Cable cylindrical insulation is first divided into sufficiently small segments whose temperatures are simulated by jointly using a finite volume method and an artificial neural network to model the thermal ageing experiment conditions. The thermal degradation of IR is then simulated by dichotomy models that randomly sample fully degraded segments based on an overall insulation (layer) ageing condition estimation and discretization models that estimate the gradual degradation of individual segments, respectively. Furthermore, uniform and non-uniform temperature profiles are incorporated into dichotomy and discretization models, respectively, for a comparison. The IR simulation results are not only compared between different models, but also discussed around the sensitivity of IR simulation to segment sizes and degradation rates. This provides cable assessment engineers with insights into model behaviour as a reference for their selection of appropriate IR degradation models.

Keywords: extruded power cables; insulation resistance; thermal ageing; dichotomy models; discretization models; insulation temperature profiles

Citation: Ge, X.; Fan, F.; Given, M.J.; Stewart, B.G. Insulation Resistance Degradation Models of Extruded Power Cables under Thermal Ageing. *Energies* **2024**, *17*, 1062. https://doi.org/10.3390/en17051062

Academic Editor: Tomasz Norbert Kołtunowicz

Received: 23 December 2023
Revised: 15 February 2024
Accepted: 17 February 2024
Published: 23 February 2024

1. Introduction

The ageing of extruded power cables refers to the gradual deterioration and alteration of their physical, chemical and electrical properties over time subject to environmental, thermal, mechanical and electrical stresses. The ageing process can result in a gradual reduction in the cable performance, including its insulation integrity, mechanical strength and overall reliability. Factors such as temperature variations, exposure to moisture, chemical interactions, mechanical stresses and electrical loading contribute to the cable ageing process, potentially diminishing cable insulation effectiveness and increasing the risk of operational failures or breakdowns [1,2].

Thermal stress is considered one of the primary stresses accelerating the ageing of extruded power cables [3]. The principal source of thermal stresses arises from the resistive heat of the heavily loaded cable core conductor. The resistive heat generated in the core conductor propagates throughout the cable insulation and might increase the insulation temperature beyond its designed limit. Furthermore, excessively high ambient temperature conditions can also contribute to the thermal ageing of cable insulation. In addition to the cable temperature itself, the thermal degradation of extruded power cables is closely related to their polymer structure and the effectiveness of antioxidants [4].

The estimation of a good electrical cable insulation condition is inherently associated with a heightened resistance to electric current flow. Insulation resistance (IR) testing is a widely acknowledged condition monitoring technique, demonstrating an ability to directly infer the current state of cables [5]. Most research related to power cable ageing has placed a notable emphasis on the increase in electrical conductivity or the decrease in electrical resistivity for the insulation materials throughout the ageing processes [6–11]. Mecheri et al. delved into the impacts of prolonged thermal ageing on dielectric and mechanical properties of cross-linked polyethylene (XLPE) insulation of high voltage (HV) cables [7]. To ascertain the degree of the XLPE material deterioration resulting from thermal ageing and prevent potential failures, an extensive examination was conducted by encompassing volumetric resistivity assessments and additional condition monitoring evaluations on dumb-bell- and circular-shaped XLPE probes over a duration of 5000 h. The findings revealed a significant reduction in resistivity, amounting to several hundredfold, caused by the concurrent decrease in polymer viscosity which was considered to increase the mobility of charge carriers within XLPE. Nedjar [8] examined the influences of thermal ageing on the electrical characteristics of XLPE employed in HV cables, demonstrating that thermal ageing induced notable modifications in the electrical properties of the polymer. In particular, elevating the thermal ageing temperature leads to a faster reduction in resistivity, aligning with the principle outlined by the Arrhenius law. With the thermal ageing process going further, molecular bonds exhibited weakening, resulting in an augmentation of free volume. This phenomenon subsequently heightened the mobility of charge carriers, accompanied by a decline of volumetric resistivity. Mecheri et al. performed a comprehensive investigation into the impacts of thermal ageing conditions on the performance of XLPE for medium voltage (MV) 18/30 kV cables [9]. XLPE specimens of the same material were subjected to thermal stresses under a controlled environment facilitated by a forced air-circulating oven. The assessments on XLPE thermal degradation were carried out through the volumetric resistivity testing under a direct current (DC) voltage of 500 V at a measurement temperature of 85 °C. After 1350 h of thermal ageing at 90 °C, XLPE resistivity was observed to significantly decrease from around 700 TΩ·cm to 2.5 TΩ·cm. Discernible resistivity reductions were further pronounced at higher ageing temperatures of 135 °C and 150 °C, where the volumetric resistivity values declined to 0.46 TΩ·cm and 0.2 TΩ·cm, respectively, after 1350 h of thermal ageing.

Zhang et al. investigated the impacts of thermal ageing temperatures and durations on the DC electrical conductivity of XLPE insulation materials for HVDC cables by using Fourier transform infrared (FTIR) spectra [10]. The experimental findings and subsequent analysis revealed a sequential pattern in the DC electrical conductivity of thermally aged XLPE insulation, which initially declined below the level observed on unaged XLPE insulation and then exhibited a gradual increase over the entire ageing period due to the combined effects of thermal decomposition, post-crosslinking and the diffusion of low molecular weight substances. In addition, it was observed that higher thermal ageing temperatures induced more pronounced changes in the electrical conductivity of XLPE insulation. When subjected to the same ageing duration of 700 h, the variations of the DC electrical conductivity of XLPE insulation with ageing temperatures complied well with the Arrhenius equation. Kang and Kim performed a comprehensive assessment on IR properties of low voltage cables, subjecting them to external flame exposure, over-current conditions and accelerated degradation trials [11]. The results revealed a notable IR reduction from a peak of 7.5 TΩ to 0.008 TΩ during direct flame contact. However, it demonstrated a complete recovery to its initial state when cooling down to room temperature. In accelerated degradation experiments simulating 10 to 30 years of cable operation, no notable IR decline was observed at room temperature. Nevertheless, upon reaching an induced ageing equivalent to 40 years, a rapid IR reduction was observed at room temperature.

A significant challenge of comprehending the dielectric insulation condition lies in developing a quantitative approach to assessing insulation degradation over time. A dichotomy model was originally proposed by Chang et al. [12] to predict the IR of rectangular (unit cube) insulation bulk. It regards an entire insulation specimen cube as a combination of multiple small sub-cubes and divides them into degraded and non-degraded sub-cubes which possess disparate resistivity. The volume ratio of the two parts is estimated from thermal ageing temperature and duration, based on which the positions of degraded sub-cubes within the specimen are randomly sampled to evaluate the overall IR. According to the original dichotomy model, the electrical resistance degradation trend of the XLPE material under thermal ageing can be classified into three phases, as shown in Figure 1 [12]:

- Phase 1: IR gradually decreases from an initial resistance R_0 since thermal ageing conditions are applied on insulation till a time t_s;
- Transition phase: IR significantly decreases within a short period from t_s to t_f due to the percolation phenomenon occurring in the insulation bulk;
- Phase 2: After t_f, IR gradually approaches toward a completely degraded resistance constant R_d.

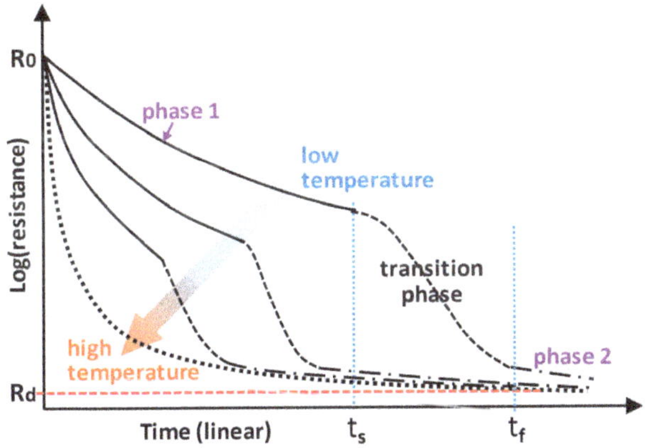

Figure 1. Temperature-dependent IR degradation of XLPE materials under thermal ageing [12].

Even though the original dichotomy model is able to describe the long-term decline of IR over thermal ageing time, it assumes that the insulation is uniform in ageing temperature and chooses the degraded sub-cubes in a random way. The contributions of this paper are to enhance the dichotomy model by simulating different degradation degrees between insulation layers under radial temperature gradients and additionally propose a discretization model that simulates the thermal degradation of individual insulation segments separately given a non-uniform temperature distribution. Furthermore, the developed IR degradation models adapt the original unit cube insulation model to a cylindrical insulation model in order to reflect the geometry of a practical cable insulation. In addition, the finite volume method (FVM) and artificial neural network (ANN) models are jointly employed to simulate the temperature distribution of cable insulation, which is then incorporated into the enhanced dichotomy model or the novel discretization model to evaluate IR degradation during Phase 1 in Figure 1. In order to perform a comparison not only between the dichotomy and discretization models but also between the uses of different temperature profiles, four IR degradation models are developed in this work:

- a dichotomy model with uniform temperature distribution;
- a dichotomy model with radial temperature gradients;
- a discretization model with uniform temperature distribution;
- a discretization model with non-uniform temperature distribution.

These models help investigate the influences of model methodologies and temperature profiles on the IR degradation simulation of power cables under thermal ageing. In addition, different segment sizes and degradation rates are applied to examine their effects on the IR degradation simulation.

The paper is structured as follows: Section 2 describes the IR degradation model development; Section 3 presents the application of the four IR degradation models with the temperature profile simulated under thermal ageing conditions; Section 4 discusses the effects of segment sizes and degradation rates on IR simulation results; and Section 5 presents conclusions and recommendations for further work.

2. IR Degradation Models

Based on the discretization methodology [13], a sample of extruded power cable insulation can be conceptualized as a cylindrical volume made up of a substantial number of small segments, as illustrated in Figure 2a, where the inner radius, outer radius and length of the cable insulation model are denoted by x, X and L, respectively, all in meters. The position of the center of each segment is represented in a cylindrical coordinate system in the form of (x_i, φ_j, l_k), as shown in Figure 2b, where x_i (m), φ_j (arc degree) or l_k (m) denotes the position of a segment along the radial, angular or longitudinal dimension, respectively.

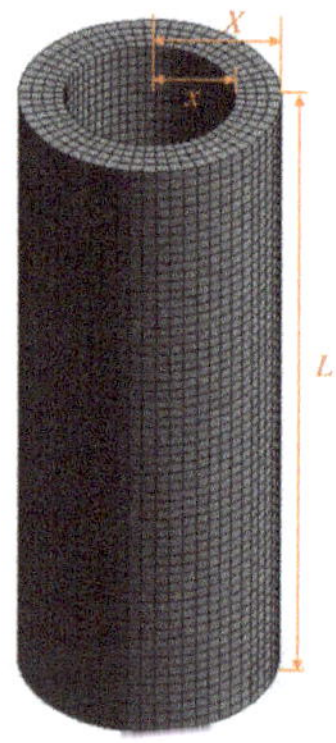

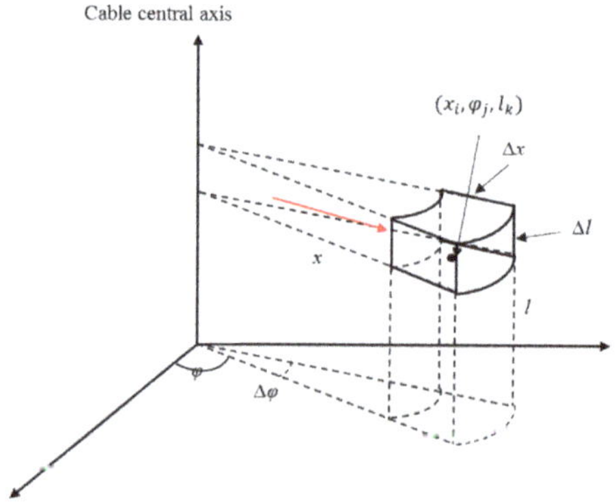

Figure 2. (**a**) Model of cable insulation divided into multiple segments; and (**b**) shape and position of a segment in the cable insulation model, with a red arrow representing the direction of the current flowing through insulation.

Denoting radial, angular and longitudinal sizes of each segment by Δx, $\Delta\varphi$ and Δl, respectively, the resulting numbers of segments along radial, angular and longitudinal dimensions are equal to $M = (X - x)/\Delta x$, $N = 360°/\Delta\varphi$ and $P = L/\Delta l$, respectively. The total number of segments in the cable insulation model then equals $N_t = M \times N \times P$.

The volume of a segment in the cable insulation model is dependent on its position along the radial axis. The segments located at outer insulation layers have larger volumes than those at inner insulation layers. The electrical resistance of any insulation material not only depends on its material resistivity (ρ in Ω-m) but also on shape and volume [14]. The integral formula of electrical resistance (R in Ω) for variable cross-section resistors with parallel curved terminals can be formulated by (1) [15]:

$$R = \frac{1}{A(l_1)} \int_{l_1}^{l_2} \frac{\rho(l)}{(1 + \kappa_{\uparrow}(l - l_1))(1 + \kappa_{\rightarrow}(l - l_1))} \quad (1)$$

where l_1 or l_2 is the arc-length parameter of each terminal measured through its normal surface ($l_2 > l_1$), $A(l_1)$ represents the cross-section area located at l_1 and $\rho(l)$ is the resistivity of the segment. The specific signed principal curvature values of $\kappa_{\uparrow}$ and $\kappa_{\rightarrow}$ for the cylindrical segment terminals are equal to 1 and 0, respectively.

It is assumed that the electric current flows through individual insulation segments along the radial axis with a uniform density (i.e., radially from the core conductor to any surrounding outer ground sheath), as indicated by the red arrow in Figure 2b. In addition, the material resistivity is assumed to be uniform within an individual insulation segment given its sufficiently small volume. For the insulation segment located at (x_i, φ_j, l_k) with resistivity $\rho(x_i, \varphi_j, l_k)$, $(i = 1, 2, \ldots, M;\ j = 1, 2, \ldots, N;\ k = 1, 2, \ldots, P)$, based on (1), its resistance $R_e(x_i, \varphi_j, l_k)$ can be formulated by (2):

$$R_e(x_i,\ \varphi_j,\ l_k) = \frac{\rho(x_i,\ \varphi_j,\ l_k)}{\Delta\varphi\Delta l}\ln\left(\frac{x_i + \frac{\Delta x}{2}}{x_i - \frac{\Delta x}{2}}\right) \quad (2)$$

The resistances of all the insulation segments are converted into the total resistance of the power cable insulation based on the equivalent resistance model for a DC circuit. The series resistance of all the segments along an individual radial column at fixed angular and longitudinal locations is first estimated, making the insulation model have resistances of $N \times P$ columns. Then, the parallel resistance of all the columns within each individual plane at every fixed longitudinal location is calculated. Finally, the total resistance R_t of cable insulation is derived from the parallel connection of all the plane resistances. In other words, all the radial columns within the insulation are parallel connected. The total IR, R_t, is formulated by:

$$R_t = \frac{1}{\sum_{k=1}^{P}\sum_{j=1}^{N}\left(\frac{1}{\sum_{i=1}^{M} R_e(x_i,\varphi_j,l_k)}\right)} \quad (3)$$

The estimation of resistivity ρ for individual insulation segments by the four different IR models is detailed in the following subsections, respectively.

2.1. Dichotomy Model with Uniform Temperature Distribution

The dichotomy models categorize insulation segments into two types, virtually non-degraded and degraded segments, which are assumed to have resistivity values of ρ_0 and ρ_d, respectively. The value of ρ_0 is considered to be much greater than ρ_d. According to (2), the resistances of non-degraded and degraded segments (denoted by R_0 and R_d) are formulated by (4) and (5), respectively:

$$R_0(x_i,\ \varphi_j,\ l_k) = \frac{\rho_0}{\Delta\varphi\Delta l}\ln\left(\frac{x_i + \frac{\Delta x}{2}}{x_i - \frac{\Delta x}{2}}\right),\quad (x_i,\ \varphi_j,\ l_k) \neq (x_l,\ \varphi_m,\ l_n) \quad (4)$$

$$R_d(x_l,\ \varphi_m,\ l_n) = \frac{\rho_d}{\Delta\varphi\Delta l}\ln\left(\frac{x_l + \frac{\Delta x}{2}}{x_l - \frac{\Delta x}{2}}\right),\quad (x_i,\ \varphi_j,\ l_k) \neq (x_l,\ \varphi_m,\ l_n) \quad (5)$$

When the ageing temperature distribution is uniform within the cable insulation and also within individual segments, the number N_d of degraded segments to be randomly sampled within the insulation model depends on the total number N_t of segments and the degradation volume ratio V_d, i.e., $N_d = N_t \times V_d$. Assuming that the degradation rate does not change with ageing time t, the value of V_d can be determined based on the cumulative distribution function (CDF) of an exponential distribution in terms of t [16]:

$$V_d(t) = 1 - e^{-\lambda(T)\cdot t} \quad (6)$$

where the degradation rate $\lambda(T)$ (in 1/h) of insulation material is a function of the ageing temperature T (in K) depending on material properties and generally obtained from the IR decay tendency rather than the measurement of chemical reaction. Considering in this work, and for this model, that the degradation of cable insulation is thermally activated, $\lambda(T)$ is presumed to follow the Arrhenius model [16] as formulated by (7):

$$\lambda(T) = \lambda_0 \cdot e^{\left(\frac{-E_a}{k_B\cdot T}\right)} \quad (7)$$

where λ_0 is a constant (1/h) obtained from experimental data, k_B is the Boltzmann constant of 1.38×10^{-23} J/K and E_a is the IR thermal degradation activation energy (J) of insulation material. Then, the degradation volume ratio and the number of degraded segments within the insulation can be calculated by (8) and (9), respectively:

$$V_d(t,T) = 1 - e^{-(\lambda_0 \cdot \exp(\frac{-E_a}{k_B \cdot T})) \cdot t} \quad (8)$$

$$N_d(t,T) = N_T \cdot V_d(t,T) \quad (9)$$

Since this model assumes a uniform temperature distribution for cable insulation, the locations of the degraded segments accompanying their resistivity ρ_d are randomly and uniformly sampled within the insulation. A flowchart showing the IR estimation by the dichotomy model with uniform temperature is presented in Figure 3.

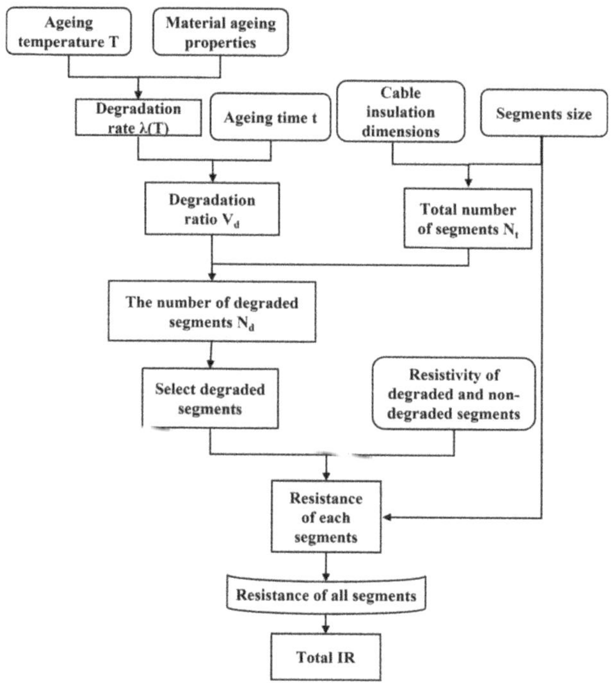

Figure 3. IR estimation process of a dichotomy model with uniform temperature.

2.2. Dichotomy Model with Radial Temperature Gradients

When considering the temperature gradients along the cable radius created by the propagation of resistive heat from the core conductors through insulation layers, outer insulation layers closer to the ambient environment and further away from the core conductors have lower temperatures than inner insulation layers, as shown in Figure 4. The introduction of insulation temperature gradients into the dichotomy model will produce more reliable IR results when power cables are carrying currents.

According to the Arrhenius model [17], the radial temperature gradients will result in inner insulation layers having higher degradation volume ratios than outer insulation layers. Given that the temperature of the *i*th insulation layer along the radial axis is fixed at T_i, its degradation volume ratio $V_{d(i)}$ and the number $N_{d(i)}$ of degraded segments at ageing time t can be calculated by (10) and (11), respectively:

$$V_{d(i)} = 1 - e^{-(\lambda_0 \cdot \exp(\frac{-E_a}{k_B \cdot T_i})) \cdot t} \quad (10)$$

$$N_{d(i)} = N_{t(i)} \cdot V_{d(i)} \tag{11}$$

where $N_{t(i)}$ is the total number of segments in the ith layer equalling $(N \times P)$. Since the model assumes a constant temperature for each individual layer, the locations of the $N_{d(i)}$ virtually degraded segments accompanying their resistivity ρ_d within each layer are randomly and uniformly selected for that layer. The IR estimation process of the dichotomy model with radial temperature gradients is shown in Figure 5.

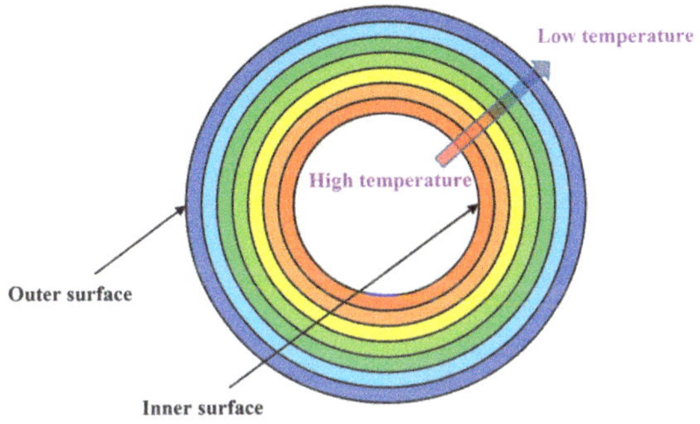

Figure 4. Schematic of insulation temperature gradients along cable radius.

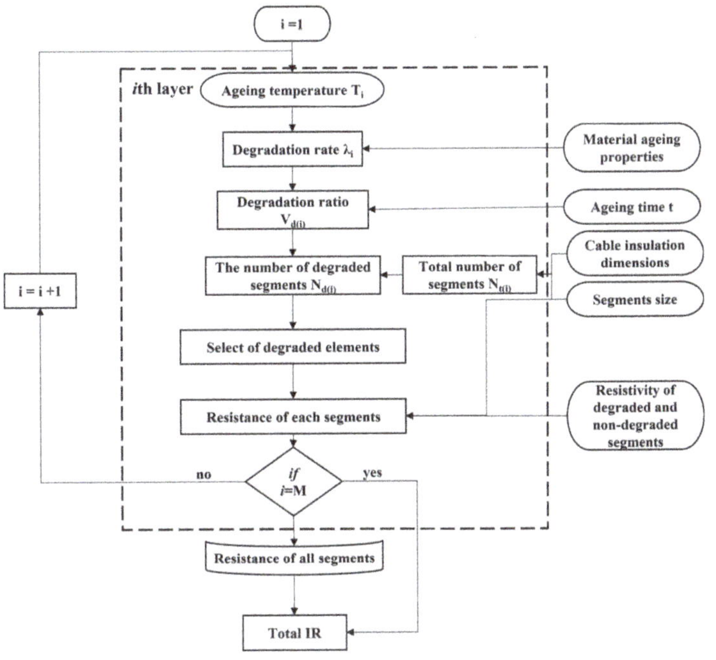

Figure 5. IR estimation process of a dichotomy model with radial temperature gradients.

2.3. Discretization Model with Uniform Temperature Distribution

Compared with the dichotomy models which differentiate the segment resistivity between non-degraded ρ_0 and fully degraded ρ_d only, the discretization models proposed here simulate the gradual resistivity degradation of individual segments separately by a function of thermal ageing time t and ageing temperature T. Assuming that insulation segments have consistent temperature, their resistivity is modelled to exponentially decline with t at the same degradation rate $\lambda(T)$:

$$\rho(t,T) = \rho_0 \cdot e^{-\lambda(T)\cdot t} \tag{12}$$

where $\lambda(T)$ complies with the Arrhenius model as formulated by (7). The resistance of each segment is then calculated by (13) based on its location along the radial axis, segment sizes and $\rho(t,T)$. The process of IR estimation by the discretization model with uniform temperature is described in Figure 6.

$$R_e(x_i,\ \varphi_j,\ l_k, t) = \frac{\rho(t,T)}{\Delta\varphi\Delta l} \ln\left(\frac{x_i + \frac{\Delta x}{2}}{x_i - \frac{\Delta x}{2}}\right) \tag{13}$$

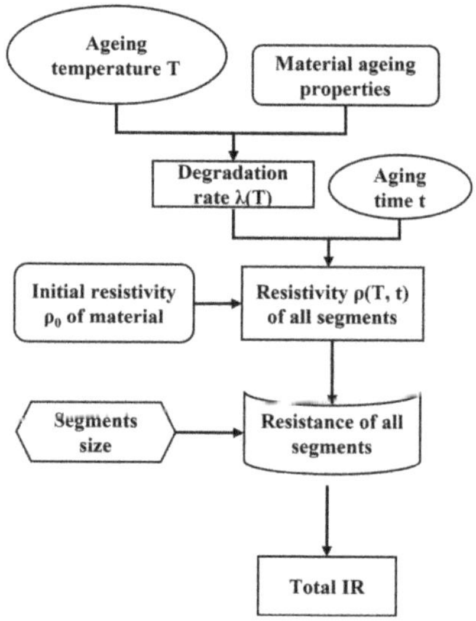

Figure 6. IR estimation process of a discretization model with uniform temperature.

2.4. Discretization Model with Non-Uniform Temperature Distribution

As it was noted in Section 2.2, the temperature distribution of cable insulation is generally non-uniform along its radial axis. The cable temperature will also vary longitudinally along the cable especially when it spans a long distance with various ambient environment such as solar irradiation, wind velocity and soil moisture content [18]. The non-uniform temperature along multiple dimensions will cause degradation rates to vary with the positions within cable insulation. Given the temperature T' of a particular insulation segment locating at (r_i, φ_j, l_k), its resistivity and resulting resistance at t can be calculated by (14) and (15), respectively. A flowchart describing the IR estimation by the discretization model with a non-uniform temperature distribution is shown in Figure 7.

$$\rho(x_i, \varphi_j, l_k, t, T') = \rho_0 \cdot e^{-\lambda(x_i,\varphi_j,l_k,T')\cdot t} \tag{14}$$

$$R_e(r_i, \varphi_j, l_k, t, T') = \frac{\rho(x_i, \varphi_j, l_k, t, T')}{\Delta\varphi\Delta l} \ln\left(\frac{x_i + \frac{\Delta x}{2}}{x_i - \frac{\Delta x}{2}}\right) \tag{15}$$

It is noted that the flow charts in Figures 3 and 5–7 are applied at each ageing time step to estimate the IR value after a particular thermal ageing period. For dichotomy models, the number of degraded segments is first determined at each individual ageing time step. Then, the locations of the degraded segments selected in the previous time step are kept, while new degraded segments are uniformly and randomly sampled from the locations of the remaining non-degraded segments. For discretization models, the resistivity of insulation segments is updated at every ageing time step.

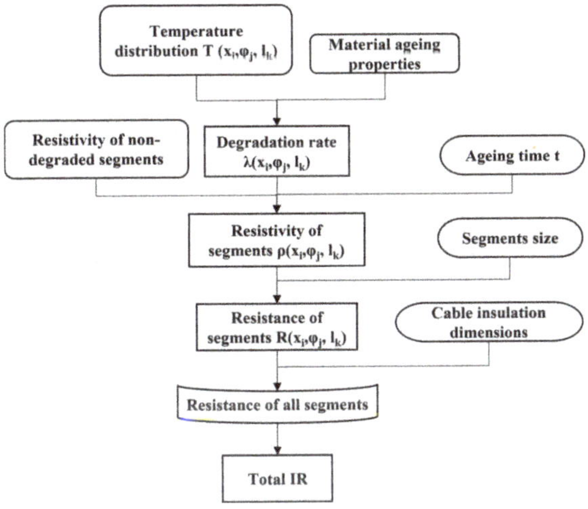

Figure 7. IR estimation process of a discretization model with non-uniform temperature.

3. Application of IR Degradation Models

3.1. Degradation Model Parameters Estimation

The parameters of the four IR degradation models are estimated based on the experimental data of XLPE insulation referenced from the research of Mecheri et al. [7] where XLPE insulation samples were thermally aged at four different temperature levels of 80 °C, 100 °C, 120 °C and 140 °C (i.e., 353.15 K, 373.15 K, 393.15 K and 413.15 K, respectively). The XLPE sample resistivity was measured through a megohmmeter by applying a DC voltage of 500 V for a duration of 10 min. Figure 8 shows the XLPE resistivity measurements under different thermal ageing temperatures, which generally declined with the thermal ageing time, with the pattern being affected by the ageing temperature.

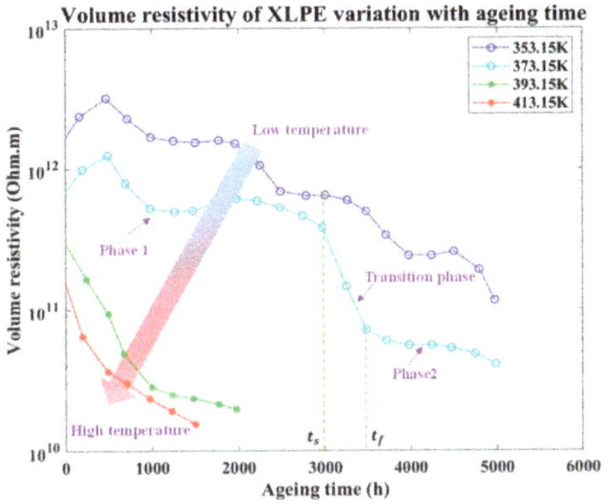

Figure 8. Volume resistivity (Ω·m) of XLPE insulation varying with ageing time and temperature [7].

To explore the dependency of the IR degradation model parameters on temperature, the values of λ, ρ_0 and ρ_d for each temperature level are first determined from the corresponding XLPE resistivity measurements in Figure 8. The software Origin 2022 is used here to estimate λ, ρ_0 and ρ_d in (12) by fitting the decay of XLPE resistivity over Phase 1 of the thermal ageing process prior to the occurrence of a percolation phenomenon in the transition phase. The results of λ, ρ_0 and ρ_d estimated for the four ageing temperature levels are listed in Table 1. Taking the natural logarithm of the Arrhenius model which links λ to T via the IR thermal ageing activation energy E_a, as formulated by (16), the parameters λ_0 and E_a are computed by fitting (16) to the degradation rate estimates in Table 1, as shown in Figure 9a, which equal 1.82×10^6 (1/h) and 64 kJ, respectively. Figure 9b,c show the linear fitting of ρ_0 and ρ_d against temperatures, which is used here to approximately describe the dependencies of ρ_0 and ρ_d on T, as formulated by (17) and (18), respectively. The statistics of the linear fitting of $\ln(\lambda)$, ρ_0 and ρ_d against T are listed in Table 2. While the linear fitting of $\ln(\lambda)$ and ρ_0 achieves an R-squared of about 0.9, the R-squared for the linear fitting of ρ_d is 0.54 only. The fitting could be improved by using a function with a higher degree of freedom, though this might cause an over-fitting problem. The data obtained from more XLPE samples and more temperature levels can be considered for a better fitting of the temperature-dependent variables.

$$\ln(\lambda) = \left(-\frac{E_a}{k_B}\right)\frac{1}{T} + \ln(\lambda_0) \quad (16)$$

$$\rho_0 = (-3.0475 \times T + 1255.52) \times 10^{10} \quad (17)$$

$$\rho_d = (0.0126 \times T - 3.4677) \times 10^{10} \quad (18)$$

Table 1. Degradation rates, non-degraded and degraded resistivity values obtained from resistivity measurements of XLPE insulation samples at different thermal ageing temperatures.

Temperature (K)	Degradation Rate λ (1/h)	ρ_0 (Ω-m)	ρ_d (Ω-m)
353.15	8.5×10^{-4}	2.0×10^{12}	1.0×10^{10}
373.15	1.0×10^{-3}	1.0×10^{12}	1.0×10^{10}
393.15	9.0×10^{-3}	3.2×10^{11}	1.9×10^{10}
413.15	1.4×10^{-2}	1.95×10^{11}	1.54×10^{10}

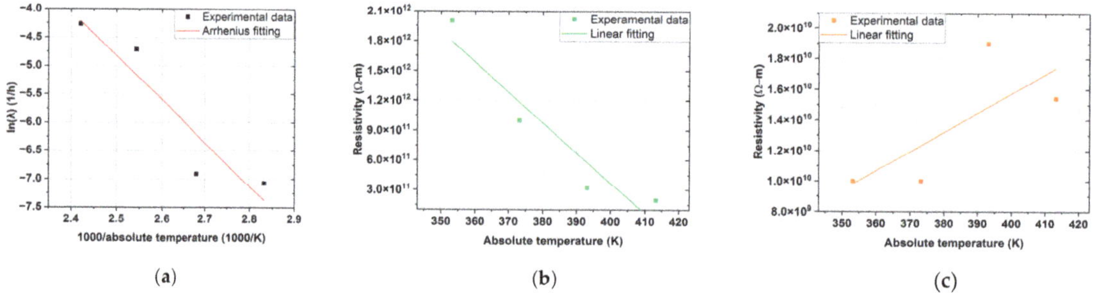

Figure 9. (**a**) The Arrhenius plot of the natural logarithm of degradation rate against the inverse of absolute temperature; and the linear fitting of (**b**) non-degraded resistivity and (**c**) degraded resistivity against absolute temperature for XLPE insulation samples.

Table 2. The statistics of linear fitting of the natural logarithm of degradation rate, non-degraded and degraded resistivity against absolute temperature for XLPE insulation samples.

Fitting Statistics		$\ln(\lambda)$	ρ_0	ρ_d
Intercept	Value	14.4153	1255.5200×10^{10}	-3.4677×10^{10}
	Std. Err.	5.3673	267.4020×10^{10}	3.1279×10^{10}
	t-Value	2.6858	4.6953	−1.1086
Slope	Value	−7.6959	-3.0475×10^{10}	0.0126×10^{10}
	Std. Err.	2.0460	0.6967×10^{10}	0.0081×10^{10}
	t-Value	−3.7615	−4.3741	1.5460
Residual Sum of Squares		0.7874	1941.6750×10^{20}	0.2657×10^{20}
R-squared		0.8762	0.9054	0.5444

3.2. Experiment Design and Temperature Simulation

The model parameters estimated in Section 3.1 are applied here to simulate the IR of an XLPE insulation sample with inner radius x of 0.0075 m, and outer radius X of 0.0115 m (i.e., a 0.004 m insulation thickness). Long term laboratory experiments have been undertaken on the surface temperature measurements along a number of 3.2 m LSXHIOE 1 × 185/35, 6.35/11 kV XLPE-insulated power cable samples in a temperature ageing enclosure running increased currents to produce cable surface temperatures of approximately 100 °C [19]. The surface temperature distribution across the length of the cables was measured using five type T temperature sensors as depicted in Figure 10. An FVM simulation model was also implemented to replicate the measurements to a reasonable degree of accuracy, as shown in Figure 11. The details are too extensive to reproduce here, but as an example an outer temperature of 100 °C was simulated in order to produce an understanding of the inner temperature distribution across the cable insulation. The example is displayed in Figure 12 which shows the temperature profile at one cross section of the cable with natural convection air flow in the ageing chamber. The purpose of this is to demonstrate the power of the FVM model in being able to provide data that can be fed to an ANN model to estimate the temperature at any position within the insulation. Different FVM simulations can be applied for different physical geometries and scenarios and this is simply one scenario to demonstrate the principles of the methodology. For the details of FVM simulation and the coefficients of cable layer and air properties, the reader is referred to [20] and [21–24], respectively.

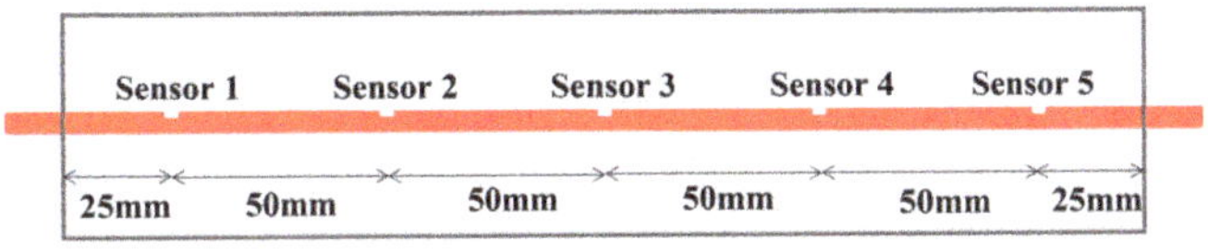

Figure 10. Positions of type T thermocouples placed along the cable sample surface.

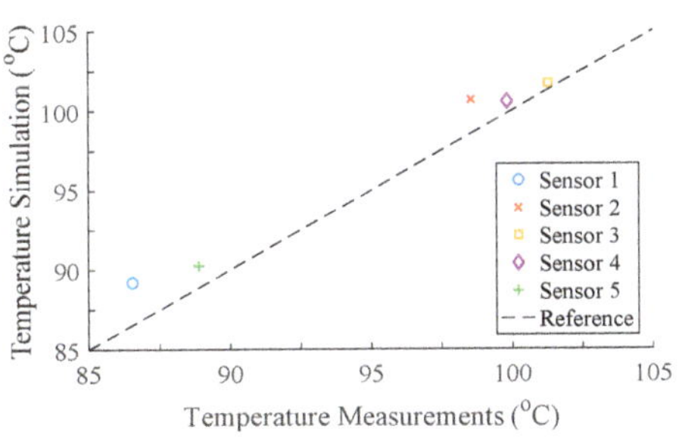

Figure 11. Measurements and simulation of cable surface temperatures (°C) at the sensor positions.

In order for the resolution of the FVM-based temperature simulation to meet the sizes of insulation segments, the FVM-based temperature simulation results were used to train the ANN model [25] which then interpolated the temperature for all the segments. The MATLAB neural network toolbox [26] was employed to construct the ANN model which was trained based on 70% of the temperature data obtained from the FVM simulation. An amount of 15% of the temperature data was designated for validation, and the remaining 15% was used for testing. The ANN models underwent training for a total of 1000 iterations, with a Pearson correlation coefficient of over 0.99 indicating an excellent performance of the ANN model. The temperature distributions estimated by the ANN model for the entire cable insulation and a selected 1 m insulation section starting from the middle of the cable sample are shown in Figures 13a and 13b, respectively. Higher temperatures generally occur at the segments closer to core conductors and/or the centre of the ageing chamber.

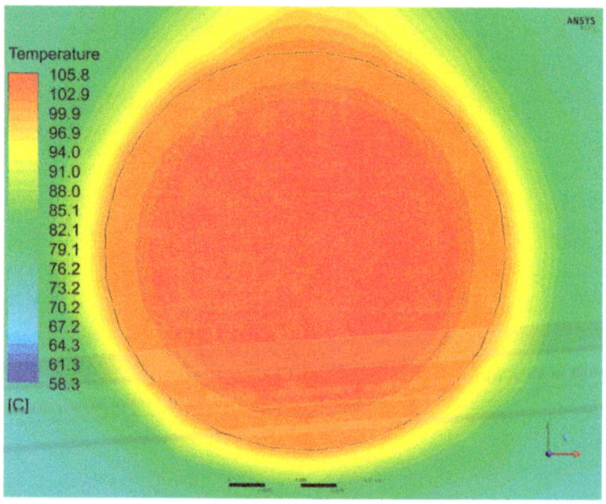

Figure 12. Example temperature profile simulation (°C) across one cross section of the cable sample.

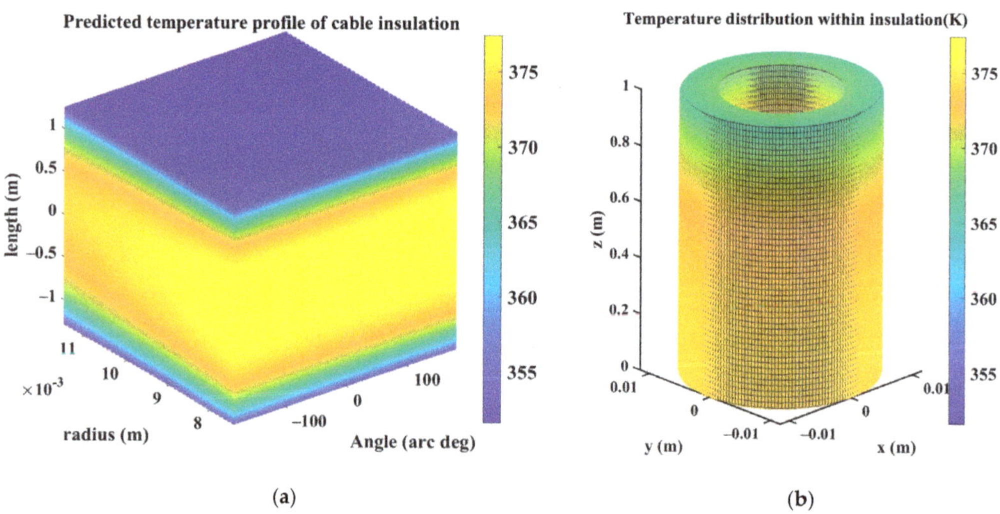

Figure 13. Example temperature profiles (K) of (**a**) the entire cable insulation and (**b**) the selected 1 m insulation section in cylindrical coordinate system.

It is noted that the insulation temperature simulation performed by the FVM depicts the temperature profiles across insulation in the thermal ageing experiment and does not reflect the local temperature elevation in partial discharge events or a highly non-uniform temperature profile in the case of a complex ambient environment. While the FVM-based simulation shows its capability of replicating the temperature profiles in the experimental environment, the FVM can be enhanced to cope with the heat exchange of cables with the surrounding environment and reflect the effects of partial discharges on local temperature under practical operation, providing reliable insulation temperature profiles for thermal degradation assessment in practice. In addition, the existence of local temperature variations means that the dichotomy models which assume a uniform temperature distribution across the entire insulation (layer) are not applicable. In such cases, the discretization models can be employed to simulate the thermal degradation of individual insulation segments under a highly non-uniform temperature distribution.

3.3. *IR Degradation Simulation Results*

The 1 m cable insulation sample with constant diameters across its length, as shown in Figure 13b, is employed here to compare the performance of the four IR degradation models and simplify the conditions for IR simulation. Given the four IR degradation models considering temperature information in different ways, the hotspot temperature of 377.26 K within the entire insulation is applied to the dichotomy or discretization model with uniform temperature, and the discretization model with non-uniform temperature uses the exact insulation temperature profile. The dichotomy model with radial temperature gradients divides the insulation sample into 10 layers along the radial dimension. Table 3 lists the hotspot temperature adopted for each layer and related degradation rate which decreases from the innermost layer #1 to the outermost layers #10. As it was noted in Sections 2.1 and 2.2, the degraded insulation segments are selected by the dichotomy models in a random way. To that end, the IR simulation of the dichotomy models is performed 50 times to ensure reliability. The 50 potential IR estimates at each thermal ageing time step are found to have a negligible standard deviation which is smaller than 0.1%. Therefore, the one-time IR calculation results of the dichotomy models will be presented in this work.

Table 3. Hotspot temperatures and associated degradation rates of 10 insulation layers in the dichotomy model with radial temperature gradients.

Layer No.	Absolute Temperature (K)	Degradation Rate (1/h)
1	377.26	0.002505
2	376.96	0.002464
3	376.66	0.002424
4	376.36	0.002385
5	376.07	0.002348
6	375.86	0.002321
7	375.59	0.002287
8	375.33	0.002256
9	375.10	0.002227
10	374.82	0.002192

3.3.1. Initial Resistance Distributions

The initial resistances of insulation segments estimated by the four IR degradation models are shown in Figure 14, respectively. In the main, the segments at inner insulation layers which have relatively smaller volumes show higher resistances than those at outer layers. Since the (non-degraded) segment resistivity at the start of the thermal ageing process depends on segment temperature only, the dichotomy and discretization models which assume uniform temperature produce the same initial resistance distributions (see Figure 14a,c) which lead to a total IR of 94.6 GΩ. When the radial temperature gradients are introduced into the dichotomy model, the segments of outer layers with lower temperature will have higher resistivity than those that are assumed to be subjected to the hotspot temperature of the entire insulation. Therefore, the dichotomy model with radial temperature gradients produces relatively higher initial resistances for segments at outer layers than the model with uniform temperature (see Figure 14a,b). This results in a higher total IR equalling around 97.18 GΩ. When the full insulation temperature profile is introduced into the discretization model, the segment resistance variations with the position-dependent temperatures are simulated along longitudinal and radial dimensions (see Figure 14d). The total IR estimated by the discretization model at the start of the thermal ageing process is around 101.32 GΩ.

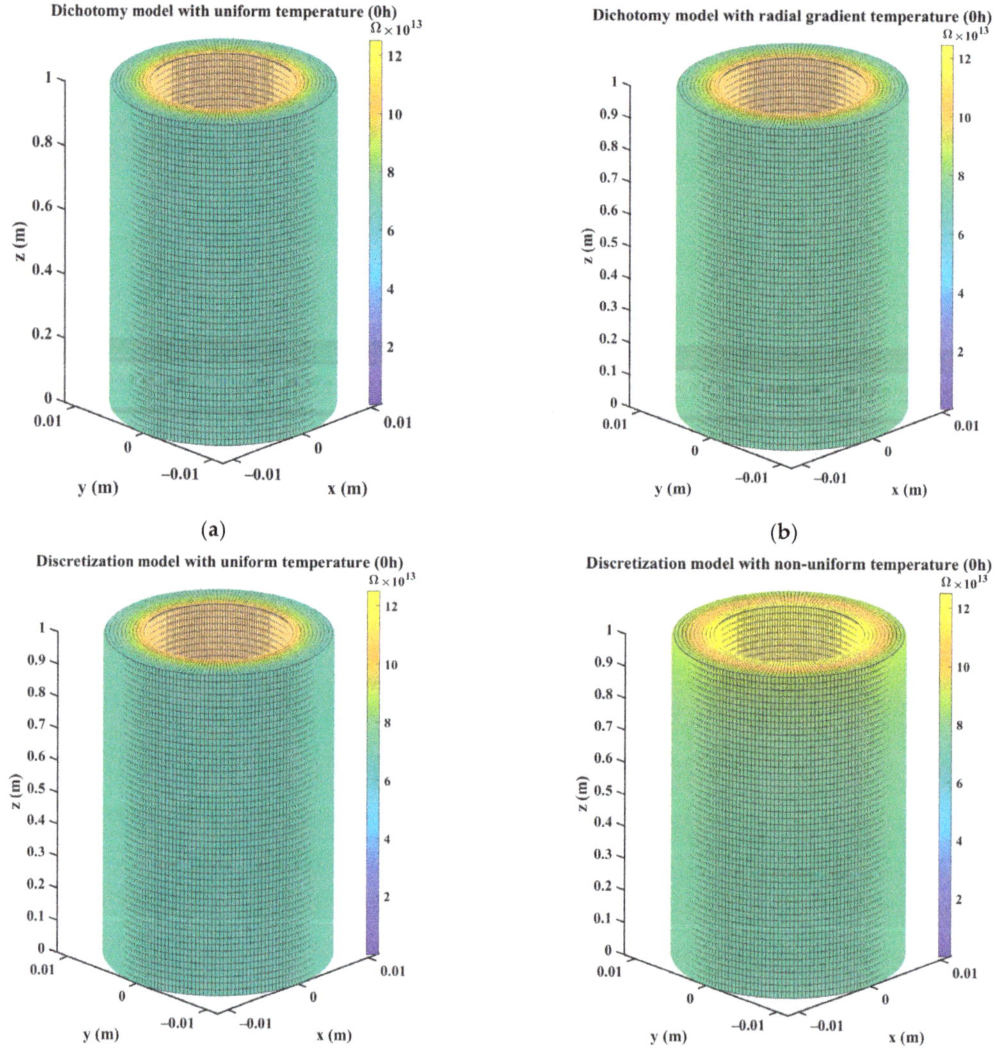

Figure 14. The segment resistances (Ω) of cable insulation at 0 h estimated by (**a**) dichotomy model with uniform temperature; (**b**) dichotomy model with radial temperature gradients; (**c**) discretization model with uniform temperature; and (**d**) discretization model with non-uniform temperature.

3.3.2. Resistance Distributions after 100 h and 300 h Thermal Ageing

Figures 15 and 16 compare the segment resistance distributions of the insulation sample modelled by the four IR degradation models for thermal ageing durations of 100 h and 300 h. Compared with the dichotomy models which assign low resistivity ρ_d to the random samples of fully degraded segments, the discretization models are shown to describe the gradual resistance degradation of individual insulation segments separately. Table 4 tabulates the degradation volume ratio and the resulting number of degraded segments that are randomly sampled for each insulation layer by the dichotomy model with radial temperature gradients. The degradation volume ratio determined from the temperature-based degradation rate not only increases with layer temperature but also with the thermal ageing duration, which is reflected in the increasing number of degraded segment samples (see Figures 15b and 16b). Since the dichotomy model with uniform temperature uses the hotspot temperature of the entire insulation, the number of degraded segment samples at the outermost layer in Figure 15a is greater than in Figure 15b which is simulated with the radial temperature gradients. In addition, the inclusion of the full insulation temperature profile allows the discretization model to describe the dependency of individual segment degradation on local temperatures which vary along both radial and longitudinal dimensions, as shown in Figures 15d and 16d.

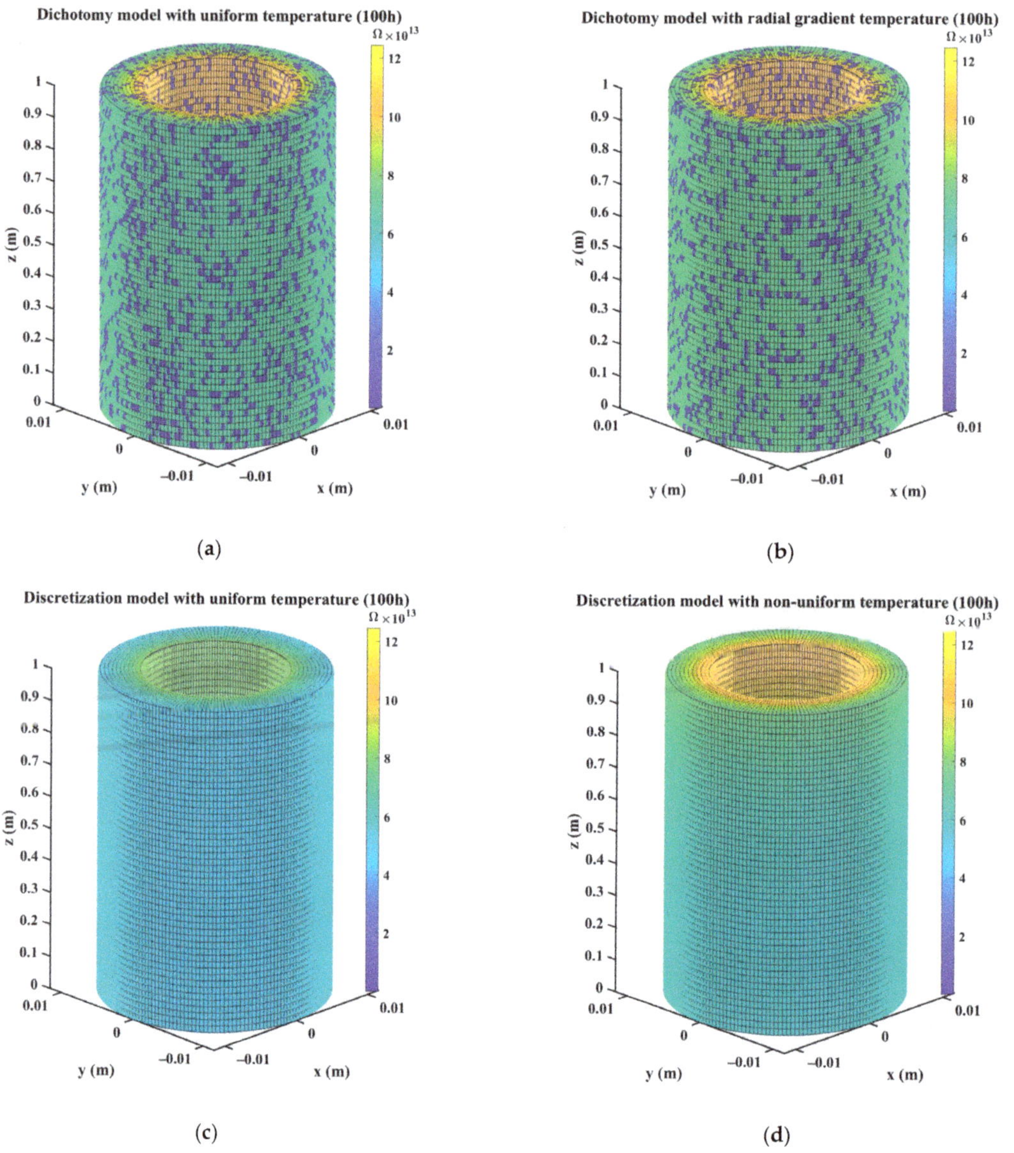

Figure 15. The segment resistances (Ω) of cable insulation at 100 h of thermal ageing estimated by (**a**) Dichotomy model with uniform temperature; (**b**) dichotomy model with radial temperature gradients; (**c**) discretization model with uniform temperature; and (**d**) discretization model with non-uniform temperature.

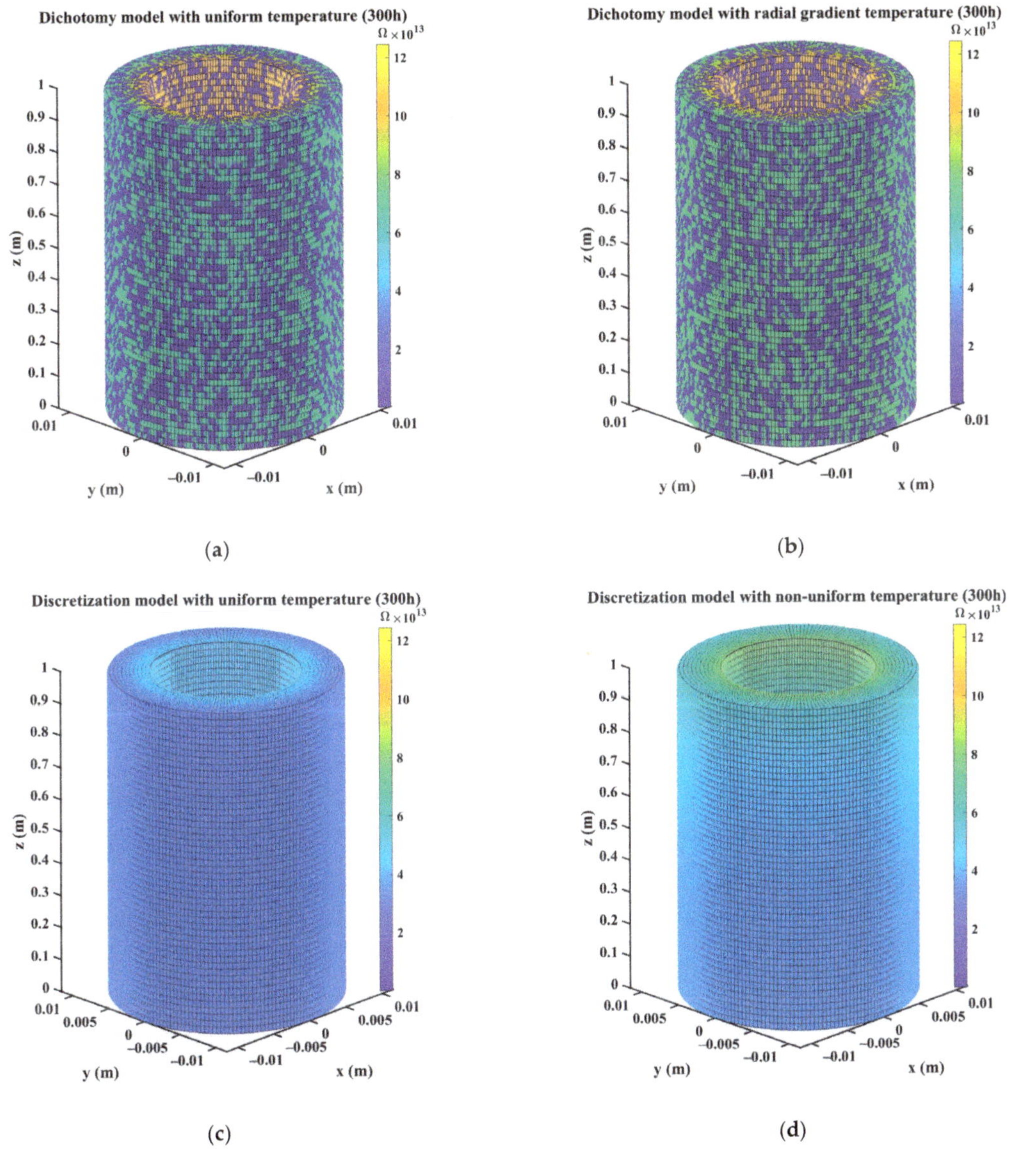

Figure 16. The segment resistances (Ω) of cable insulation at 300 h of thermal ageing estimated by (**a**) Dichotomy model with uniform temperature; (**b**) dichotomy model with radial temperature gradients; (**c**) discretization model with uniform temperature; and (**d**) discretization model with non-uniform temperature.

Table 4. Degradation volume ratios and numbers of degraded segments estimated for 10 insulation layers by the dichotomy model with radial temperature gradients after 100 h and 300 h of thermal ageing.

Layer	Thermal Ageing for 100 h		Thermal Ageing for 300 h	
	Degradation Ratio	**Number of Degraded Segments**	**Degradation Volume Ratio**	**Number of Degraded Segments**
1	22.158%	1994	52.833%	4755
2	21.836%	1965	52.245%	4702
3	21.526%	1937	51.674%	4651
4	21.220%	1910	51.106%	4600
5	20.927%	1883	50.559%	4550
6	20.715%	1864	50.161%	4514
7	20.443%	1840	49.645%	4468
8	20.194%	1817	49.171%	4425
9	19.964%	1797	48.731%	4386
10	19.687%	1772	48.197%	4338

3.3.3. IR Degradation Curves in Short Term of Thermal Ageing

To examine the impacts of model assumptions on IR degradation simulation, the IR degradation curves generated by the four models over the initial short term (0–300 h) of thermal ageing are compared in Figure 17. Compared with the dichotomy model which uses the hotspot temperature of the entire insulation, the adoption of hotspot temperature for each of the 10 insulation layers under radial temperature gradients reduces the degradation rates and numbers of degraded segments at outer layers. This not only produces relatively greater IR simulation results but also slightly reduces the speed of the simulated IR degradation. In addition, different IR simulation results between dichotomy and discretization models which use the same uniform temperature reveal the inherent differences between the two models. The discretization model simulates the gradual IR degradation of individual segments separately, while the dichotomy model assesses the overall degradation condition of the insulation sample through a degradation volume ratio based on which the fully degraded segments are randomly located within the insulation sample. With the degradation volume ratio increasing over the thermal ageing process, the radial column consisting of series connected segments along the radial dimension becomes more likely to have multiple degraded segments, which largely reduces the column resistance and causes more degradation in the total IR. Therefore, the random selection of degraded segments by the dichotomy model leads to lower IR values than the discretization model, as shown in Figure 17. Furthermore, the discretization model with non-uniform temperature distribution is shown to generate the highest IR estimates throughout the duration of interest due to the use of a detailed insulation temperature profile instead of the hotspot temperature(s) of cable insulation.

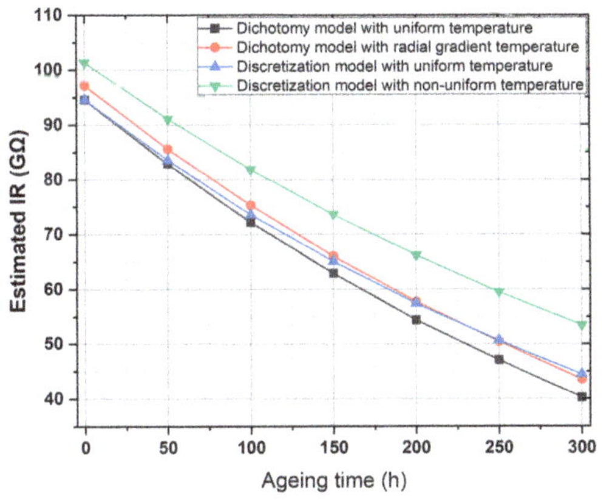

Figure 17. IR degradation curves (GΩ) estimated by different models within 300 h of thermal ageing.

4. Sensitivity Analysis

4.1. Influence of Segment Size

To examine the influence of the segment size adopted for the IR degradation simulation, the IR simulation results of the four IR models are simulated based on four different segments sizes, respectively, as listed in Table 5 where the size #2 is different from the size #1, #3 or #4 along the radial, longitudinal or angular dimension, respectively. The material properties and related parameters are the same as those that were employed in Section 3. To explore the IR degradation trend over a longer thermal ageing period, the IR results are simulated at every 50 h during 2000 h of thermal ageing, as shown in Figure 18.

Table 5. Insulation segment sizes and resulting numbers of segments for comparison.

Size Index	Δx (m)	$\Delta\varphi$ (°)	Δl (m)	M	N	P	N_t
1	2×10^{-4}	2	0.02	20	180	50	1.8×10^5
2	1×10^{-4}	2	0.02	40	180	50	3.6×10^5
3	1×10^{-4}	2	0.01	40	180	100	7.2×10^5
4	1×10^{-4}	1	0.02	40	360	50	7.2×10^5

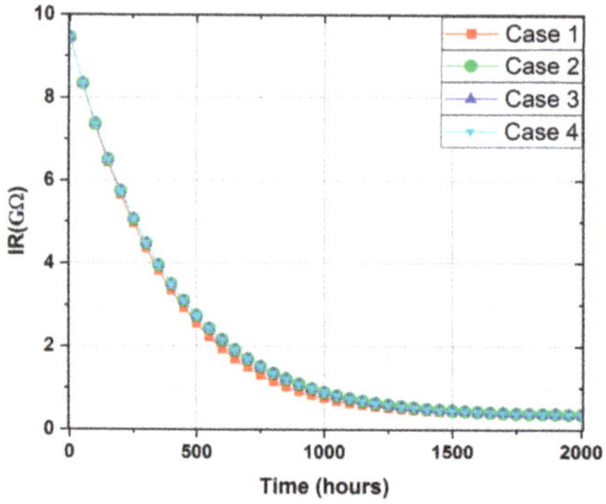

(a)

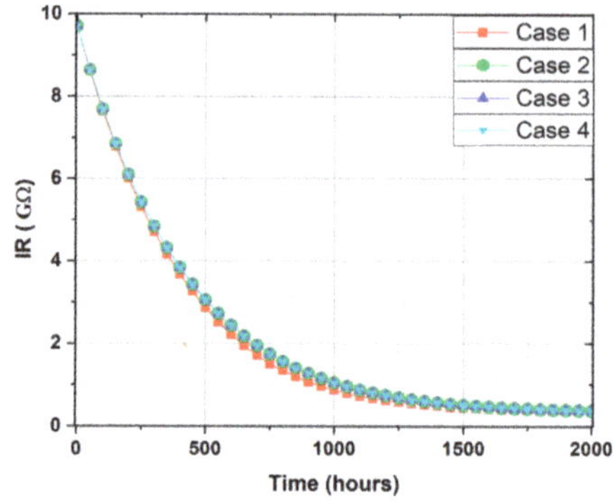

(b)

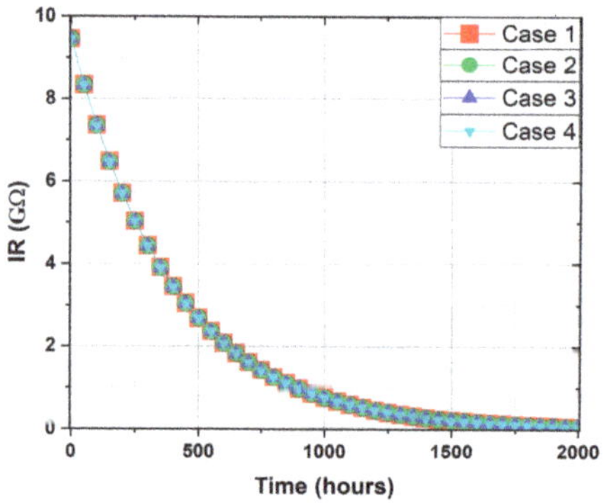

(c)

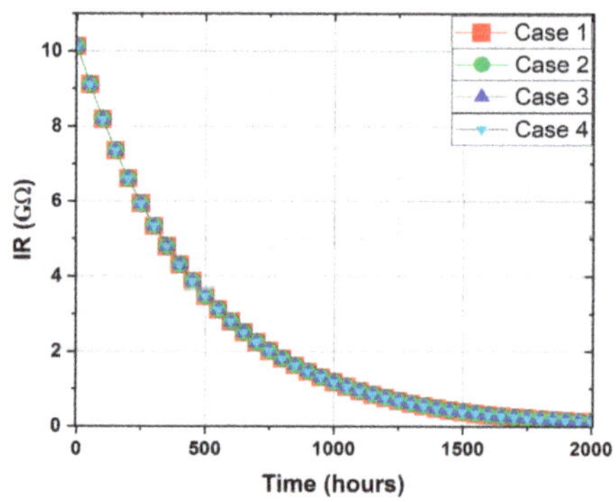

(d)

Figure 18. IR degradation trends (Ω) over 2000 h of thermal ageing simulated based on different segment sizes by (**a**) dichotomy model with uniform temperature; (**b**) dichotomy model with radial temperature gradients; (**c**) discretization model with uniform temperature; and (**d**) discretization model with non-uniform temperature.

Figure 18a,b show that the adoption of sizes #2, #3 and #4 in the dichotomy models almost produces the same IR results, meaning that the segment size change along the angular or longitudinal dimension examined in this work has negligible impacts on the IR simulation. When the size #1 is used by the dichotomy models, a degraded segment with a radial size $\Delta x = 0.2$ mm can be regarded as two degraded segments of $\Delta x = 0.1$ mm that are series connected in a column along the radial dimension. This increases the probability of a single radial column having more degraded parts and thus reducing the total IR. Therefore, compared with the sizes #2, #3 and #4 of $\Delta x = 0.1$ mm, the size #1 of $\Delta x = 0.2$ mm results in the dichotomy models producing slightly lower IR values, especially during 200–1200 h of thermal ageing. When the thermal ageing process approaches 2000 h, the effect of Δx on IR simulation becomes negligible due to a sufficient degradation volume ratio which makes the two different Δx values result in similar degradation parts of single radial columns. For the discretization models which simulate the individual segment degradation rather than rely on the random selection of fully degraded segments, the use of the four different segment sizes produces very similar IR results, as shown in Figure 18c,d. This illustrates that the segment size variations assessed in this work have little impact on the IR simulation of discretization models.

4.2. Influence of Degradation Rate

The degradation rates employed in the simulation above are determined through the Arrhenius fitting of experimental data sourced from Mecheri et al. [7]. Considering the possible inherent uncertainties, a percentage error of $\pm 10\%$ is introduced in the degradation rate, denoted by 0.9λ and 1.1λ. With other model parameters remaining fixed, the IR degradation trends simulated by the four models using the smaller segment size #3 combined with the three degradation rates are compared in Figure 19a–d, respectively.

Even though the uncertainties of degradation rates do not affect the IR degradation pattern, it is estimated that the IR percentage errors are less sensitive to the λ error of +10% and vary with thermal ageing time in different ways between the dichotomy and discretization models. The IR percentage error amplitudes of dichotomy models increase to a maximum of about 22% for 0.9λ or 16% for 1.1λ at around 900–1000 h and then relatively slowly drop with the ageing time going further. This might be because the dichotomy models randomly assign a constant fully degraded resistivity to insulation segments based on the number of degraded segments jointly determined by the degradation rate and ageing time. The variation in the number of degraded segments induced by the λ error of $\pm 10\%$ increases at the beginning of the thermal ageing process and then decreases at a certain ageing time when the number of degraded segments approaches the total number of insulation segments. In an extreme case where all the segments are fully degraded after a sufficiently long thermal ageing process under any degradation rate, the dichotomy models will always give the same IR results independent of degradation rates. For the discretization models which simulate the gradual degradation of individual segments, the IR percentage error amplitudes increase with the ageing time throughout the thermal ageing process and approximately reach 60% for 0.9λ or 40% for 1.1λ at the end of 2000 h. However, the corresponding IR absolute errors at the end of 2000 h are insignificant due to the fact that the IR results become very small after a long thermal ageing time compared with initial IR levels. Since the uncertainty propagating from degradation rates to IR values can be magnified especially when using the discretization models, further exploration should reflect the confidence bounds for IR degradation estimation subject to various sources of uncertainty.

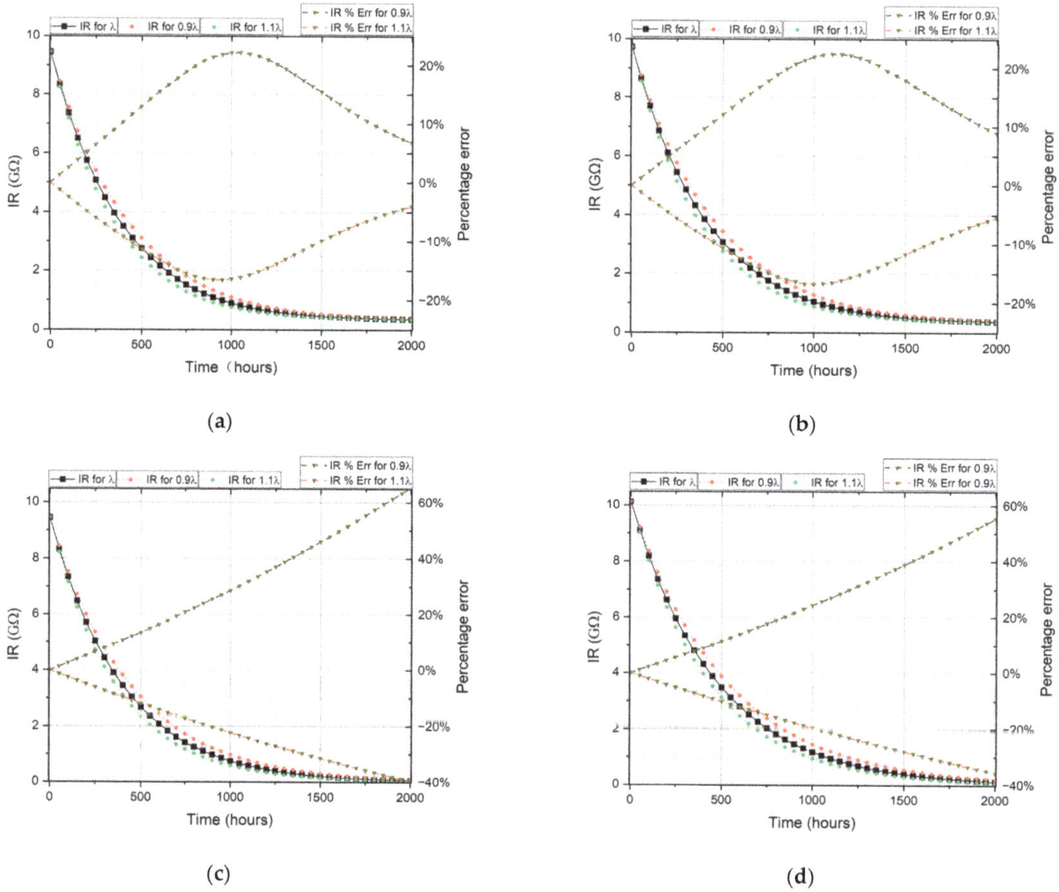

Figure 19. IR degradation trends (Ω) with percentage errors over 2000 h of thermal ageing simulated under uncertain degradation rates by (**a**) dichotomy model with uniform temperature; (**b**) dichotomy model with radial temperature gradients; (**c**) discretization model with uniform temperature; and (**d**) discretization model with non-uniform temperature.

5. Conclusions and Future Work

The insulation of heavily loaded power cables is subjected to varying extents of thermal ageing that causes the degradation of electrical dielectric properties including insulation resistance (IR). To identify an appropriate approach for the IR degradation simulation under thermal ageing, this paper has developed four IR degradation models for cross-linked polyethylene-insulated power cables and compared the IR simulations between different methodologies and also between the uses of uniform and non-uniform temperature profiles. A cylindrical cable insulation sample has been modelled in this work and divided into multiple sufficiently small segments, enabling the random selection of fully degraded segments by dichotomy models and the estimation of individual segment degradation processes by discretization models. The insulation temperature profile has been estimated by modelling the thermal ageing experiment via a finite volume method (FVM) and then refined by an artificial neural network (ANN) to match profile resolutions with the insulation segment sizes.

When assuming a uniform insulation temperature equal to the hotspot temperature of the entire insulation, the dichotomy model has generated lower IR values than the discretization model, especially for a longer thermal ageing duration due to an increased probability of assigning the fully degraded insulation resistivity to multiple segments that are series connected in the same radial column. Although the incorporation of radial temperature gradients into the dichotomy model has mitigated the overestimation of IR degradation, the resulting IR levels are lower than those that are produced by the discretization model with the exact temperature profile. This is because the discretization model has additionally considered the temperature variation along the longitudinal dimension and simulated the gradual degradation of individual segments separately rather than randomly sampling fully degraded segments from the insulation (layers). In addition, the IR degradation simulation has been performed based on different segment sizes and degradation rates for a sensitivity analysis, suggesting that IR results are more sensitive to the segment size along the radial dimension and to the underestimation of degradation rates. Furthermore, the degree of sensitivity not only varies with the thermal ageing time but also depends on the IR degradation model adopted.

It is noted that the IR degradation models developed here consider thermal ageing effects only. Based on the present work, the proposed discretization model with the full temperature profile should be enhanced to simulate the joint effects of thermal ageing and annealing by additionally considering the diffusion of chemical components from semicon shields which may increase trap density and promote hopping conduction within cable insulation [27]. The cumulation and exhaustion of the chemical components could result in the IR trend exhibiting a U-shape variation at the initial stage of the heating process [28]. Furthermore, the enhanced discretization model will be fitted to laboratory IR measurements to understand the conduction mechanisms in insulation with the presence of semicon layers over the heating process. Moreover, different sources of uncertainty will be translated into the confidence bounds of an IR estimate by simulating their propagation through IR degradation models. In addition, the FVM will be further developed to simulate highly non-uniform insulation temperature profiles which may be induced by local partial discharges and/or complex ambient environments, providing reliable temperature profiles to discretization models for resistivity degradation estimation. The low-resistance current path generated during partial discharges will also be considered in future model development. In the case of overhead cables that are exposed to air, the insulation degradation accelerated by solar radiation should also be explored.

Author Contributions: Conceptualization, B.G.S. and M.J.G.; methodology, X.G. and B.G.S.; software, X.G.; validation, X.G., F.F. and B.G.S.; formal analysis, X.G., F.F. and B.G.S.; investigation, X.G., F.F. and B.G.S.; resources, X.G. and B.G.S.; data curation, X.G. and F.F.; writing—original draft preparation, X.G. and F.F.; writing—review and editing, B.G.S.; visualization, X.G.; supervision, B.G.S. and M.J.G.; project administration, B.G.S.; funding acquisition, B.G.S. All authors have read and agreed to the published version of the manuscript.

Funding: This research was funded by the UK Engineering and Physical Sciences Research Council grant EP-T001445-1.

Data Availability Statement: Data are contained within the article.

Conflicts of Interest: The authors declare no conflicts of interest.

References

1. Klauck, M.; Nobrega, G.; Reinecke, E.A.; Bentaib, A.; Maas, L.; Chaumeix, N.; Allelein, H.J. Experimental investigation on the impact of cable fire products from flame-retardant cables on catalysts used in passive auto-catalytic recombiners. *Prog. Nucl. Energy* **2022**, *152*, 104365. [CrossRef]
2. *IEC Std 60505*; Evaluation and Qualification of Electrical Insulation Systems. IEC Standard: Geneva, Switzerland, 2011.
3. *IEEE Std 1064-1991*; IEEE Guide for Multifactor Stress Functional Testing of Electrical Insulation Systems. IEEE Standard: New York, NY, USA, 1991.
4. Densley, J. Ageing and diagnostics in extruded insulations for power cables. In Proceedings of the 1995 IEEE 5th International Conference on Conduction and Breakdown in Solid Dielectrics, Leicester, UK, 10–13 July 1995; pp. 1–15.
5. Zhou, C.; Yi, H.; Dong, X. Review of recent research towards power cable life cycle management. *IET High Volt.* **2017**, *2*, 179–187. [CrossRef]

6. Boukezzi, L.; Boubakeur, A. Effect of thermal aging on the electrical characteristics of XLPE for HV cables. *Trans. Electr. Electron. Mater.* **2018**, *19*, 344–351. [CrossRef]
7. Mecheri, Y.; Boukezzi, L.; Boubakeur, A.; Lallouani, M. Dielectric and mechanical behavior of cross-linked polyethylene under thermal aging. In Proceedings of the 2000 Annual Report Conference on Electrical Insulation and Dielectric Phenomena, Victoria, BC, Canada, 15–18 October 2000; Volume 2, pp. 560–563.
8. Nedjar, M. Effect of thermal aging on the electrical properties of crosslinked polyethylene. *J. Appl. Polym. Sci.* **2009**, *111*, 1985–1990. [CrossRef]
9. Mecheri, Y.; Bouazabia, S.; Boubakeur, A.; Lallouani, M. Effect of thermal ageing on the properties of XLPE as an insulating material for HV cables. In Proceedings of the International Electrical Insulation Conference, IET Centre, Birmingham, UK, 29–31 May 2013; pp. 29–31.
10. Zhang, C.C.; Li, Y.F.; Hu, M.Y.; Ma, F.L.; Zhao, H.; Han, B.Z. Conductivity properties of XLPE insulation used for HVDC cable after accelerated thermal ageing. In Proceedings of the 2018 12th International Conference on the Properties and Applications of Dielectric Materials (ICPADM), Xi'an, China, 20–24 May 2018; pp. 500–503.
11. Kang, S.D.; Kim, J.H. Investigation on the insulation resistance characteristics of low voltage cable. *Energies* **2020**, *13*, 3611. [CrossRef]
12. Chang, Y.S.; Mosleh, A. Predictive Model on the Degradation of the Electrical Resistance of Cable Insulation. In Proceedings of the Probabilistic Safety Assessment and Management PSAM, Los Angeles, CA, USA, 14 September 2018.
13. DeCarlo, R.A. *Linear Systems: A State Variable Approach with Numerical Implementation*; Prentice-Hall, Inc.: Hoboken, NJ, USA, 1989.
14. Serway, R.A.; Jewett, J.W. Current and resistance. In *Physics for Scientists and Engineers with Modern Physics*, 10th ed.; Cengage: Boston, MA, USA, 2018; pp. 691–712.
15. Rodríguez-Valverde, M.Á.; Tirado-Miranda, M. A simpler derivation of the integral formula of electrical resistance. *Eur. J. Phys.* **2009**, *30*, L47. [CrossRef]
16. Mosleh, A.; Al-Sheikhly, M.; Chang, Y.S.; Reister, R. *Physics-Based Probabilistic Model of the Effects of Ionizing Radiation on Polymeric Insulators of Electric Cables Used in Nuclear Power Plants*; No. 15-8258; University of California: Los Angeles, CA, USA, 2019.
17. Navidi, W.C. *Statistics for Engineers and Scientists*; McGraw-Hill: New York, NY, USA, 2006; p. 2.
18. Stojanović, M.; Klimenta, J.; Panić, M.; Klimenta, D.; Tasić, D.; Milovanović, M.; Perović, B. Thermal aging management of underground power cables in electricity distribution networks: A FEM-based Arrhenius analysis of the hot spot effect. *Electr. Eng.* **2023**, *105*, 647–662. [CrossRef]
19. *IEEE 1407-2007*; IEEE Guide for Accelerated Aging Tests for Medium-Voltage (5kV–35kV) Extruded Electric Power Cables Using Water-Filled Tanks. Power Engineering Society: New York, NY, USA, 2007; Sponsored by the Insulated Conductors Committee.
20. Meng, X.K.; Wang, Z.Q.; Li, G.F. Dynamic analysis of core temperature of low-voltage power cable based on thermal conductivity. *Can. J. Electr. Comput. Eng.* **2016**, *39*, 59–65. [CrossRef]
21. Lee, K.Y.; Yang, J.S.; Choi, Y.S.; Park, D.H. Specific heat and thermal conductivity measurement of XLPE insulator and semi-conducting materials. In Proceedings of the 2006 IEEE 8th International Conference on Properties & Applications of Dielectric Materials, Bali, Indonesia, 26–30 June 2006; pp. 805–809.
22. Jia, Y.; Mao, Z.; Huang, W.; Zhang, J. Effect of temperature and crystallinity on the thermal conductivity of semi-crystalline polymers: A case study of polyethylene. *Mater. Chem. Phys.* **2022**, *287*, 126325. [CrossRef]
23. Gray, D.D.; Giorgini, A. The validity of the Boussinesq approximation for liquids and gases. *Int. J. Heat Mass Transf.* **1976**, *19*, 545–551. [CrossRef]
24. The Engineering Toolbox. Air—Density, Specific Weight and Thermal Expansion Coefficient vs. Temperature and Pressure. Available online: https://www.engineeringtoolbox.com/air-density-specific-weight-d_600.html (accessed on 16 November 2021).
25. Ge, X.; Given, M.J.; Stewart, B.G. Determining accelerated aging power cable spatial temperature profiles using Artificial Neural Networks. In Proceedings of the 2022 IEEE International Conference on High Voltage Engineering and Applications (ICHVE), Chongqing, China, 25–29 September 2022; pp. 1–4.
26. Beale, M.H.; Hagan, M.T.; Demuth, H.B. *Deep Learning Toolbox*; User's Guide; The MathWorks, Inc.: Natick, MA, USA, 2018.
27. Diego, J.A.; Belana, J.; Orrit, J.; Cañadas, J.C.; Mudarra, M.; Frutos, F.; Acedo, M. Annealing effect on the conductivity of XLPE insulation in power cable. *IEEE Trans. Dielectr. Electr. Insul.* **2011**, *18*, 1554–1561. [CrossRef]
28. Ge, X.; Fan, F.; Given, M.J.; Stewart, B.G. XLPE cable insulation resistance modelling under annealing and thermal ageing effects. In Proceedings of the 98th IEEE Conference on Electrical Insulation and Dielectric Phenomena, East Rutherford, NJ, USA, 15–19 October 2023.

MDPI AG
Grosspeteranlage 5
4052 Basel
Switzerland
Tel.: +41 61 683 77 34

Energies Editorial Office
E-mail: energies@mdpi.com
www.mdpi.com/journal/energies

www.ingramcontent.com/pod-product-compliance
Lightning Source LLC
Chambersburg PA
CBHW040752250926
51850CB00007BB/1472
9783725819737